Microwave Techniques in Superconducting Quantum Computers

Alan Salari

ARTECH HOUSE
BOSTON | LONDON
artechhouse.com

Library of Congress Cataloging-in-Publication Data
A catalog record for this book is available from the U.S. Library of Congress.

Cover design by Joi Garron

ISBN-13: 978-1-63081-987-3

Accompanying appendices can be found at: htttps://quaxys.com/book.

© 2024 Artech House
685 Canton Street
Norwood, MA 02062

All rights reserved. Printed and bound in the United States of America. No part of this book may be reproduced or utilized in any form or by any means, electronic or mechanical, including photocopying, recording, or by any information storage and retrieval system, without permission in writing from the publisher.

All terms mentioned in this book that are known to be trademarks or service marks have been appropriately capitalized. Artech House cannot attest to the accuracy of this information. Use of a term in this book should not be regarded as affecting the validity of any trademark or service mark.

10 9 8 7 6 5 4 3 2 1

To my beloved parents, whose unwavering guidance and wisdom have illuminated my path and shaped me into the person I am today.

To the educators who have inspired, challenged, and supported me throughout my academic and professional journeys.

Contents

Foreword xi
Preface xiii
Acknowledgments xv

CHAPTER 1
Introduction to Quantum Physics 1

1.1 A Brief History of Quantum Mechanics 1
1.2 Quantum Versus Classical Mechanics 3
1.3 Schrödinger Equation 4
1.4 The Machinery of Quantum Calculations 5
1.5 Solving the Schrödinger Equation 6
    1.5.1 Time-Independent Schrödinger Equation 7
    1.5.2 Standard Hamiltonians 8
1.6 Quantum Measurement 17
    1.6.1 Collapse of the Wave Function 17
    1.6.2 Expectation Value 18
    1.6.3 Variance or Uncertainty 19
    1.6.4 Uncertainty Principle 20
    1.6.5 Heisenberg's Picture 23
    1.6.6 Quantum Coherence 24
1.7 Quantum Entanglement 27
    References 29

CHAPTER 2
Introduction to Quantum Computing 31

2.1 Quantum Computing 31
    2.1.1 The Power of Quantum Computing 31
    2.1.2 DiVincenzo Criteria 32
    2.1.3 Applications of a Quantum Computer 33
2.2 Quantum Information Processing 34
    2.2.1 Single-Qubit Gates 34
    2.2.2 Two-Qubit Gates 37
    2.2.3 Gate Fidelity 39
    2.2.4 Quantum Circuits 41
    2.2.5 Quantum Algorithm 42
    2.2.6 Quantum Error Correction 46
    2.2.7 Quantum Supremacy 46

2.3	Quantum Computing Platforms	47
	2.3.1 Ions	47
	2.3.2 Neutral Atoms	47
	2.3.3 Semiconductor Qubits	48
2.4	Challenges and Opportunities in Quantum Computing	51
	2.4.1 Technical Challenges of Scaling	52
	2.4.2 Skillsets for Quantum Hardware Engineers	54
	References	55

CHAPTER 3
Superconducting Qubits — 57

3.1	Introduction to Superconductivity	57
	3.1.1 Cooper Pairs	61
	3.1.2 Types of Superconductors	64
	3.1.3 Josephson Junction	66
3.2	Superconducting Qubit	69
	3.2.1 Artificial Atom	69
	3.2.2 Cooper Pair Box	71
	3.2.3 Transmon Qubit	73
	3.2.4 Qubit Coherence Time Scales	78
3.3	Qubit Control and Readout	86
	3.3.1 Qubit Control	86
	3.3.2 Qubit Readout	93
	3.3.3 Spectroscopic Measurement Methods	112
	3.3.4 Equivalent Circuit of Qubit-Cavity Coupling	113
	3.3.5 Qubit Control and Readout in Practice	115
3.4	Two-Qubit System	116
	3.4.1 Dispersive Two-Qubit Interactions	120
3.5	Calibration of Single-Qubit Operations	121
3.6	Testing the Performance of a Quantum Processor	122
	References	122

CHAPTER 4
Microwave Systems — 125

4.1	A Brief History of Microwave Engineering	125
4.2	Microwave Engineering	127
4.3	Microwave System Analysis	128
	4.3.1 Microwave Link	129
	4.3.2 Signal Degradation Factors	131
	4.3.3 Nonlinear Effects in Microwave Systems	149
	4.3.4 Dynamic Range	161
	4.3.5 Error Vector Magnitude	162
	References	164

CHAPTER 5
Microwave Components — 167

- 5.1 Microwave Component Analysis — 167
  - 5.1.1 Tools for the Analysis of Microwave Components — 167
- 5.2 Signal Generation — 177
- 5.3 Signal Transmission — 183
  - 5.3.1 TEM-Mode Transmission Lines — 185
  - 5.3.2 Non-TEM Transmission Lines — 196
  - 5.3.3 Types of Transmission Lines — 200
  - 5.3.4 Microwave Connectors — 207
- 5.4 Signal Processing — 209
  - 5.4.1 Performance Specifications of Microwave Components — 209
  - 5.4.2 Amplitude Manipulation — 209
  - 5.4.3 Frequency Manipulation — 229
  - 5.4.4 Phase Manipulation — 244
- 5.5 Signal Detection — 247
  - 5.5.1 Homodyne Detection — 250
  - 5.5.2 Superheterodyne Detection — 251
  - 5.5.3 Direct RF Sampling — 251
  - References — 255

CHAPTER 6
Principles of Electromagnetic Compatibility — 257

- 6.1 Signal Integrity — 257
- 6.2 EMC — 258
  - 6.2.1 Interaction of an Electronic System with the Environment — 259
  - 6.2.2 Interference Sources — 260
  - 6.2.3 Crosstalk — 261
- 6.3 Electromagnetic Shielding — 261
  - 6.3.1 Shielding Effectiveness — 262
  - 6.3.2 Effect of Penetration in the Shield — 265
  - 6.3.3 Effect of Grounding on the Shield — 267
  - 6.3.4 Shielding Techniques for Qubits — 268
- 6.4 Filtering — 271
- 6.5 Grounding — 272
  - 6.5.1 Grounding of Wires — 274
  - References — 275

CHAPTER 7
Control Hardware for Superconducting Qubits — 277

- 7.1 High-Level Description of the Setup — 277
- 7.2 Low-Level Description of the Setup — 279
- 7.3 Room-Temperature Setup — 279
  - 7.3.1 Signal Generation — 279
  - 7.3.2 Signal Processing — 285

	7.3.3 Further Considerations at Room Temperature	288
7.4	Cryogenic Setup	290
	7.4.1 DC Wiring	290
	7.4.2 Heat Loads	290
	7.4.3 Noise-Suppression Techniques	291
	7.4.4 Signal Amplification	307
	7.4.5 Further Considerations for the Cryogenic Setup	310
	7.4.6 Vibrational Damping and Decoupling	311
7.5	Future Directions for the Control Hardware	314
	7.5.1 Integrated Hardware	314
	7.5.2 Cryogenic CMOS Chips	314
7.6	Automation and Control of the Experiment	315
	References	316

CHAPTER 8
Principles of Cryogenics — 317

8.1	Introduction	317
8.2	An Overview of Cooling Techniques	318
8.3	Cryogens	319
	8.3.1 Cooling Mechanisms	319
	8.3.2 Storage and Transportation of Cryogens	321
8.4	Mechanical Refrigerators	322
8.5	Pumped-Helium Refrigerators	325
8.6	Dilution Refrigerator	326
	8.6.1 Principle of Dilution Refrigeration	327
	8.6.2 Components of a Dilution Refrigerator	332
	8.6.3 Dry and Wet Dilution Fridge	337
8.7	Cryogenic Thermometry	338
	8.7.1 Cryogenic Temperature Measurements	338
	8.7.2 Installation Considerations	341
8.8	Materials in Cryogenic Environments	346
	References	348

About the Author — 351

Index — 353

Foreword

Despite the almost infinite supply of information at our disposal today, we still need books. The author of a scientific book becomes a curator of knowledge, pulling together that which is important for their subject. Such curation is certainly not possible in the first moment after discovery; a process must take place in which knowledge, which has been gained often haphazardly and with great struggle, acquires some recognizable structure.

Quantum computers have been with us, intellectually speaking, for several decades now. But their physical realization has been slow and painful, and so the pedagogy of quantum computers has been, necessarily, also slow to emerge. It has taken decades for us to know what we need to know.

Alan Salari's book thus arrives, in a pioneering spirit, with an effort to define the intellectual landscape of the subject. It concerns itself with just one part of the vast subject, namely superconducting quantum computing. This is unquestionably a major branch of the subject, one which has a chance of being the dominant branch—maybe all the quantum computers of the future will be of this form. This is by no means certain, but it is best to be prepared for this outcome.

While some researchers 25 years ago understood that the superconducting device could be the realization of a qubit, they had no inkling of what would have to surround these qubits to make the quantum computer function. About 15 years of exploratory work followed. In the last 10 years, a consensus has gradually been appearing of how the superconducting quantum processor will really be built.

The big message from this consensus is that the superconducting quantum computer needs a lot of advanced techniques from cryogenic microwave engineering. This is the premise that drives Alan Salari's book, and is the featured item in the landscape that he depicts. We see here a cornucopia of microwave components, functions, and methodologies. Salari has covered in great detail all the microwave art that is seen as essential for the functioning of the quantum computer, and some art that is perhaps not yet fully in use—the state of system engineering is still in flux in this young field.

Why is such a large collection of high-performance, high-precision microwave techniques needed for the job? As an answer, I would like to offer an overall assessment of what must be done, by quoting from the opening pages of the book by theoretical physicist N. David Mermin, *Quantum Computer Science: An Introduction* (Cambridge University Press, 2007):

> It is tempting to say that a quantum computer is one whose operation is governed by the laws of quantum mechanics. But since the laws of

quantum mechanics govern the behavior of all physical phenomena, this temptation must be resisted. Your laptop operates under the laws of quantum mechanics, but it is not a quantum computer. A quantum computer is one whose operation exploits certain very special transformations of its internal state, whose description is the primary subject of this book. The laws of quantum mechanics allow these peculiar transformations to take place under very carefully controlled conditions.

In a quantum computer the physical systems that encode the individual logical bits must have *no physical interactions whatever that are not under the [almost] complete control of the program.*

I feel that this statement provides an excellent guide to the formidable job that the microwave engineer must perform in the constructing of the quantum computer. The insertion of the word "almost" is mine, but I think that it is in the spirit of the serious error analysis that Mermin provides later in his book, and as summarized also by Salari. Without "almost," there is no engineering job to be done, as complete control is a clearly impossible task and not worth attempting. But "almost complete control," the degree of which is made precise by the quantum theory, is what drives the unheard-of precision and performance requirements that the microwave system must achieve.

I hope that some of you readers, especially from the engineering disciplines, will be infected by the pioneering spirit, and use the knowledge gained to push the frontiers further toward a fully functioning quantum computer.

David P. DiVincenzo
January 2024

Preface

The first quantum revolution, which took place in the early twentieth century, transformed our understanding of the fundamental nature of matter and energy. This profound shift in our knowledge paved the way for developing groundbreaking technologies, such as transistors, atomic clocks, and lasers.

Today, we are witnessing the second quantum revolution, which seeks to push the boundaries of what we can achieve in controlling and manipulating quantum systems. The second quantum revolution aims to achieve unprecedented levels of precision and capabilities, leading to revolutionary applications in areas such as quantum computing, communication, sensing, and simulation. These remarkable advancements are poised to transform various fields, from telecommunications and cryptography to machine learning, finance, materials science, chemistry, and biology.

Quantum computing is one of the most exciting and challenging fields in the realm of quantum technologies. Among the various approaches to quantum computing, superconducting quantum computers have emerged as a promising technology. Superconducting quantum computing is a rapidly developing field that requires a multidisciplinary approach bringing together expertise in quantum physics, microwave engineering, cryogenic engineering, nanofabrication, material science, and software development.

The talent gap in quantum technologies has been challenging, requiring more educational resources, investments, and public awareness. While there has been much focus on training for developing quantum algorithms and software, there is a shortage of dedicated resources for understanding quantum hardware. This shortage motivated me to write this book to bridge the gap in the literature and to provide a resource for physicists and microwave engineers covering the theoretical principles and practical implementation of hardware for superconducting quantum processors.

This book covers a wide range of topics, from the basic principles of quantum mechanics, quantum computing, and superconducting qubits to the analysis of microwave systems and components, principles of cryogenics, and hardware implementation of superconducting quantum processors. From the outset, striking a balance between content that is accessible to both physicists and microwave engineers has been a significant challenge. In many cases, I have had to provide details that may seem obvious to one group but not the other. Despite these difficulties, I have strived to present the material in a clear and engaging way for all readers. Depending on the student's background and the instructor's preference, the material presented in this book can be used in both advanced undergraduate and graduate-level courses.

Chapters 1 and 2 introduce the principles of quantum mechanics and quantum computing, which are essential for understanding the operation of superconducting qubits. Chapter 3 delves into the details of superconducting qubits, discussing superconductivity, the principles of superconducting qubits, and control and readout techniques for single- and two-qubit gates. Chapters 4–6 cover the principles of microwave systems, microwave components, and electromagnetic compatibility, which are crucial for designing and optimizing the performance of superconducting quantum computers. Chapter 7 details cryogenic and room-temperature control hardware, collecting the knowledge from Chapters 1–6 to understand the hardware setup for the control and readout of superconducting qubits. Chapter 8 covers the principles of cryogenics, focusing on the dilution refrigeration technique.

A set of problems with a solutions manual, as well as a set of PowerPoint slides for instructors, is available through the following website: https://quaxys.com/book.

The second quantum revolution promises to unlock entirely new capabilities and drive transformative breakthroughs that were once thought impossible. I sincerely hope this book will provide a valuable resource for students, researchers, and professionals seeking to understand the intricacies of quantum hardware and, ultimately, help accelerate the development and adoption of this revolutionary technology.

Acknowledgments

I am deeply grateful for the unwavering support and invaluable assistance provided by numerous individuals throughout this book's writing, reviewing, and production. I would especially like to extend my heartfelt appreciation to David DiVincenzo for his overarching insight and thoughtful foreword.

Furthermore, I would like to acknowledge the contributions of the following individuals who meticulously reviewed my work:

Ankur Agrawal (Amazon Web Services);

Marc Almendros (Keysight Technologies);

Andy Cobin (Quantum Microwave);

David DiVincenzo [Forschungszentrum Jülich, Institute for Quantum Information (IQI) RWTH-Aachen];

Danilo Erricolo (University of Illinois, Chicago);

Yale Eckert (Rohde and Schwarz);

Giuseppe Esposito [Italian National Research Council (CNR); Institute for Electromagnetic Sensing of the Environment (IREA)];

Allen Fernandez (Jacobs);

Steven Frankel (Technion-Israel Institute of Technology);

Dean Friesen (Boeing);

Eric Holland (Keysight Technologies);

Alex Khan (ZebraKet);

Saesun Kim (NASA, Jet Propulsion Laboratory);

Piotr Kulczakowicz (Quantum Startup Foundry, University of Maryland, College Park);

James Liu (Intellian Technologies);

Dennis Lucarelli (Error Corp.);

Luke Mauritsen (Montana Instruments, member of the National Quantum Initiative Advisory Committee to the White House);

Brian McMenamin (Rohde and Schwarz);

Kevin Osborne (University of Maryland, College Park);

Matej Stefanac (Yazaki Group Europe);

Sara Sussman (Princeton University);

Oscar Viyuela (MIT, Harvard);

Matija Zesko (McKinsey);

Silvia Zorzetti (Fermi National Laboratory).

I am beyond grateful for their support and willingness to share their knowledge and insight.

Last, I am much obliged to Natalie McGregor, David Michelson, Julia Sayger, and Isabel Fowler, whose expert management and guidance were instrumental in the successful publication of this book.

CHAPTER 1
Introduction to Quantum Physics

> What we observe is not nature itself, but nature exposed to our method of questioning.
>
> —*Werner Heisenberg*

This chapter focuses on the principles of quantum mechanics essential for understanding the fundamental concepts of quantum computing and the operation of superconducting qubits. First, Section 1.1 offers a brief historical overview. Next, the chapter explores the Schrödinger equation and its applications in solving basic yet insightful quantum mechanical problems, such as the quantum harmonic oscillator, a two-level quantum system, and the quantum potential well. The chapter also delves into important quantum mechanics concepts including Heisenberg's uncertainty principle, entanglement, coherence, and quantum measurement.

1.1 A Brief History of Quantum Mechanics

The theory of quantum mechanics emerged in the 1920s to explain the behavior of the smallest particles in nature, such as electrons, atoms, and protons. At the quantum level, the laws of nature vastly differ from those that govern our everyday, macroscopic world. Quantum mechanics introduces concepts such as wave-particle duality, superposition, and entanglement, which have no classical counterparts. Despite these differences, classical mechanics and quantum mechanics are not entirely disconnected. According to the correspondence principle, quantum mechanics converges toward classical mechanics when the quantum numbers involved become large. For instance, in the case of the quantum harmonic oscillator, we can observe classical harmonic oscillator behavior for large quantum numbers, as discussed in Section 1.5.2.3.

The quantization of electromagnetic energy was crucial to the development of quantum mechanics. In the late nineteenth century, classical physics predicted the Rayleigh-Jeans catastrophe, which referred to an ideal black body in thermal equilibrium emitting an unbounded amount of energy as the wavelength approached the ultraviolet range. Planck resolved this issue by proposing that electromagnetic radiation could only be emitted or absorbed in discrete energy packets [1]. Each packet carries an energy E, which is given by $E = hf$, where h represents Planck's constant, and f represents the frequency of the electromagnetic wave. Planck's quantum of electromagnetic radiation was later named the photon by Einstein, who postulated that a photon was a physical particle. This assumption laid the groundwork for the wave-particle duality of electromagnetic radiation, whose wave nature had already been proved by Maxwell.

In 1913, Bohr made a significant breakthrough with his atomic model, which helped explain the Rydberg formula for the spectral emission lines of the hydrogen atom. To develop this model, Bohr proposed that the angular momentum L of the atom was quantized and could only take on values that were integer multiples of $\hbar = h/2\pi$, written as $L = n\hbar$ [2]. The discovery of the angular momentum's quantization was further reinforced by the Stern-Gerlach experiment of 1922, which demonstrated that electron spin was also quantized. This experiment provided additional evidence for the quantized nature of atomic particles [3].

In 1924, Louis de Broglie proposed the theory that matter has both particle and wave-like behavior, holding that particles with mass m and momentum p also behave as a wave with wavelength $\lambda = h/p$. Figure 1.1(a) shows the wave-like nature of an electron. These matter waves are a manifestation of the wave-particle duality in quantum mechanics. De Broglie connected the requirement that the angular momentum of an electron be an integer multiple of $\hbar$ to a standing-wave condition. In this interpretation, the electron is represented by a wave, and the number of wavelengths that fit along the circle of the electron's orbit must be an integer. The dual nature of matter was experimentally verified by Davisson and Germer, who showed that electrons scattered by the surface of a nickel's crystal displayed a diffraction pattern, a wave phenomenon. Later, in 1957, Claus Jönsson demonstrated the wave behavior of electrons using the double-slit experiment depicted in Figure 1.1(b).

In 1925, Werner Heisenberg, Max Born, and Pascual Jordan introduced matrix mechanics, which was the first formulation of quantum mechanics that was conceptually self-contained and mathematically consistent. In their theory, the physical properties of particles were represented as matrices that evolve over time. Around the same time, Erwin Schrödinger developed wave mechanics to describe the state of a quantum system by a wave function. Born interpreted the wave function as the probability density of finding a particle in a particular position at a particular time. Both matrix and wave mechanics were successful in reproducing the outcomes of previous theories, including Bohr's atomic model.

In 1928, Paul Dirac accomplished a major achievement by developing a theory combining quantum mechanics and special relativity. The Dirac equation is

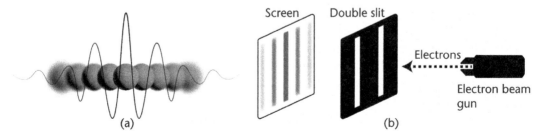

Figure 1.1 (a) The wave-like nature of electrons in quantum mechanics allows them to exist simultaneously at several locations. (b) Double slit experiment with electrons: An electron gun generates a stream of electrons, passing through the slits and interfering with one another to create an interference pattern on the screen.

a relativistic wave equation that describes the behavior of spin-1/2 particles, such as electrons and quarks, in their free form or under electromagnetic interactions. This equation was instrumental in explaining the electron gyromagnetic ratio and calculating the atom's spontaneous emission coefficients.

Through the end of the 1940s, advancements in microwave technology enabled more accurate measurements of the electron magnetic moment and the shift of the hydrogen atom's levels, known as the Lamb shift. These measurements revealed inconsistencies that Dirac's theory could not explain. To address these issues, Shinichiro Tomonaga, Julian Schwinger, and Richard Feynman developed quantum electrodynamics (QED), a relativistic quantum field theory of electrodynamics that describes how light and matter interact. QED, the first theory to fully reconcile special relativity with quantum mechanics, has made incredibly precise predictions, such as the electron's anomalous magnetic moment and the Lamb shift of hydrogen's energy levels. Feynman called QED "the jewel of physics" for its remarkable accuracy [4]. As discussed in subsequent chapters, the foundation of quantum computing is the interaction between light and matter, where laser light or microwave radiation interacts with atoms, ions, or quantum circuits to perform quantum computation.

A fundamental grasp of linear algebra and differential equations is essential to comprehending the material presented in this chapter.

1.2 Quantum Versus Classical Mechanics

The key distinction between classical and quantum mechanics lies in the inherently probabilistic nature of quantum mechanics. While statistical approaches are employed in classical mechanics due to incomplete knowledge of all degrees of freedom in a system, in quantum mechanics, the probabilistic nature is inherent and not just a reflection of our limited information about the system. In other words, the statistical nature of a quantum system persists even if we have precise information about all the system's degrees of freedom. This probabilistic nature is encoded in the wave function, which is the solution to the Schrödinger equation. The wave function enables us to calculate probabilistic quantities, such as expectation values, and variances of physically measurable quantities, such as position and momentum.

The principle of reversibility of physics laws is crucial to understanding the differences between the probabilistic characteristics of quantum and classical physics. This principle asserts that if a physical system is allowed to evolve from its initial state and then reversed precisely for the same amount of time, the system will return to its initial state. However, in classical systems where probability is involved, the system may not revert to its initial state. If we possess complete information about all the degrees of freedom, we can reverse every evolution of a classical system.

In quantum systems, the laws of physics are reversible, even in the presence of probability. Unlike in classical systems, this probability is inherent and does not result from a lack of information about the system. However, there is a crucial condition for the reversibility of the laws of quantum mechanics. Reversibility is possible only if we do not perform any measurement on the quantum system, allowing it to travel backward in time without any intervention. Following this condition, the quantum system will eventually return to its initial state. The measurement concept is another key distinction between classical and quantum physics, as discussed in Section 1.6. When a quantum system is measured, its state changes, leading to the loss of information through wave-function collapse. Therefore, we must avoid measuring the quantum system while checking for its reversibility.

Let us consider an example to better understand the fundamental concept of reversibility. Consider a system oscillating between two states, such as a binary pendulum with states L and R. If we start from state R, the system will return to R after two oscillations (RLRLR). However, suppose there is a small probability that the system randomly stays at state L. In this case, if the system stays at L during the first oscillation, we get RLLRL. When we run the system backward in time, assuming that the event of staying at L does not occur, we get LRLRL and end up in L state instead of returning to the initial state R. This example shows that the reversibility of physical laws does not apply to classical systems when probabilities are involved.

The conservation of information is crucial for the reversibility of physical processes. Chapter 2 covers quantum algorithms, which consist of a series of time-evolving quantum processes known as quantum gates. The reversibility of these gates is a critical requirement in quantum computing, ensuring that any operation on a normalized quantum state preserves the sum of probabilities of all possible outcomes to 1 to prevent the loss of quantum information. As discussed in Section 2.1.2, the reversibility in quantum mechanics is mathematically expressed through unitary transformations.

1.3 Schrödinger Equation

The laws of physics describe how the state of a system evolves over time. The state of a system can be characterized by various physical quantities, such as position, momentum, and energy. If the initial state of a system is known, we can use the appropriate laws to predict its future state. For instance, in classical mechanics, Newton's second law can be used to predict the position $x(t)$ of a particle at any time, given the initial conditions and the force acting on the particle. From the position $x(t)$, other quantities like velocity $v = dx/dt$, momentum $p = mv$, and kinetic energy $T = mv^2/2$ can be determined.

The Schrödinger equation provides a mathematical framework for quantum systems to calculate measurable quantities such as position, momentum, and energy. In the Schrödinger picture, the state of a quantum system is represented by the wave function $\Psi(x, t)$, which encodes the probabilities of all the possible outcomes. The wave function is used to calculate measurable physical quantities, such as position and momentum, which are known as observables. The Schrödinger equation is written as follows [5]:

$$\left[-\frac{\hbar^2}{2m}\frac{\partial^2}{\partial x^2} + V(x,t)\right]\Psi(x,t) = i\hbar\frac{\partial}{\partial t}\Psi(x,t) \qquad (1.1)$$

where m represents the mass of the particle, $V(x, t)$ is the potential, and $\hbar$ is the reduced Planck's constant. Alternatively, (1.1) can also be written in terms of Hamiltonian operator $H = -(\hbar^2/2m)(\partial^2/\partial x^2) + V(x,t)$ as $H\Psi(x,t) = i\hbar(\partial/\partial t)\Psi(x,t)$. Initially, the interpretation of the wave function $\Psi(x, t)$ was unclear. It was later proposed by Born that the square of the absolute value of the wave function, $|\Psi(x, t)|^2$, can be interpreted as the probability density of finding the particle at a particular location at a particular time.

The probability that a particle with wave function $\Psi(x, t)$ will be found between positions x and $x + dx$ at the time t is given by $|\Psi(x, t)|^2 dx$.

The wave function is more than just a probability function. Its phase is critical in enabling interference effects, a wave phenomenon, so referring to it as a wave function rather than a probability function is appropriate. Interference effects are essential for quantum algorithms to arrive at the correct answer, as illustrated in Section 2.2.5. It is important to note that the absolute phase of a wave function cannot be measured in experiments. However, the relative phase of two probability amplitudes can be determined by observing interference effects.

For a state to be physically realizable, the wave function must be square-integrable to ensure that the probability of finding a particle at any possible location adds up to one. This condition is known as normalization and is expressed mathematically as $\int_{-\infty}^{\infty} |\Psi(x,t)|^2 dx = 1$. In addition to being normalized, the wave function needs to fulfill the following mathematical properties [4]:

- The wave function must be continuous and single-valued everywhere.
- Derivative of the wave function is always continuous and single-valued except at a boundary where the potential $V(x, t)$ is infinite.
- The wave function must be normalizable. This requires $|\Psi(x, t)|^2$ to approach zero fast enough as $x \to \pm\infty$ so that $\int_{-\infty}^{\infty} |\Psi(x,t)|^2 dx$ remains finite.

1.4 The Machinery of Quantum Calculations

Before delving further into the mathematics of quantum mechanics, it is insightful to examine the structure of a quantum mechanical calculation. Compared to classical equations such as Maxwell's or Newton's equations, the way the Schrödinger equation is applied to calculations on observables, such as position and momentum, is different due to the probabilistic nature of quantum mechanics.

Figure 1.2 illustrates the process of quantum calculations. The first step is solving the Schrödinger equation to obtain the wave function. To accomplish this, we need to know the initial conditions, boundary conditions, and the potential function $V(x, t)$. It is worth noting that the Schrödinger equation can only be solved analytically for a limited number of cases. Therefore, numerical methods are commonly used to solve the Schrödinger equation for real-world problems. It is essential to mention that a wave function has a mathematical structure that falls under the concept of vector space.

In quantum mechanics, observables such as position and momentum are represented by operators. The operator and wave function is used to calculate the expectation value of the observable, indicating the most likely outcome when

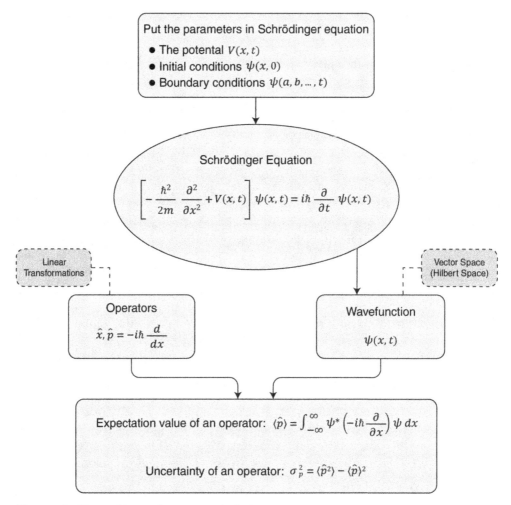

Figure 1.2 The machinery of quantum calculations.

we measure it. Due to the probabilistic nature of quantum mechanics, we work with expectation values instead of the deterministic outcomes known in classical mechanics. Additionally, we can calculate the uncertainty, which is the variance of the probability distribution. We will discuss these concepts in more depth later in Sections 1.6.2 and 1.6.3.

1.5 Solving the Schrödinger Equation

This section covers the time-independent Schrödinger equation and standard Hamiltonians, such as the potential well, potential barrier, and free particle. This section concludes with an examination of the quantum harmonic oscillator and two-level quantum systems, also known as spin 1/2 or qubits, essential for understanding the superconducting qubits.

1.5.1 Time-Independent Schrödinger Equation

If the potential function in the Schrödinger equation is time-independent, the wave function can be expressed as a product of a spatial function $\psi(x)$ and a temporal function $f(t)$ [i.e., $\Psi(x, t) = \psi(x)f(t)$]. This method is called the separation of variables. If we substitute the solution $\Psi(x, t)$ back in the Schrödinger equation (1.1), we obtain $\partial\Psi/\partial t = \psi(df/dt)$, $(\partial^2\Psi/\partial x^2) = (d^2\psi/dx^2)f$. Now the Schrödinger equation reads

$$i\hbar \frac{1}{f}\frac{df}{dt} = -\frac{\hbar^2}{2m}\frac{d^2}{dx^2} + V \tag{1.2}$$

The left side of the (1.2) is only a function of t, and the right is only a function of x. This is only possible if both sides are constant, where we call this constant E, then $i\hbar(1/f)(df/dt) = E$, where E represents the energy of the system. The solution of this ordinary differential equation is straightforward $f(t) = Ce^{-iEt/\hbar}$

We can absorb the constant C into ψ and only use the $e^{-iEt/\hbar}$ term as the time solution. If we substitute $i\hbar(1/f)(df/dt) = E$ back in (1.2) and multiply both sides by ψ, we obtain the time-independent Schrödinger equation with Hamiltonian H

$$H\psi = E\psi \tag{1.3}$$

The $\psi(x)$ function can be obtained by solving the time-independent Schrödinger equation. The states given by solving (1.3) are called stationary states. The general solution can be written as a product of the stationary states $\psi(x)$ and the time evolution function as $\Psi(x, t) = \psi(x)e^{-iEt/\hbar}$.

Note that the probability density is time-independent and depends only on the stationary states $|\Psi(x, t)|^2 = \Psi^* \Psi = \psi^* e^{+iEt/\hbar}\psi e^{-iEt/\hbar} = |\psi(x)|^2$.

The general solution to the Schrödinger equation consists of a linear combination of stationary states, where the time evolution of each state is dependent on the energy of that state, as can be seen in the exponent of the temporal function $\Psi(x, t) = \sum_{n=1}^{\infty} c_n \psi_n(x) e^{-iE_n t/\hbar}$.

This profound result reveals that a quantum state can be expressed as a linear combination, or superposition, of multiple states. In other words, a quantum system can exist in multiple states simultaneously. The concept of superposition is one of the fundamental features of quantum mechanics, which makes quantum computing so powerful. The ability to exist in multiple states simultaneously allows quantum computers to perform computations in parallel, as explored in Chapter 2.

The relative phase of the wave functions plays a crucial role in interference effects, which is utilized in quantum algorithms to obtain the correct answer. To illustrate this, let us consider an example. As we will see later, the state of a quantum system can be represented using the Dirac notation, also known as bra-ket notation. Suppose that the state $|\psi_1\rangle$ is an equal superposition of the $|0\rangle$ and $|1\rangle$ qubit states $|\psi_1\rangle = (|0\rangle + |1\rangle)/\sqrt{2}$. And suppose that $|\psi_2\rangle$ is also an equal superposition of $|0\rangle$ and $|1\rangle$ states but that the two states are out-of-phase $|\psi_2\rangle = (|0\rangle - |1\rangle)/\sqrt{2}$. When we examine the states $|\psi_1\rangle$ and $|\psi_2\rangle$ individually, the probabilities of both states being in |0⟩ and |1⟩ are equal. As a result, the relative phase between the superimposed states does not matter. However, if we create a new superposition $|\psi_3\rangle = (|\psi_1\rangle + |\psi_2\rangle)/\sqrt{2}$, where the system has an equal probability of being in either $|\psi_1\rangle$ or $|\psi_2\rangle$ state, the relative phase between the two superposed states becomes crucial. This is because the phases of the two states can interfere constructively or destructively, leading to different outcomes when the system is measured. We can see that by substituting equations $|\psi_1\rangle$ and $|\psi_2\rangle$ into superposition $|\psi_3\rangle = (|\psi_1\rangle + |\psi_2\rangle)/\sqrt{2}$, we get $|\psi_3\rangle = |0\rangle$. This means that the $|1\rangle$ state vanished due to the destructive interference between the two states.

Section 1.5.2 examines some standard Hamiltonians to gain more insight into the solutions of the Schrödinger equation.

1.5.2 Standard Hamiltonians

While the Schrödinger equation can only be solved analytically for a few cases, these solutions offer valuable insights. This section explores the solutions for the free particle, potential well, potential barrier, harmonic oscillator, and two-level quantum systems.

1.5.2.1 Free Particle

For a free particle, the potential $V(x,t) = 0$. The Schrödinger equation reads $-(\hbar^2/2m)(d^2\psi/dx^2) = E\psi$ or $d^2\psi/dx^2 = -k^2\psi$, where $k \equiv \sqrt{2mE}/\hbar$. The general solution of this equation consists of two traveling waves $\psi(x) = Ae^{ikx} + Be^{-ikx}$. The first term in $\psi(x) = Ae^{ikx} + Be^{-ikx}$ represents a wave traveling to the right, and the second represents a wave traveling to the left. There is no constraint on the values of k since no boundary conditions are associated with the free particle. The energy of a free particle can take any value, and its energy spectrum is continuous. A superposition of plane waves describes the time-dependent solution $\Psi(x,t) = \psi(x)e^{-iEt/\hbar} = Ae^{ik(x-(\hbar k/2m)t)} + Be^{-ik(x+(\hbar k/2m)t)}$. The wave function of a free particle cannot be normalized,

$$\int_{-\infty}^{+\infty} \Psi_k^* \Psi_k \, dx = |A|^2 \int_{-\infty}^{+\infty} 1 \, dx = |A|^2 (\infty) \tag{1.4}$$

This proves that there is no such thing as a free particle with definite energy. However, we can obtain a meaningful physical interpretation by using a linear combination of solutions, as described by the superposition principle, where the resulting wave function can be normalized

$$\Psi(x,t) = \frac{1}{\sqrt{2\pi}} \int_{-\infty}^{+\infty} \phi(k) e^{i\left(kx - \frac{\hbar k^2}{2m}t\right)} dk \tag{1.5}$$

Equation (1.5) bears a resemblance to the Fourier transform, where the coefficients $\phi(k)$ can be computed using the inverse Fourier transform

$$\phi(k) = \frac{1}{\sqrt{2\pi}} \int_{-\infty}^{+\infty} \Psi(x,0) e^{-ikx}\, dx \tag{1.6}$$

Equation (1.6) represents a range of energies that correspond to each value of k. This results in a localized and coherent superposition of plane waves, which is known as a wave packet [see Figure 1.3(a)]. The size and shape of the wave packet change as it propagates [Figure 1.3(b)].

As shown in Figure 1.3(a), two velocities are associated with a wave packet, the group velocity v_g and the phase velocity v_p. The phase velocity is related to the velocity of the ripples given by $v_{\text{phase}} = \omega/k$. The group velocity is related to the velocity of the envelope. The group velocity can be approximated as follows if the k has a small spread[1] $v_{\text{group}} = d\omega/dk$. The dispersion relation ($\omega - k$ relation) for a free particle is given by $\omega = (\hbar k^2/2m)$. The relations for the dispersion, group, and phase velocity lead to $v_{\text{classical}} = v_{\text{group}} = 2v_{\text{phase}}$.

1.5.2.2 Potential Well

Let us dive into the fascinating topic of potential wells. In classical mechanics, potentials can be engineered to manipulate the movement of particles, creating bound states or controlling scattering states. The same holds for quantum particles, but the possibilities are even broader in the quantum realm. For instance, unlike classical particles, a quantum particle can tunnel through a finite potential barrier, meaning that it can escape even if it lacks the required energy. This tunneling phenomenon allows for exciting applications in nanoelectronics, quantum computing, and even nuclear fusion.

An infinite square well with the following potential function is shown in Figure 1.4(a)

$$V(x) = \begin{cases} 0, & 0 \leq x \leq L \\ \infty, & \text{otherwise} \end{cases} \tag{1.7}$$

This potential function results in the following wave function after solving the Schrödinger equation [6] $\psi_n(x) = \sqrt{2/L} \sin(n\pi x/L)$ with energy levels $E_n = n^2\pi^2\hbar^2/2mL^2$.

In an infinite square potential well, wave functions exhibit a simple sinusoidal pattern: standing waves. These states are known as bound states since each state's energy is lower than that of the potential well, which, in this case, is infinite. This property makes potential wells ideal for quantum mechanical systems, as they can create stable and controllable environments for experiments in fields like quantum information processing and quantum simulations. One of the most exciting applications of potential wells is particle trapping. For example, single electrons

---

1. For a large range of k's, the wave packet's shape changes rapidly since different wave components travel at a different speed; therefore, we're not dealing anymore with a well-defined group of waves with a well-defined velocity.

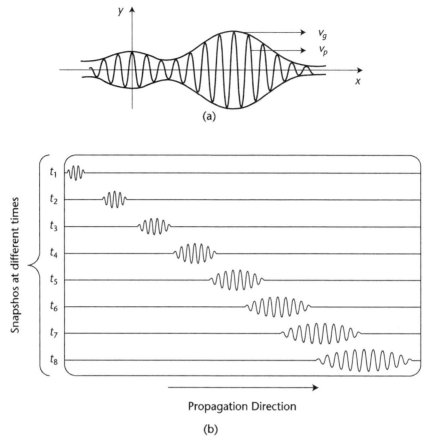

Figure 1.3 (a) A wave packet, where the envelope travels with a group velocity, whereas the ripples travel with the phase velocity, and (b) the spread of a wave packet; the shape and size of the wave packet change as it propagates.

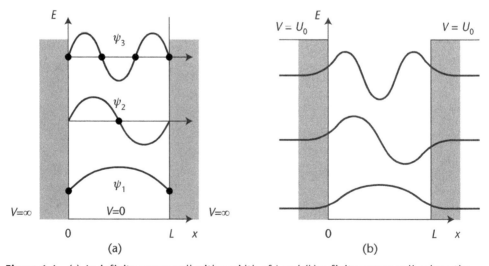

Figure 1.4 (a) An infinite square well with a width of L and (b) a finite square well, where the wave functions decay exponentially outside the well.

can be trapped in a spin qubit, or neutral atoms can be trapped using an optical potential well.

An interesting phenomenon called tunneling can occur in a finite square well, as shown in Figure 1.4(b). This is a characteristic of quantum mechanics, and it means that even when a particle's energy is lower than the surrounding potential, there is still a nonzero probability of finding the particle outside the well.

Quantum tunneling has many applications in semiconductor devices, including tunneling diodes and imaging techniques, such as the scanning tunneling microscope (STM). In addition, the tunneling effect is used for the spin-charge conversion readout of spin qubits. Josephson junctions, critical components in superconducting qubits, also exhibit quantum tunneling, as discussed in Chapter 3.

1.5.2.3 Harmonic Oscillator

The quantum harmonic oscillator is a fundamental concept in understanding and modeling quantum mechanical systems. It is used to model a wide range of physical systems, including electromagnetic waves, resonators, qubits, and molecular potentials near their minimum. In classical mechanics, a harmonic oscillator is a mass m attached to a spring with a force constant k. The motion of the system is governed by Hooke's law $F = -kx = md^2x/dt^2$, where the potential energy is given by $V(x) = kx^2/2$.

The equivalent problem in quantum mechanics is to solve the Schrödinger equation for the following potential $V(x) = m\omega^2 x^2/2$, where ω is the angular frequency of the oscillator. To determine the system's energy levels, it is sufficient to solve the time-independent Schrödinger equation.

$$\left(-\frac{\hbar^2}{2m}\frac{d^2}{dx^2} + \frac{1}{2}m\omega^2 x^2\right)\psi = E\psi \tag{1.8}$$

We do not delve into the mathematical details of the solution, as it is readily available in practically all introductory books on quantum mechanics [5]. However, it is crucial to understand each state's energy levels and the wave functions of the harmonic oscillator. The solution to the Schrödinger equation is given by

$$\Psi_n(x) = \sqrt{\frac{1}{2^n n!}} \cdot \left(\frac{m\omega}{\pi\hbar}\right)^{\frac{1}{4}} \cdot e^{-\frac{m\omega x^2}{2\hbar}} \cdot H_n\left(\sqrt{\frac{m\omega}{\hbar}}x\right) \tag{1.9}$$

$H_n(\sqrt{m\omega/\hbar}x)$ are Hermite polynomials. The Hermite polynomials and, consequently, the wave function is an odd function for odd values of n and an even function for even values of n, as shown in Figure 1.5(a). The energy of each state is given by

$$E_n = \left(n + \frac{1}{2}\right)\hbar\omega, \quad \text{for } n = 0,1,2,\ldots \tag{1.10}$$

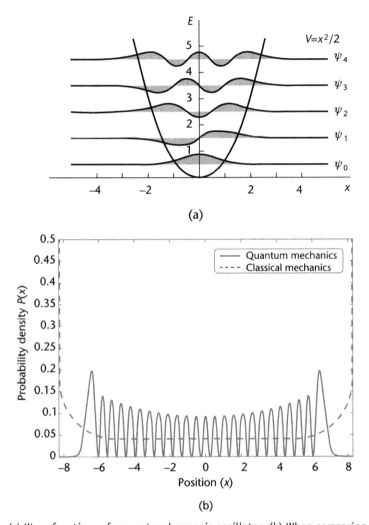

Figure 1.5 (a) Wave functions of a quantum harmonic oscillator. (b) When comparing a quantum harmonic oscillator to a classical one, we observe that the behavior of the quantum system becomes increasingly like that of a classical harmonic oscillator as the quantum numbers become large. The wave function of a quantum harmonic oscillator for n = 100 indicates this convergence.

There are several notable differences between quantum and classical harmonic oscillators, including the following.

- The lowest allowed harmonic oscillator energy $E_0 = \hbar\omega/2$ is not zero.
- The probability of finding the particle outside the classically allowed range is not zero [see Figure 1.5(a)].

Moreover, it is worth noting the following.

- The ground state of a harmonic oscillator is a minimum uncertainty state (see Section 1.6.3 for the concept of uncertainty), resulting in a Gaussian position-space wave function.

- All energy eigenstates of the harmonic oscillator obey the virial theorem. This theorem relates the kinetic energy T and potential energy V of the system through their expectation values, [i.e., $\langle T \rangle$ and $\langle V \rangle$, respectively]. (See Section 1.6.2 for the concept of expectation value.) $\langle T \rangle = \langle V \rangle = E_n/2$.

Despite all the differences between the classical and quantum harmonic oscillators, they behave similarly for large quantum numbers, as illustrated in Figure 1.5(b). This is an example of the correspondence principle introduced in Section 1.1.

1.5.2.4 Ladder Operators

Sometimes, instead of being interested in the wave function corresponding to each state, we seek to understand the transitions between these states. For instance, in a harmonic oscillator, we can assign a number to each state, such as $|0\rangle$, $|1\rangle$, $|2\rangle$, …, making a ladder of states [7]. Then, we can define ladder operators, which consist of raising and lowering operators acting on a state, moving it up or down the ladder of states. Before delving into this concept, let's have a quick look at the Dirac notation. The bra-ket or Dirac notation is widely used in quantum mechanics. The bra vector $\langle A|$ and the ket vector $|A\rangle$ are defined as $\langle A| = (A_1^* \; A_2^* \; \ldots \; A_N^*)$ and $|A\rangle = (A_1 \; A_2 \; \ldots \; A_N)^T$, respectively, where A_N^* is the complex conjugate of A_N. In quantum mechanics, the dagger symbol (†) denotes the adjoint or Hermitian conjugate of an operator or a state vector. The Hermitian conjugate, which is the transposed conjugate of a bra, is the corresponding ket, and vice versa (i.e., $|A\rangle^\dagger = \langle A|$ and $\langle A|^\dagger = |A\rangle$). Moreover, $|\phi\rangle = A|\psi\rangle$ if and only if $\langle \phi| = \langle \psi|A^\dagger$.

Now, let us explore how the ladder operators are constructed. Recall that the Hamiltonian of a harmonic oscillator is given by

$$\hat{H} = \frac{\hat{p}^2}{2m} + \frac{1}{2}m\omega^2 \hat{x}^2 \tag{1.11}$$

With a clever variable change, the Hamiltonian can be written in a more straightforward and useful form. We define $\hat{a}$ and $\hat{a}^\dagger$ operators as follows:

$$\hat{a} = \sqrt{\frac{m\omega}{2\hbar}}\left(\hat{x} + \frac{i}{m\omega}\hat{p}\right) \tag{1.12}$$

$$\hat{a}^\dagger = \sqrt{\frac{m\omega}{2\hbar}}\left(\hat{x} - \frac{i}{m\omega}\hat{p}\right) \tag{1.13}$$

where a and $a^\dagger$ operators are called lowering and raising operators, respectively. The Hamiltonian in (1.11) can be expressed in terms of raising and lowering operators $H = (\hbar\omega/2)(a + a^\dagger)$. The following relations show how these operators act on the state $|n\rangle$

$$\hat{a}|n\rangle = \sqrt{n}|n-1\rangle, \; \hat{a}^\dagger|n\rangle = \sqrt{n+1}|n+1\rangle \tag{1.14}$$

So when $\hat{a}$ acts on the state $|n\rangle$, it brings it down by one level to the state $|n-1\rangle$. For the raising operator, it brings the state $|n\rangle$ one level up to $|n+1\rangle$. Section 1.5.2.4 use the concept of raising and lowering operators to show the transition between the ground and excited states of the qubit. The commutation relation between the raising and lowering operators is $[a, a^\dagger] = 1$.

The matrix form of the raising and lowering operators can be obtained by using $a^\dagger_{ij} = \langle \psi_i | a^\dagger | \psi_j \rangle$ and $a_{ij} = \langle \psi_i | a | \psi_j \rangle$. The eigenvectors ψ_i are those of the harmonic oscillator expressed in the number basis. The matrix form of the eigenstates of the harmonic oscillator $|n\rangle$ is given by $|0\rangle = [1\ 0\ \ldots\ 0]$, $|1\rangle = [0\ 1\ \ldots\ 0]$, and so on.

1.5.2.5 Quantum LC Oscillator

The concept of a quantum harmonic oscillator can be used to model a quantum LC oscillator, as depicted in Figure 1.6(a) [8, 9]. A qubit is an example of a quantum anharmonic oscillator due to the nonlinear term in its potential, as discussed in Chapter 3. Despite this, the harmonic oscillator can still accurately approximate the anharmonic oscillator at low energy levels, as demonstrated in Figure 1.6(b).

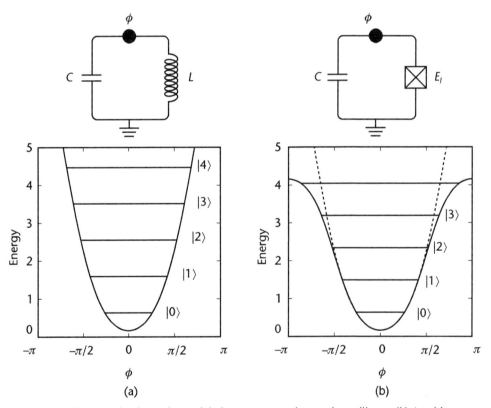

Figure 1.6 (a) An LC circuit can be modeled as a quantum harmonic oscillator. (b) A qubit consisting of a nonlinear element called a Josephson junction parallel to a capacitor is an anharmonic oscillator. However, a qubit can be approximated as a harmonic oscillator at low energy levels.

1.5 Solving the Schrödinger Equation

The Hamiltonian of an LC oscillator has a similar form to the quantum harmonic oscillator introduced

$$H = \frac{1}{2C}Q^2 + \frac{\varphi^2}{2L} \tag{1.15}$$

where C and L are the capacitance and inductance of the circuit, Q is the charge on the capacitor plates, and φ is the flux through the inductor. Table 1.1 compares the parameters of quantum electrical and mechanical oscillators.

The coordinate Φ and its conjugate momentum Q can be promoted to quantum operators obeying the canonical commutation relation $[\hat{Q},\hat{\Phi}] = -i\hbar$ [8, 9]. We can rewrite the Hamiltonian in terms of lowering and raising operators.

$$H = \frac{\hbar\Omega}{2}\left(a^\dagger a + aa^\dagger\right) = \hbar\Omega\left(a^\dagger a + \frac{1}{2}\right) \tag{1.16}$$

where $\Omega = 1/\sqrt{LC}$ is the resonant frequency and the raising $a^\dagger$, and lowering a operators can be expressed as

$$a = +i\frac{1}{\sqrt{2L\hbar\Omega}}\hat{\Phi} + \frac{1}{\sqrt{2C\hbar\Omega}}\hat{Q}, \; a^\dagger = -i\frac{1}{\sqrt{2L\hbar\Omega}}\hat{\Phi} + \frac{1}{\sqrt{2C\hbar\Omega}}\hat{Q} \tag{1.17}$$

which obeys $[a, a^\dagger] = 1$. Chapter 3 discusses the two coupled LC oscillators to model the qubit-cavity coupling.

Table 1.1 Comparison of Quantum Electrical and Mechanical Oscillators

Electrical Oscillator	Mechanical Oscillator
Φ	x
Q	p
C	m
L^{-1}	k
$V = \dot{\Phi} = \frac{Q}{C}$	$v = \dot{x} = \frac{p}{m}$
$I = \dot{Q} = -\frac{1}{L}\Phi$	$F = \dot{p} = -kx$
$\omega_0 = \frac{1}{\sqrt{LC}}$	$\omega_0 = \sqrt{\frac{k}{m}}$
$H = \frac{Q^2}{2C} + \frac{\Phi^2}{2L}$	$H = \frac{p^2}{2m} + \frac{kx^2}{2}$
$\left[\hat{\Phi},\hat{Q}\right] = i\hbar$	$\left[\hat{x},\hat{p}\right] = i\hbar$

1.5.2.6 Two-Level Quantum System: Qubit

Assume an isolated two-level quantum system, also called a qubit, with the ground state $|g\rangle \equiv [1\ 0]^T$, the excited state $|e\rangle \equiv [0\ 1]^T$, and energy difference of $E = E_e - E_g = \hbar\omega_{eg}$. As the ground and excited states are orthogonal, we have $\langle e|g\rangle = 0$ and $\langle g|g\rangle = \langle e|e\rangle = 1$. The Hamiltonian of the qubit $\hat{H}_q$ is given by

$$\hat{H}_q = \frac{\hbar\omega_{eg}}{2}(|e\rangle\langle e| - |g\rangle\langle g|) = \frac{\hbar\omega_{eg}}{2}\sigma_z \tag{1.18}$$

where $\sigma_z = \begin{pmatrix} 1 & 0 \\ 0 & -1 \end{pmatrix}$ is the Pauli-Z matrix.

When this Hamiltonian acts on the ground state $|g\rangle$, it returns $\hat{H}_q|g\rangle = (\hbar\omega_{eg}/2)(|e\rangle\langle e|g\rangle - |g\rangle\langle g|g\rangle) = -(\hbar\omega_{eg}/2)|g\rangle$, and when it acts on the excited state $|e\rangle$, it returns $\hat{H}_q|e\rangle = (\hbar\omega_{eg}/2)|e\rangle$.

Like a quantum harmonic oscillator, lowering and raising operators can be defined for a qubit. The raising operator $\sigma^+ = |e\rangle\langle g|$ acting on the ground state brings the qubit to the excited state, and the lowering operator $\sigma^- = |g\rangle\langle e|$ acting on the excited state brings the qubit to the ground state. For a qubit, it is more common to represent the ground and excited states by $|0\rangle$ and $|1\rangle$, respectively.

1.5.2.7 Bloch Sphere

A two-level system's pure state $|\psi\rangle$ can be expressed as a superposition of basis vectors $|0\rangle \equiv [1\ 0]^T$ and $|1\rangle \equiv [0\ 1]^T$ (i.e., $|\psi\rangle = \alpha|0\rangle + \beta|1\rangle$), where the coefficients α and β are complex numbers and normalized $|\alpha|^2 + |\beta|^2 = 1$.

The Bloch sphere shown in Figure 1.7(a) is a helpful tool to visualize the state space of a qubit and has no equivalent for more than one qubit. Figure 1.7 compares the state space of a classical bit and a qubit. While a bit has only two possible states, a qubit can have any arbitrary state on the Bloch sphere.

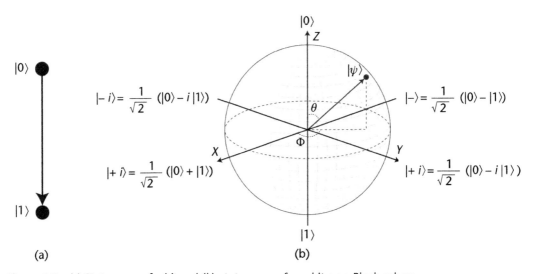

Figure 1.7 (a) State space of a bit and (b) state space of a qubit on a Bloch sphere.

We can associate a vector with the state of a qubit on a Bloch sphere. This vector extends from the origin to the surface of the Bloch sphere, which has a unit radius. Its orientation is determined by the ϕ and θ angles shown in Figure 1.7(a). Therefore, an arbitrary state $|\psi\rangle$ on a Bloch sphere can be expressed as follows [5]:

$$|\psi\rangle = \cos\left(\frac{\theta}{2}\right)|0\rangle + e^{i\phi}\sin\left(\frac{\theta}{2}\right)|1\rangle \text{ where } 0 \leq \theta \leq \pi \text{ and } 0 \leq \phi \leq 2\pi \qquad (1.19)$$

The relative phase between the basis states is encoded in the ϕ angle. As shown in Figure 1.7(a), the intersection of the Bloch sphere with the x and y axes represents various superposition states, while the intersection with the z-axis represents the $|0\rangle$ and $|1\rangle$ states.

1.6 Quantum Measurement

Early in the development of quantum mechanics, physicists and philosophers struggled with misunderstandings stemming from the theory's inherent probabilistic character, measurement process, and interpretation of measured data. In discussing the many interpretations of quantum mechanics and their consequences beyond the scope of this book, we cover one widely accepted and experimentally confirmed interpretation: the Copenhagen interpretation, associated with Bohr and proponents of this interpretation. According to this interpretation, the act of measurement results in the collapse of the wave function. Through measurement, we force the system to choose one state among all possible states and remain in that state afterward. The collapse of the wave function implies that passive observation is impossible, as the act of observing may alter what is being observed.

1.6.1 Collapse of the Wave Function

The concept of quantum measurement and the collapse of the wave function is depicted in Figure 1.8. The amplitude of the $|\Psi|^2$ function at each point represents the probability of finding the particle at that point. As shown in Figure 1.8, when a measurement is taken, the wave function collapses, and the particle is found at one single position and remains in that position afterward. Therefore, if we measure the particle after the collapse of the wave function, we will find the particle at the same position, regardless of how many times we repeat the measurement.

The question is how to perform a meaningful measurement of position or any other observable if the measurement results might change every time owing to the system's probabilistic character. The solution is to determine the expectation value of an observable. For instance, when measuring the position, we can prepare n identical quantum systems in the same state Ψ, as shown in Figure 1.8, and measure their positions. By taking an average of this position data, we can determine the position where we expect to find the particle the most. Naturally, the more measurement points we have, the closer we can get to the expectation value. Section 1.6.2 discusses how to calculate the expectation value mathematically.

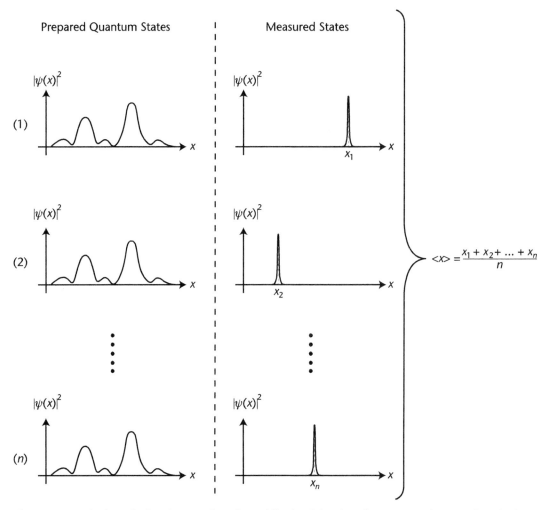

Figure 1.8 Left plots display the wave function, while the right plots demonstrate the wave function's collapse following measurement. This illustration presents the concept of expectation value in quantum measurements. We prepare 'n' identical quantum systems in the same quantum state and conduct measurements on each system. As a result of the probabilistic nature of quantum mechanics, every measurement yields a different outcome. The expectation value is defined as the average of these measured values. Consequently, we anticipate the particle to predominantly occupy the position ⟨x⟩.

The measured position values depicted in Figure 1.8 exhibit a certain distribution. This distribution can be created by plotting the measured positions on the x-axis and the frequency of each position on the y-axis. The standard deviation measures how much the values in a distribution deviate or spread out from the mean value, and the variance is referred to as the uncertainty of the position operator. Sections 1.6.2 and 1.6.3 cover the mathematical calculation of the expectation and uncertainty values.

1.6.2 Expectation Value

Recall from Section 1.4 that an operator in quantum mechanics represents an observable. The expectation value of an observable A with the operator $\hat{A}$ is given by

1.6 Quantum Measurement

$$\langle A \rangle = \int_{-\infty}^{\infty} \psi^*(x,t) \hat{A} \psi(x,t) \, dx \tag{1.20}$$

If $\Phi(p, t)$ represents the momentum-space wave function, we have, in general,

$$\langle Q(x,p,t) \rangle = \begin{cases} \int \Psi^* \hat{Q}\left(x, \frac{\hbar}{i}\frac{\partial}{\partial x}, t\right) \Psi \, dx, & \text{in position space;} \\ \int \Phi^* \hat{Q}\left(-\frac{\hbar}{i}\frac{\partial}{\partial p}, p, t\right) \Phi \, dp, & \text{in momentum space.} \end{cases} \tag{1.21}$$

As shown in Section 1.5.2.2, the wave functions are $\psi_n(x) = \sqrt{2/L}\sin(n\pi x/L)$. We can use (1.22) to calculate the expectation value of a particle's position in an infinite square well in the ground state ($n = 1$).

$$\langle x \rangle = \int_0^L \sqrt{\frac{2}{L}} \sin\left(\frac{\pi x}{L}\right) x \sqrt{\frac{2}{L}} \sin\left(\frac{\pi x}{L}\right) dx = \frac{L}{2} \tag{1.22}$$

where the $u = \pi x/L$ change of variable is used to solve the above integral. It shows that we usually expect to find the particle in the center of the well for the ground state.

Let us calculate the momentum expectation value for a particle in its ground state in an infinite square well.

$$\langle p \rangle = \int_0^L \left(\sqrt{\frac{2}{L}} \sin\left(\frac{\pi x}{L}\right)\right)\left(-i\hbar \frac{d}{dx}\left(\sqrt{\frac{2}{L}} \sin\left(\frac{\pi x}{L}\right)\right)\right) dx = -i\hbar \frac{L}{\pi}(0) = 0 \tag{1.23}$$

The expectation value of the momentum of a particle in an infinite square well is zero due to the standing waves that fit in the well. Standing waves oscillate in a place like a guitar string, meaning there is no net energy flow or momentum in either direction.

To investigate whether there is a connection between expectation values in quantum mechanics and classical mechanics, we can turn to Ehrenfest's theorem. This theorem states that, similar to Newton's second law, the time rate of change of momentum (i.e., the force) is equal to the gradient of the potential $d\langle p \rangle/dt = \langle -\partial V/\partial x \rangle$.

1.6.3 Variance or Uncertainty

The probabilistic nature of quantum systems results in measured values having a certain distribution. The standard deviation is used to express the spread of this distribution. The standard deviation of the position operator is defined as $\sigma_x = \sqrt{\langle x^2 \rangle - \langle x \rangle^2}$ [5]. The variance or σ_x^2 is referred to as the uncertainty of an operator. Note that every calculation of the standard deviation σ_x or expectation value is done for a specific eigenstate.

Let us calculate the standard deviation for the position of a particle in the ground state of an infinite square well

$$\langle x^2 \rangle = \int_0^L \left(\sqrt{\frac{2}{L}} \sin\left(\frac{\pi x}{L}\right) \right) x^2 \left(\sqrt{\frac{2}{L}} \sin\left(\frac{\pi x}{L}\right) \right) dx \qquad (1.24)$$

By using the change of variable $u = \pi x/L$ we obtain $\langle x^2 \rangle = 0.283 L^2$, and therefore $\sigma_x = \sqrt{\langle x^2 \rangle - \langle x \rangle^2} = \sqrt{0.283 L^2 - (0.5L)^2} = 0.182 L$.

The standard deviation of the momentum of a particle in the ground state of an infinite square well is calculated as follows $\sigma_p = \sqrt{(h^2/4L^2) - (0)^2} = h/2L$.

Section 1.6.4 explores the relationship between the uncertainties in position and momentum, as described by Heisenberg's uncertainty principle.

1.6.4 Uncertainty Principle

Recall from Section 1.1 that wave-particle duality is a significant property of quantum mechanics. Both wave and particle pictures can represent quantum physics processes equally well. However, there is a limit to the application of the particle picture to quantum systems dictated by the uncertainty principle. This implies that as one concept, such as the position, approaches the particle picture, the other, such as the velocity, becomes fuzzy and distant from its initial classical meaning. Let us consider an example to clarify this.

As explained in Section 1.5.2.1, a wave packet can be used to visualize the position of a quantum particle spread over the distance Δx, implying that the particle can be anywhere in the Δx range. The electron's velocity is associated with the group velocity of the wave packet, as mentioned earlier. Due to dispersion, the size and shape of the wave packet change as it moves; therefore, the group velocity cannot be precisely defined. As a result, the velocity also spreads over a range defined by Δv, which implies that the particle's velocity has a value that belongs to the Δv range. It is important to note that this uncertainty should be viewed as a fundamental aspect of the electron rather than a sign that the wave picture is not applicable.

The wave packet can be written as a superposition of plane sinusoidal waves with wavelengths centered around λ_0 with a spread of $\Delta \lambda$ [6]. The number of crests that fall within the Δx boundary is approximately $n = \Delta x / \lambda_0$. Outside the Δx boundary, the superposition of the waves must have a canceling effect. This is possible if, and only if, some of the component waves contain at least $n + 1$ waves that fall in the critical range. This phenomenon results in an interference pattern, where the waves constructively add up within the Δx boundary and destructively outside it. This is expressed mathematically as [10]

$$\frac{\Delta x}{\lambda_0 - \Delta \lambda} \geq n + 1 \qquad (1.25)$$

If we substitute $n = \Delta x / \lambda_0$ in (1.25), then we obtain $(\Delta x / \lambda_0 - \Delta \lambda) \geq (\Delta x / \lambda_0) + 1$. After doing some algebra, we obtain $\Delta x \Delta \lambda / \lambda_0 (\lambda_0 - \Delta \lambda) \geq 1$. Since $\lambda_0 \gg \Delta \lambda$, the following approximation can be applied $\lambda_0(\lambda_0 - \Delta\lambda) \approx \lambda_0^2$, so we arrive at $\Delta x \Delta\lambda / \lambda_0^2 \geq 1$.

De Broglie's formula relates the group velocity of the wave packet v_g to the wavelength and mass m of the particle $v_g = h/m\lambda_0$. The range of velocities that determine the spreading of the packet is given by $\Delta v_g = h\Delta\lambda/m\lambda_0^2$

We use the definition of the momentum $\Delta p = m\Delta v_g$ and $\Delta v_g = h\Delta\lambda/m\lambda_0^2$ to obtain Heisenberg's uncertainty principle $\Delta x \Delta p \geq h$. Another form of the uncertainty principle relates to the standard deviation of the position σ_x and the standard deviation of momentum σ_p. This form is expressed as $\sigma_x \sigma_p \geq \hbar/2$.

Any usage of the terms "position" and "velocity" that exceeds the precision provided by $\sigma_x \sigma_p \geq \hbar/2$ is equally meaningless to the use of undefined words. This means that if we want to apply classical concepts such as position and velocity to a quantum system, we must acknowledge that there is a limit to how precisely these variables can be defined in the quantum context. As one variable becomes more particle-like, the other necessarily becomes less so.

Let us take a numerical example to understand how the uncertainty relation works. Suppose we can determine the position of an electron with an uncertainty of 10 nm. Using the uncertainty principle, we can find the minimum uncertainty in the electron's velocity as $m\Delta v \geq h/\Delta x$. The mass of an electron is $m_e = 9.10 \times 10^{-31}$ kg. Therefore, the minimum uncertainty in the velocity of the electron is $\Delta v \geq 72.74 \times 10^3$ m/s.

As discussed in Section 1.6.4.1, the uncertainty principle implies that no quantum state can simultaneously be a position and a momentum eigenstate. In other words, the wave function for position and momentum cannot be identical. However, this is not the case when the observables are compatible or commutable.

1.6.4.1 Uncertainty Principle and Fourier Transform

The Fourier transform is a powerful mathematical tool that finds applications in diverse fields of science and engineering. It represents a method for reconstructing a function by combining sine waves of different frequencies with appropriate amplitude and phase.

The uncertainty principle can be expressed using the language of the Fourier transform. A wave function can be represented either as a position-space wave function $\psi(x)$ or a momentum-space wave function $\phi(p)$.[2]

$$\psi(x) = \frac{1}{\sqrt{2\pi\hbar}} \int_{-\infty}^{\infty} e^{\frac{ipx}{\hbar}} \phi(p)\, dp \qquad (1.26)$$

$$\phi(p) = \frac{1}{\sqrt{2\pi\hbar}} \int_{-\infty}^{\infty} e^{-\frac{ipx}{\hbar}} \psi(x)\, dx \qquad (1.27)$$

It is clear that (1.26) and (1.27) (apart from the factors of $\hbar$) constitute a Fourier transform pair between the "position-space" and momentum space wave

---

2. This is analogous to the familiar time-domain and frequency-domain representations of a signal.

functions.[3] The Fourier transform bridges these two representations, allowing us to transform a wave function from one space to another.

The scaling property of the Fourier transform $\psi(ax) \overset{FT}{\leftrightarrow} (1/|a|)\phi(p/a)$ shows that if we scale the ψ wave function by the a factor ($a > 1$ corresponding to compression and $a < 1$ corresponding to the expansion of the wave function), then its Fourier transform ϕ is scaled by the $1/a$ factor. (See Figure 1.9.) Expansion (compression) in the position domain corresponds to compression (expansion) in the momentum domain. This behavior is related to the uncertainty principle, whereby the more compressed and less uncertain the position wave function becomes, the more extended and uncertain the momentum wave function becomes.

The only function with the same Fourier transform pair is the Gaussian wave function, which has the minimum uncertainty.

1.6.4.2 Uncertainty Principle and Commutation Relations

The uncertainty principle can also be expressed using a mathematical operator known as the commutator. The commutator of two operators $\hat{A}$ and $\hat{B}$ is defined as follows $[\hat{A}, \hat{B}] \equiv \hat{A}\hat{B} - \hat{B}\hat{A}$.

An operator can be viewed as a matrix, and matrix multiplication is generally noncommutative. Therefore, if we have two operators, $\hat{A}$ and $\hat{B}$, in general, $AB \neq BA$. The uncertainty principle can be expressed in terms of commutators as follows: For two physical quantities to be concurrently observable, their operator representations must commute (that is, $[\hat{A}, \hat{B}] = 0$). In other words, the order of measurement does not matter, and the result of the second measurement does

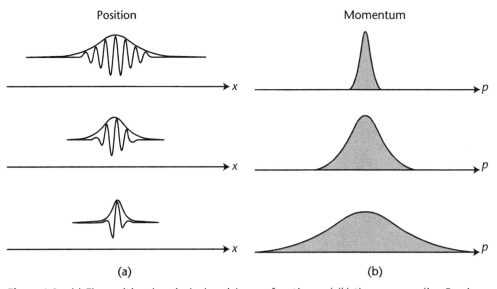

Figure 1.9 (a) The position domain (x-domain) wave function and (b) the corresponding Fourier transform in the momentum domain (p-domain). The scaling property of the Fourier transform shows that an expansion of the wavefunction in the x-domain results in the compression in the p-domain.

---

3. The probability of obtaining momentum in the range dp is $|\Phi(p, t)|^2 dp$.

not depend on the outcome of the first measurement (i.e., the two measurements commute). However, if two operators do not commute, the order of measurement matters, and the result of the second measurement depends on the outcome of the first measurement.

The commutation relation between the position and momentum is given by $[\hat{x},\hat{p}] = i\hbar$.

Since position and momentum do not commute, there is uncertainty when both observables are measured simultaneously.

The uncertainty principle in its most general form for two operators $\hat{A}$ and $\hat{B}$ can be expressed as follows $\sigma_A^2 \sigma_B^2 \geq \left(1/2i\langle[\hat{A},\hat{B}]\rangle\right)^2$.

So, if we substitute $[\hat{x},\hat{p}] = i\hbar$ in $\sigma_A^2 \sigma_B^2 \geq \left(1/2i\langle[\hat{A},\hat{B}]\rangle\right)^2$, we can easily obtain the uncertainty relation for the position and momentum.

1.6.5 Heisenberg's Picture

In the Schrödinger picture, the operators are constant, and the states evolve over time. For the time-independent Hamiltonian, the time evolution operator $U(t) = e^{-i\hat{H}t/\hbar}$ acts on an initial state $\psi(t_1)$, to generate $|\psi(t_2)\rangle = U(t)|\psi(t_1)\rangle$.

In Heisenberg's picture, the operators (observables) are time-dependent, but the state vectors are time-independent. The two pictures only differ by a basis change with respect to the time dependency.

In Heisenberg's equation of motion, the time derivative of the expectation value of an observable, $\hat{Q}(x,p,t)$ is a measure of how fast the system is changing. It can be shown that the time derivative of the expectation value of the observable $\hat{Q}$ is given by

$$\frac{d}{dt}\langle\hat{Q}\rangle = \frac{i}{\hbar}\langle[\hat{H},\hat{Q}]\rangle + \left\langle\frac{\partial\hat{Q}}{\partial t}\right\rangle \quad (1.28)$$

In the normal case, the operator $\hat{Q}$ is not explicitly dependent on t, so $\partial\hat{Q}/\partial t = 0$. Therefore, the operator's commutator with the Hamiltonian $[\hat{H},\hat{Q}]$ determines the time rate of change of the expectation value of $\hat{Q}$ given by $d\langle\hat{Q}\rangle/dt$. If $\hat{Q}$ commutes with $\hat{H}$ (i.e., $[\hat{H},\hat{Q}] = 0$), then $\langle Q\rangle$ must be a constant and conserved quantity.

Let us see how Heisenberg's picture helps to understand the uncertainty relation between the time $\hat{t}$ and energy $\hat{E} = i\hbar d/dt$ operators given by $\Delta E \Delta t \geq \hbar/2$ where Δt is the amount of time it takes for the expectation value of the operator $\hat{Q}$ to change by one standard deviation σ_Q, and where σ_Q is given by

$$\sigma_Q = \left|\frac{d\langle\hat{Q}\rangle}{dt}\right|\Delta t \quad (1.29)$$

The Δt entirely relies on the observable $\hat{Q}$ because one observable may vary quickly while another may change slowly. For small values of ΔE, the time rate of change of the observable must be very gradual. If an observable changes rapidly with time, the uncertainty in the energy must be large.

1.6.6 Quantum Coherence

Recall that passive observation is impossible in quantum mechanics, because the act of observing a quantum system can alter what is being observed. Any disturbance caused by the environment acting on a quantum system is equivalent to measuring or observing the system. As a result, when the environment interacts with a quantum system, a process called decoherence occurs, eventually causing the wave function to collapse [11].

Figure 1.10(a) illustrates the decoherence process, where a two-level system in spin-up and spin-down superposition gradually collapses to spin-up due to noise acting on it. As discussed in Chapters 2 and 3, decoherence presents a significant obstacle to quantum computing. Later, Section 2.1.2 demonstrates that minimizing decoherence in a quantum computing system is crucial for prolonging the maintenance of quantum information. Therefore, it is necessary to decouple quantum systems from the environment to minimize the impact of uncontrolled and unwanted measurements on the system.

Now, let's explore the meaning of a coherent superposition of quantum states. Constructive and destructive interference effects occur when two or more waves

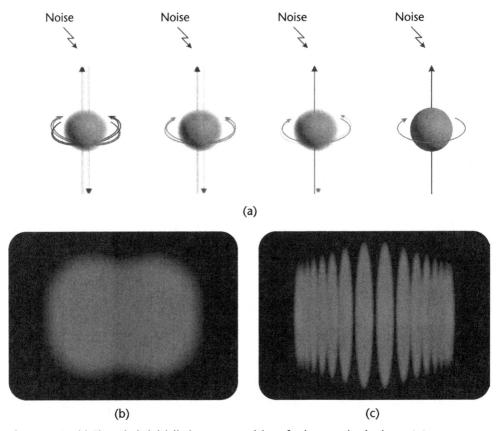

Figure 1.10 (a) The spin is initially in a superposition of spin-up and spin-down states. Decoherence occurs when environment-related disturbances such as noise act on a quantum system resulting in a gradual wave function collapse. (b) Double-slit experiment with an incoherent light source and (c) a coherent light source.

overlap, and their amplitudes add up or cancel out. To observe clear interference patterns, the phases of the interfering waves need to be well-defined and not randomly change with time. Figure 1.10(b, c) demonstrates the interference patterns in a double-slit experiment with incoherent and coherent light sources. No clear interference effect is observed with an incoherent light source.

Coherent superposition of wave functions leads to interesting quantum phenomena such as macroscopic effects in superconductors and lasers. In quantum computing, interference is the underlying principle that leads to constructive interference where the solution lies and destructive interference elsewhere, as explained in Chapter 2.

Chapter 2 further demonstrates that qubits must retain quantum information for a sufficient duration to complete the qubit operations. The coherence time, or the duration that a qubit can remain coherent and retain quantum information, determines the number of quantum gates or qubit operations that can be applied to the qubit. A fault-tolerant quantum computer aims to reach a billion to a trillion quantum gate operations. Single-qubit gates typically take tens of nanoseconds, and two-qubit gates take a few hundred nanoseconds, depending on the hardware. Currently, superconducting qubits have already reached millisecond coherence times, which allows for the execution of a few thousand quantum gates before the qubit loses quantum information.

Nonideal properties in the qubit, such as quasi-particle tunneling, critical current noise, and nonideal control signals and noise in its environment, can result in the loss of quantum information and cause decoherence, as shown in Chapter 3.

1.6.6.1 Quantum Coherence and Energy Uncertainty

Heisenberg's uncertainty principle, $\Delta E \Delta t \geq \hbar/2$, can be utilized to determine the uncertainty in the energy levels of a qubit. As we learned in Section 1.6.5, in Heisenberg's picture, Δt represents the amount of time it takes for the expectation value of the operator $\hat{Q}$ to change by one standard deviation. This operator can be the charge operator corresponding to a charge qubit such as a Cooper pair box (see Chapter 3). The longer it takes for the charge of the qubit to change, the less uncertain the qubit energy will be. This implies that the more resilient the qubit is to noise-induced changes in the charge, the more well-defined the qubit transition frequency will be. Figure 1.11(a) illustrates a qubit with well-defined energy levels, while Figure 1.11(b) displays a noisy qubit with broadened energy levels. Thus, in this case, the transition could occur at a range of frequencies instead of a single sharp frequency.

As illustrated in Figure 1.11(c), the uncertainty in energy levels leads to line broadening, where the transition happens at a range of frequencies with a bandwidth determined as follows: $E = \hbar \omega$, and using Heisenberg's uncertainty principle results in $\Delta E = \hbar \Delta \omega \rightarrow \Delta \omega \propto 1/\Delta t$. The line has a "Lorentz shape" with a sharper peak and wider wings than a Gaussian profile.

$$\phi(\omega) = \frac{\Delta \omega / 2\pi}{\left(\omega - \omega_0\right)^2 + \left(\Delta \omega / 2\pi\right)^2} \quad (1.30)$$

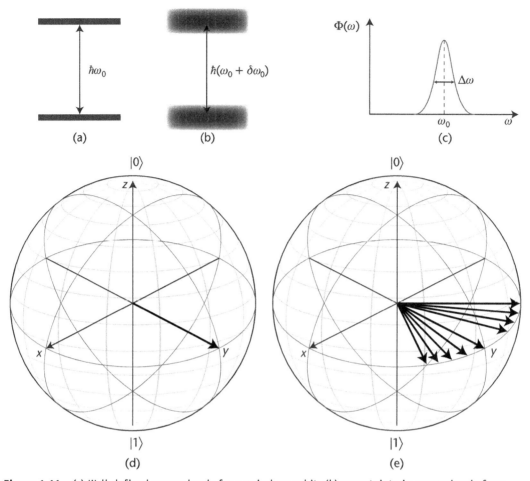

Figure 1.11 (a) Well-defined energy levels for a noiseless qubit, (b) uncertainty in energy levels for a noisy qubit, (c) line broadening due to uncertain energy levels, (d) qubit prepared in a superposition state, and (e) qubit starting to precess around the Z axis at f_{10} frequency. When considering an ensemble of experiments, frequency noise causes the Bloch vectors of each experiment to spread out from the average as some rotate with greater frequency and some with lower.

But how does energy uncertainty lead to decoherence? Let us use the Bloch sphere to explain this concept. Assume that the qubit is in the superposition state and located at $|+\rangle = (|0\rangle + |1\rangle)/\sqrt{2}$ on the Bloch sphere, as shown in Figure 1.11(d). When the qubit is allowed to evolve over time, it starts to precess around the Z-axis at transition frequency f_{01}. Therefore, the qubit phase ϕ evolves in time according to the equation $\phi(t) = 2\pi f_{10} t$. Any noise-induced fluctuations in frequency δf will create a phase offset $\delta \phi = \delta f t$.[4] Consequently, frequency noise ($\delta f(t)$) over time results in random phase fluctuations ($\delta \phi$), leading to decoherence.

---

[4]. If the frequency offset is constant, it is possible to eliminate the phase offset by measuring the bare frequency.

1.6.6.2 Quantum Coherence and Density Matrix

According to decoherence theories, a nonisolated quantum system develops entanglement with other systems, including measurement apparatuses, eventually resulting in decoherence. The density matrix is a helpful tool for describing decoherence processes. If we define the state vector $|\psi\rangle$ as a superposition of the $|0\rangle$ and $|1\rangle$ states $|\psi\rangle = (|0\rangle + e^{i\theta}|1\rangle)/\sqrt{2}$. Then the density matrix ρ is obtained by

$$\rho = |\psi\rangle\langle\psi| = \begin{pmatrix} \frac{1}{2} & \frac{1}{2}e^{-i\theta} \\ \frac{1}{2}e^{i\theta} & \frac{1}{2} \end{pmatrix} \quad (1.31)$$

As shown in Section 1.6.6.1, as a quantum system interacts with the environment, the wave function's phase undergoes random changes resulting in decoherence. The off-diagonal entries gradually average to zero with time, and the density matrix reduces to $\begin{pmatrix} 1/2 & 0 \\ 0 & 1/2 \end{pmatrix}$, where no wave function in a superposition state can be associated with this density matrix. So, this density matrix has been reduced to a classical probability function.

The density matrix of a single-qubit state can be expanded in the following form

$$\begin{aligned} \rho &= \frac{1}{2}\left(\hat{I} + \vec{r}\cdot\hat{\vec{\sigma}}\right) = \frac{1}{2}\hat{I} + \frac{1}{2}r_x\hat{\sigma}_x + \frac{1}{2}r_y\hat{\sigma}_y + \frac{1}{2}r_z\hat{\sigma}_z \\ &= \frac{1}{2}\begin{bmatrix} 1+r_z & r_x - ir_y \\ r_x + ir_y & 1-r_z \end{bmatrix} \end{aligned} \quad (1.32)$$

where $\hat{I}$ is the identity matrix, and $\hat{\sigma}_x, \hat{\sigma}_y, \hat{\sigma}_z$ the X, Y, Z Pauli matrices, respectively

$$\sigma_x \equiv X \equiv \begin{pmatrix} 0 & 1 \\ 1 & 0 \end{pmatrix}, \sigma_y \equiv Y \equiv \begin{pmatrix} 0 & -i \\ -i & 0 \end{pmatrix},$$
$$\sigma_z \equiv Z \equiv \begin{pmatrix} 1 & 0 \\ 0 & -1 \end{pmatrix}, I \equiv \begin{pmatrix} 1 & 0 \\ 0 & 1 \end{pmatrix} \quad (1.33)$$

The r_x, r_y and r_z coefficients correspond to the components of the Bloch vector $\vec{r}$.

The time evolution of the density matrix can be expressed using Heisenberg's equation $i\hbar\dot{\rho} = [H, \rho]$. Quantum state tomography allows the determination of the density matrix elements through measurements.

1.7 Quantum Entanglement

So far, we have focused on single-particle quantum systems that exhibit superposition. This property is unique to quantum mechanics and has no classical equivalent. The intriguing quantum mechanical phenomenon of entanglement emerges in two-particle quantum systems. Like superposition, entanglement has no classical counterpart. Chapter 2 explains how these two phenomena underlie the computational power of quantum computers. We include this information at this point in the text to maintain the logical flow of the contents.

In a classical system with two interacting particles, such as two masses coupled with a spring, it is possible to determine the state of each mass (i.e., its position) by solving a system of coupled differential equations. However, in a quantum mechanical system with two interacting entangled particles, it is impossible to know the state of any individual entangled particle. Therefore, the entire system must be described as a whole.

Assume two qubits represented by $|\psi_1\rangle = a_1|0\rangle + b_1|1\rangle$ and $|\psi_2\rangle = a_2|0\rangle + b_2|1\rangle$. The state of these two unentangled qubits $|\Psi\rangle$ is given by the tensor product $\otimes$ of the individual single-qubit states [12]

$$|\Psi\rangle = |\psi_1\rangle \otimes |\psi_2\rangle = |\psi_1 \psi_2\rangle \tag{1.34}$$

If we replace each state in (1.34), we obtain

$$\begin{aligned}|\Psi\rangle &= (a_1|0\rangle + b_1|1\rangle) \otimes (a_2|0\rangle + b_2|1\rangle) \\ &= a_1 a_2 |00\rangle + a_1 b_2 |01\rangle + a_2 b_1 |10\rangle + b_1 b_2 |11\rangle\end{aligned} \tag{1.35}$$

So, a two-qubit state is a four-dimensional state vector expressed as

$$|\psi_1 \psi_2\rangle = \begin{pmatrix} a_1 a_2 \\ a_1 b_2 \\ a_2 b_1 \\ b_1 b_2 \end{pmatrix} \tag{1.36}$$

Not all two-qubit states can be represented as a tensor product of individual states. Consider the following state called the Bell state

$$|\Phi^+\rangle = \frac{1}{\sqrt{2}} (|00\rangle + |11\rangle) \tag{1.37}$$

Let us try to decompose the Bell state into a tensor product of two single-qubit states as follows

$$\begin{aligned}|\Phi^+\rangle &= (a_1|0\rangle + b_1|1\rangle) \otimes (a_2|0\rangle + b_2|1\rangle) \\ &= a_1 a_2 |00\rangle + a_1 b_2 |01\rangle + a_2 b_1 |10\rangle + b_1 b_2 |11\rangle\end{aligned} \tag{1.38}$$

From (1.37) and (1.38), we need to have $a_1 b_2 = a_2 b_1 = 0$, which implies that either $a_1 a_2 = 0$ or $b_1 b_2 = 0$. But, from (1.37), both $a_1 a_2 \neq 0$ and $b_1 b_2 \neq 0$. This proves that the Bell state $|\Phi^+\rangle$ cannot be decomposed into the tensor product of two single-qubit states. The only way to describe the Bell state is by describing the whole state, which is the property of an entangled state.

When measuring the state of a particle in an entangled state, any changes made to its state will instantaneously affect the state of the second particle. For example, if we measure the spin of one of the entangled particles and find it up, the other particle will immediately change to the spin-down state. Entanglement plays a central role in many quantum algorithms, as seen in Chapter 2.

References

[1] Cohen-Tannoudji, C., B. Diu, and F. Laloë, *Quantum Mechanics* (Second Edition), Vol. 1, Paris, France: Hermann, 1997.

[2] Foot, C. J., *Atomic Physics*, Oxford, U.K.: Oxford University Press, 2005.

[3] Sakurai, J. J., and J. Napolitano, *Modern Quantum Mechanics* (Second Edition), Boston, MA: Addison-Wesley, 2010.

[4] Haroche, S., *Exploring the Quantum: Atoms, Cavities, and Photons*, New York, NY: Oxford University Press, 2006.

[5] Griffiths, D. J., and D. F. Schroeter, *Introduction to Quantum Mechanics*, Cambridge, UK: Cambridge University Press, 2018.

[6] Demtröder, W., *Experimentalphysik 3, Atome, Moleküle, und Festkörper*, Berlin, Germany: Springer, 2010.

[7] Lancaster, T., *Quantum Field Theory for the Gifted Amateur Illustrated Edition*, Oxford, UK: Oxford University Press, 2014.

[8] Girvin, S., *Circuit QED: Superconducting Qubits Coupled to Microwave Photons*, Oxford University Press, 2012.

[9] Vool, U., and M. Devoret, "Introduction to Quantum Electromagnetic Circuits," https://arxiv.org/pdf/1610.03438.pdf.

[10] Heisenberg, W., *Die Physikalischen Prinzipien der Quantentheorie*, Leipzig, Germany: S. Hirzel Verlag, 1930.

[11] Zagoskin, A. M., *Quantum Engineering: Theory and Design Of Quantum Coherent Systems*, Cambridge University Press, 2011.

[12] Lang, S., *Linear Algebra* (Third ed.), New York, NY: Springer, 1987.

CHAPTER 2
Introduction to Quantum Computing

> The measure of greatness in a scientific idea is the extent to which it stimulates thought and opens up new lines of research.
> —*Paul Dirac*

This chapter provides an overview of quantum computing to supply readers with the information they need to implement superconducting qubit hardware.

To this end, the chapter examines a generic quantum computing architecture based on the DiVincenzo criteria and delves into the concepts of quantum gates, quantum algorithms, and quantum error correction. Additionally, the chapter explores various quantum computing platforms and addresses the technical challenges associated with scaling. The chapter concludes by examining the practical skills required for constructing superconducting quantum computers.

2.1 Quantum Computing

Quantum mechanics, with its counterintuitive concepts such as superposition and entanglement, has revolutionized our understanding of nature. Feynman and Manin first conceived of using quantum effects to perform computations with quantum objects [1]. In 1982, Feynman postulated the idea of a universal quantum simulator, which Seth Lloyd later demonstrated in 1996 to be a quantum computer. This means that a quantum computer can be programmed to simulate the behavior of arbitrary quantum systems whose dynamics are determined by local interactions [2].

The following sections examine the architecture of a quantum computer, explore the diverse applications that quantum computers can address, and explain quantum information processing using single-qubit and two-qubit gates.

2.1.1 The Power of Quantum Computing

Chapter 1 introduced the concept of superposition and demonstrated how microscopic objects, such as atoms, can exist in multiple states simultaneously. The superposition principle enables each qubit to exist in two states simultaneously. Consequently, having n qubits provides a quantum computer with 2^n computational resources. This exponential growth in computational resources with the number of qubits distinguishes quantum computers from classical computers, where the computational resources increase linearly with the number of bits.

One may wonder how superposition can be utilized to speed up calculations. Imagine, for instance, maze-goers trying to find their way out. The search process would be much easier if they could create multiple duplicates of themselves. Being in a superposition state allows the search process to be parallelized. Section 2.2.5 examines how quantum parallelism works in a quantum algorithm.

Entanglement is another fascinating quantum mechanical phenomenon that has no classical counterpart. Recall from Chapter 1 that when two or more quantum systems are allowed to interact, they can be in an entangled state where the only way to describe the system is to describe the entire system. In other words, the quantum system with entanglement cannot be decomposed and characterized by describing its individual subsystems. Moreover, the entangled particles, such as entangled qubits, arc correlated in such a way that if we measure one qubit and find it in $|0\rangle$ state, we can be certain that the other qubit will be in $|1\rangle$ state. This correlation between entangled particles can be utilized in quantum computers, quantum communication, and quantum imaging systems. Section 2.2.2 introduces two-qubit entangling gates necessary to build a universal set of quantum gates to perform quantum algorithms.

The superposition and entanglement phenomena, along with the interference effect, can be utilized to create quantum processors that can perform specific computational tasks exponentially faster than classical processors. Quantum computers are primarily helpful for large-scale simulations and optimizations, factoring, and cryptography problems. It should be emphasized that quantum computers are not necessarily faster and more efficient than classical computers in all types of computations. For instance, adding two integers on a quantum computer is incredibly sluggish and inefficient. If you ask a quantum computer what the outcome of 1 + 1 is, it will respond with a probability of 80% that the answer is 2, which is not helpful for everyday use cases.

2.1.2 DiVincenzo Criteria

David DiVincenzo has outlined five requirements that a physical system must satisfy to be a viable platform for quantum computation [3]. According to these criteria, we need the following:

1. We need a scalable physical system with well-characterized two-level quantum systems or qubits. An atom can act as a qubit, where the atom's ground state is represented by $|0\rangle$ and the excited state by $|1\rangle$. Excitation with the right energy causes the atom to go from $|0\rangle$ to $|1\rangle$. The atom eventually relaxes to the ground state and emits a photon. Section 2.3 demonstrates that qubits can be realized using both natural and synthetic physical systems. These qubits can be manifested through various physical properties, such as the charge of a superconducting circuit, the polarization of a photon, or the spin of an electron, all of which can represent the distinct states of the qubit. Multiple qubits make up a quantum register.
2. We need a quantum register that can be initialized to a known value before

the start of computation. The initialization requirement is also essential for quantum error correction.
3. In classical computers, algorithms can be expressed in terms of Boolean functions. Any arbitrary Boolean function can be implemented using a single type of gate known as a universal logic gate. Examples of universal logic gates are NAND and NOR gates. A quantum algorithm is a sequence of unitary transformations $U_1 = e^{iH_1 t/\hbar}$, $U_2 = e^{iH_2 t/\hbar}$ ... with the Hamiltonians $H_1, H_2, \ldots$ that generate these transformations (see Chapter 1 for unitary transformation). The unitary transformations are called quantum logic gates or quantum gates. As the classical logic gates operate on one or two classical bits, the most common quantum gates operate on spaces of one or two qubits. This means that 2×2 or 4×4 unitary matrices can describe single-qubit and two-qubit quantum gates, respectively.

Like classical computers, we need a set of universal quantum gates to which any transformation possible on a qubit can be reduced (i.e., any other unitary operation can be expressed as a finite sequence of gates from the set). A universal set of quantum operations consists of arbitrary single-qubit operations and an entangling two-qubit operation, typically a controlled-NOT (CNOT) gate.
4. Finally, we need to be able to read out the result of a computation.
5. All these tasks must be done with high fidelity and sufficient speed such that decoherence processes do not destroy the quantum information.

Once such a set of criteria is met in a system, it can be used to implement every conceivable quantum algorithm. The implementation of these criteria might differ significantly for each platform. Section 2.3 discusses different quantum computing platforms.

In summary, to implement a quantum computer in a scalable system (a system with many qubits), it is necessary to combine access to qubits (initialization, control, readout) with a high degree of isolation (coherence).

We see that a quantum computer must fulfill contradictory requirements. On the one hand, it must be controlled (i.e., coupled with the environment). On the other hand, it must be decoupled from the environment as it deteriorates the system's quantum behavior due to decoherence. See [3] for two further criteria DiVincenzo also proposed for implementing quantum communication.

2.1.3 Applications of a Quantum Computer

The next question is: what can be accomplished with a quantum computer now that we understand the prerequisites for building one? This section provides a brief overview of the potential applications of quantum computers.

2.1.3.1 Cybersecurity

Rivest–Shamir–Adleman (RSA) is a widely used encryption standard for secure data transmission. The security of RSA encryption relies on the exponential complexity of factoring the product of two large prime numbers. In 1994, Peter Shor showed that using a quantum computer reduces factorization complexity to a polynomial, which makes it possible to break the RSA encryption [4].

Quantum computers could also enhance current security protocols. One potential future application of qubits in cybersecurity is quantum key distribution (QKD). This method utilizes the observer effect[1] to create a theoretically impossible transmission to intercept [5].

2.1.3.2 Chemistry, Material Science, and Drug Development

Many accounts from the history of material discovery and drug development emphasize the importance of luck and serendipity in such discoveries. Due to the vast number of atom combinations and bonding configurations, combinatorics is crucial to creating new and useful compounds. This is where quantum computers can be helpful in simulating complex molecule structures.

2.1.3.3 Banking and Finance

The complexity and diversity of financial factors make it challenging for financial institutions to accurately address the problem at hand, often leading to oversimplified solutions that compromise the accuracy of the results. Quantum computing can solve more complex problems by relaxing the constraints and allowing for more outcomes to be explored.

2.2 Quantum Information Processing

A universal set of quantum gates consisting of single-qubit and two-qubit entangling gates is needed to implement any quantum algorithm, as discussed in the DiVincenzo criteria. We will discuss single- and two-qubit gates in this section. The section concludes with an examination of the architecture of a quantum algorithm, quantum error correction, and quantum supremacy.

2.2.1 Single-Qubit Gates

This section introduces single-qubit gates and how applying these gates to a quantum state leads to rotations of the states on the Bloch sphere.

All single-qubit gates can be represented by 2×2 unitary matrices. These gates receive the state of a qubit as input and generate the qubit's new state as output. Any 2×2 unitary matrix or, in other words, any single-qubit gate, can be written as a linear combination of the identity matrix I and the three Pauli matrices given in (1.33). In other words, the space of 2×2 unitary matrices is spanned by three Pauli matrices and the identity matrix I.

Any single-qubit gate can be expressed as a combination of rotations around the Bloch sphere's x-, y-, and z-axes. The rotation around different axes $R_i(\theta)$ ($i = x, y, z$) is given by

---

1. The observer effect refers to the phenomenon wherein the act of measuring or observing a quantum system leads to an alteration of its quantum state.

2.2 Quantum Information Processing

$$R_x(\theta) \equiv X_\theta \equiv e^{-\frac{i\theta\sigma_x}{2}} = \begin{pmatrix} \cos\frac{\theta}{2} & -i\sin\frac{\theta}{2} \\ -i\sin\frac{\theta}{2} & \cos\frac{\theta}{2} \end{pmatrix} \tag{2.1}$$

$$R_y(\theta) \equiv Y_\theta \equiv e^{-\frac{i\theta\sigma_y}{2}} = \begin{pmatrix} \cos\frac{\theta}{2} & -\sin\frac{\theta}{2} \\ \sin\frac{\theta}{2} & \cos\frac{\theta}{2} \end{pmatrix} \tag{2.2}$$

$$R_z(\theta) \equiv Z_\theta \equiv e^{-\frac{i\theta\sigma_z}{2}} = \begin{pmatrix} e^{-\frac{i\theta}{2}} & 0 \\ 0 & e^{\frac{i\theta}{2}} \end{pmatrix} \tag{2.3}$$

Chapter 3 discusses how the rotation can be done using modulated microwave pulses with appropriate amplitude, phase, and width. Some pulses have a particular name, such as the π-pulse and $\pi/2$-pulse corresponding to the rotation angle on the Bloch sphere. If we start from the $|0\rangle$ state, a π-pulse brings the state from $|0\rangle$ to $|1\rangle$, and a $\pi/2$-pulse creates the superposition.

The identity matrix and Pauli operators can be combined to form a generalized rotation about any axis represented by the unit vector $\hat{n} = (n_x\hat{a}_x + n_y\hat{a}_y + n_z\hat{a}_z)$. The rotation $R_{\hat{n}}(\theta)$ about the $\hat{n}$ axis is given by

$$R_{\hat{n}}(\theta) \equiv \exp\left(-i\theta\hat{n}\cdot\frac{\vec{\sigma}}{2}\right) = \cos\frac{\theta}{2}I - i\sin\frac{\theta}{2}(n_x\sigma_x + n_y\sigma_y + n_z\sigma_z) \tag{2.4}$$

The various types of single-qubit gates are discussed as follows:

- The *bit-flip gate* or *X-gate*, also referred to as the NOT gate, transforms the basis states as $X|0\rangle = |1\rangle$ *and* $X|1\rangle = |0\rangle$. The X-gate is given by $X = \begin{pmatrix} 0 & 1 \\ 1 & 0 \end{pmatrix}$.

  This gate is equivalent to the σ_x Pauli operator, which is a π rotation around the x-axis, as shown in Figure 2.1(a).

  The time it takes for the trajectory traveled on a Bloch sphere to reach the final state is called the gate operation time. Depending on the quantum computing platform and hardware implementation, the gate operation time can vary from tens of nanoseconds for superconducting qubits to microseconds for ion traps [6].

  Also note that the trajectory traveled on a Bloch sphere looks more like Figure 2.1(b) in reality, which is not as smooth and straight as implied in Figure 2.1(a).

- The *Hadamard gate* generates the superposition state, which is the essential starting point for many quantum algorithms. The Hadamard gate H is given by $H = \frac{1}{\sqrt{2}}\begin{pmatrix} 1 & 1 \\ 1 & -1 \end{pmatrix}$.

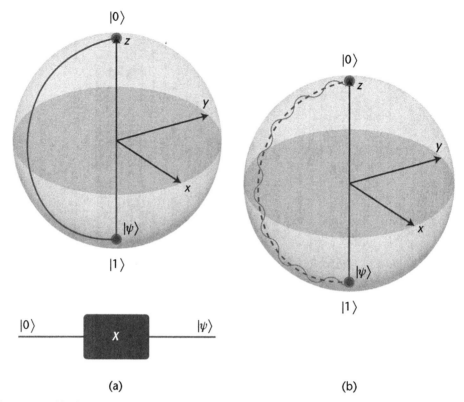

Figure 2.1 (a) The trajectory traveled on a Bloch sphere for an X-gate acting on the |0⟩ state and (b) the trajectory of a real-world X-gate on the Bloch sphere.

When the Hadamard gate acts on the |0⟩ state the output is an equal superposition of |0⟩ and |1⟩ states $H|0\rangle = (|0\rangle + |1\rangle)/\sqrt{2}$. In terms of rotations, a Hadamard gate reflects a π rotation around an axis in the x-z plane of the Bloch sphere that lies halfway between the sphere's poles, as shown in Figure 2.2(a).

- The *phase-flip gate* or *Z-gate* is equivalent to σ_Z Pauli operator with an azimuthal rotation of the qubit's state by π. The Z-gate is given by $Z = \begin{pmatrix} 1 & 0 \\ 0 & -1 \end{pmatrix}$.

  Z-gate acting on a superposition state $(|0\rangle + |1\rangle)/\sqrt{2}$ results in $Z((|0\rangle + |1\rangle)/\sqrt{2}) = (|0\rangle - |1\rangle)/\sqrt{2}$ [i.e., a π rotation in the azimuthal direction bringing the qubit to the opposite side of the x-axis, as shown in Figure 2.2(b)].

- The *phase gate* S can be seen as the square root of the Z-gate, as $S^2 = Z$. This is equivalent to an azimuthal rotation by $\pi/2$ and is given by $S = \begin{pmatrix} 1 & 0 \\ 0 & i \end{pmatrix}$.

- The $\pi/8$-gate or *T-gate* can be seen as the square root of the phase gate, as $T^2 = S$. This is equivalent to an azimuthal rotation by $\pi/4$, which is not what the gate's name conveys. The T-gate is given by $T = \begin{pmatrix} 1 & 0 \\ 0 & \exp\left(\dfrac{i\pi}{4}\right) \end{pmatrix}$.

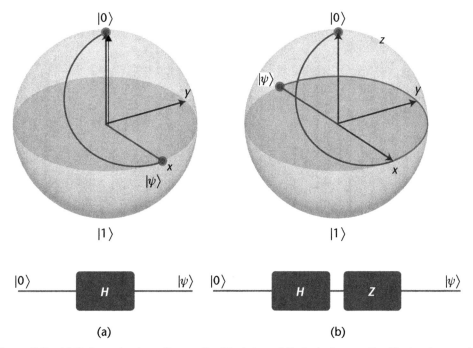

Figure 2.2 (a) Hadamard gate acting on the |0⟩ state and its trajectory on the Bloch sphere and (b) Hadamard gate followed by a Z-gate acting on the |0⟩ state and its trajectory on the Bloch sphere.

2.2.2 Two-Qubit Gates

As mentioned in Section 2.1.2, single-qubit and entangling two-qubit gates are necessary to create a universal set of quantum gates. As discussed in Chapter 1, not all quantum states can be expressed as a tensor product of two individual quantum states. These undecomposable states, such as the Bell state $|\Phi^+\rangle = (|00\rangle + |11\rangle)/\sqrt{2}$, are referred to as entangled states. Entangled states cannot be decomposed into their constituent states, and the only way to describe them is by describing the entire system as a whole [5].

Now let us see how the single-qubit gates act on two-qubit states. Assume Hadamard gate H operates on the first qubit $|\psi_1\rangle$ and phase gate S operates on the second qubit $|\psi_2\rangle$; then we can rewrite their simultaneous effect on both qubits using their tensor product $H|\psi_1\rangle \otimes S|\psi_2\rangle = (H \otimes S)|\psi_1\psi_2\rangle$. The tensor product of the Hadamard H and S gates is given by

$$H \otimes S = \frac{1}{\sqrt{2}} \begin{pmatrix} 1 \times \begin{pmatrix} 1 & 0 \\ 0 & i \end{pmatrix} & 1 \times \begin{pmatrix} 1 & 0 \\ 0 & i \end{pmatrix} \\ 1 \times \begin{pmatrix} 1 & 0 \\ 0 & i \end{pmatrix} & -1 \times \begin{pmatrix} 1 & 0 \\ 0 & i \end{pmatrix} \end{pmatrix} = \frac{1}{\sqrt{2}} \begin{pmatrix} 1 & 0 & 1 & 0 \\ 0 & i & 0 & i \\ 1 & 0 & -1 & 0 \\ 0 & i & 0 & -i \end{pmatrix} \quad (2.5)$$

This 4×4 unitary matrix operates on a four-dimensional state vector given in (1.36).

One class of entangled gates is controlled operations, where one qubit, labeled as the control qubit, determines when the gate acts on the target qubit.

The controlled-NOT or CNOT gate is the canonical two-qubit entangled gate represented by

$$\text{CNOT} = |0\rangle\langle 0| \otimes 1 + |1\rangle\langle 1| \otimes X = \begin{pmatrix} 1 & 0 & 0 & 0 \\ 0 & 1 & 0 & 0 \\ 0 & 0 & 0 & 1 \\ 0 & 0 & 1 & 0 \end{pmatrix} \quad (2.6)$$

The schematic of a CNOT gate is shown in Figure 2.3(a). If the control qubit is $|1\rangle$, the NOT operation is applied to the target qubit. Therefore, the CNOT gate maps $|10\rangle$ to $|11\rangle$ and $|11\rangle$ to $|10\rangle$, while it leaves the states $|00\rangle$ and $|01\rangle$ unchanged.

Another two-qubit gate is the controlled-phase (CPHASE) gate shown in Figure 2.3(b). This gate applies a Z-gate to the target qubit if the control qubit is in the $|1\rangle$ state. The CPHASE gate is defined as follows

$$\text{CPHASE} = |0\rangle\langle 0| \otimes 1 + |1\rangle\langle 1| \otimes Z = \begin{pmatrix} 1 & 0 & 0 & 0 \\ 0 & 1 & 0 & 0 \\ 0 & 0 & 1 & 0 \\ 0 & 0 & 0 & -1 \end{pmatrix} \quad (2.7)$$

Applying the iSWAP gate results in swapping an excitation between the two qubits and adding a phase of $i = e^{i\pi/2}$. This gate can be implemented in a superconducting qubit by turning on a capacitive coupling between the qubits for a certain amount of time. An iSWAP gate is given by

$$i\text{SWAP} = \begin{bmatrix} 1 & 0 & 0 & 0 \\ 0 & 0 & -i & 0 \\ 0 & -i & 0 & 0 \\ 0 & 0 & 0 & 1 \end{bmatrix} \quad (2.8)$$

It can be shown that the CNOT gate and single-qubit gates can be used to build the iSWAP gate [6].

The $\sqrt{i\text{SWAP}}$ gate called the square root-iSWAP gate can generate Bell-like superposition states, such as $|01\rangle + i|10\rangle$. The $\sqrt{i\text{SWAP}}$ gate is given by

$$\sqrt{i\text{SWAP}} = \begin{bmatrix} 1 & 0 & 0 & 0 \\ 0 & 1/\sqrt{2} & -i/\sqrt{2} & 0 \\ 0 & -i/\sqrt{2} & 1/\sqrt{2} & 0 \\ 0 & 0 & 0 & 1 \end{bmatrix} \quad (2.9)$$

Alongside two-qubit gates, three-qubit gates such as Toffoli and CSWAP execute more complex multiple qubits operations. These gates find application in quantum error–correction schemes, among others.

We have reviewed the standard single and two-qubit gates. Certain gate sets are preferred over others depending on the qubit platform, architecture, and the

2.2 Quantum Information Processing

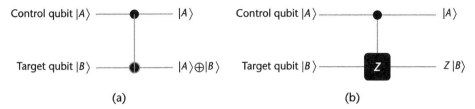

Figure 2.3 Circuit representation for (a) the CNOT gate and (b) the controlled-phase gate.

gate implementation complexity. One choice for a universal set of quantum gates is G = {X_θ, Y_θ, Z_θ, Ph_θ, CNOT} consisting of three single-qubit rotations around the three axes of the Bloch sphere, the $PH_\theta = e^{i\theta}\mathbb{1}$ gate that applies an overall phase θ to a single qubit and a two-qubit CNOT gate. Another possible set of universal quantum gates is G = {H, S, T, CNOT} [6].

2.2.3 Gate Fidelity

Quantum circuits, which are constructed by combining single-qubit and two-qubit gates, enable us to implement quantum algorithms. The computational outcome of quantum algorithms depends on the quality of the implemented quantum gates, which is determined by how closely the resulting state aligns with the ideal quantum gates. Factors such as noise and imperfections in qubit control can introduce a discrepancy between a quantum gate's actual and ideal outputs, quantified by the gate fidelity.

Achieving high-fidelity gates is crucial for implementing quantum-error correction, an essential component in realizing a fault-tolerant quantum computer. Section 2.2.6 demonstrates that for each quantum error–correction algorithm, the gate fidelity needs to surpass a certain fault-tolerance threshold [7]. For example, the threshold for implementing the surface code, a well-known quantum error–correction algorithm, requires gate fidelities exceeding 99% [8].

Suppose a quantum system is in the state $|\psi\rangle$ and some physical process with dynamics ε occurs, transforming the quantum system to the state $\varepsilon(|\psi\rangle\langle\psi|)$ [5]. The gate applied on the $|\psi\rangle$ state is shown by a unitary operation U. If $\varepsilon(|\psi\rangle\langle\psi|)$ is applied to different pure states $|\psi\rangle$, the average probability of ending up with the ideal target state can be estimated. This idea can be used to calculate the average gate fidelity as follows

$$F_{avg}(\varepsilon, U) = \int \langle\psi|U^\dagger \varepsilon(|\psi\rangle\langle\psi|) U|\psi\rangle d\psi \qquad (2.10)$$

The calculation of this integral involves averaging over all possible pure states sampled using a uniform distribution [5]. The fidelity ranges from 0 to 1, where a fidelity of 1 indicates a perfect gate operation that produces the desired target state.

The coherence time, or the duration that a qubit can remain coherent and retain quantum information, determines the number of quantum gates or qubit operations that can be applied to the qubit. Therefore, short gate times are desirable since the gate fidelity is bounded from above by the coherence times of the qubits. The ultimate goal in a fault-tolerant quantum computer is to reach a billion to a trillion quantum gate operations.

2.2.3.1 Evaluating the Performance of a Quantum Processor

This section discusses three methods for evaluating gate fidelity: the double-π metric, quantum-process tomography (QPT), and randomized benchmarking. These approaches play a crucial role in assessing the accuracy and performance of quantum gates, enabling quantum engineers to gain valuable insights into gate operations' quality.

2.2.3.1.1 Double-π Metric

The double-π ($\pi - \pi$) metric involves the application of two π pulses in succession, which ideally results in the identity operation $\mathbb{1}$. Deviations from $\mathbb{1}$ can be determined by measuring the residual population of the excited state following the pulses [9]. This simple test helps detect the presence of levels beyond a two-level Hilbert space and the effects of qubit relaxation. However, it can only provide a rough estimate of the actual gate fidelity because it lacks information on all possible errors. Additionally, the double-π metric does not effectively account for errors that exclusively affect the eigenstates of σ_x and σ_y, or variations in the rotation angle from π [9].

2.2.3.1.2 QPT

QPT aims to determine the process matrix χ. For this purpose, we apply the first pulse, chosen from $\{\mathbb{1}, R_x(\pi), R_x(\pi/2), R_y(\pi/2)\}$ to prepare the four linearly independent input states $|0\rangle$, $|1\rangle$, $(|0\rangle + i|1\rangle)/\sqrt{2}$, and $(|0\rangle - |0\rangle)/\sqrt{2}$, whose projectors[2] span the space of 2×2 density matrices ρ. The second pulse is chosen from $\{\mathbb{1}, R_x(\pi/2), R_y(\pi/2)\}$ and corresponds to the process we use to determine χ. By rotating the measurement axis using the final pulse ($\{\mathbb{1}, R_x(\pi), R_x(\pi/2), R_y(\pi/2)\}$), the state tomography can be performed on the state resulting from the first two pulses [9]. Figure 2.4 shows the QPT for a $\pi/2$ phase gate.

The number of measurements required for QPT increases exponentially with the number of qubits. Additionally, imperfect measurements make it challenging to attribute the results obtained from QPT to a single gate error. The randomized benchmarking introduced in Section 2.2.3.1.3 addresses these issues.

2.2.3.1.3 Randomized Benchmarking

Every qubit operation involves three steps: preparing the qubit in a known quantum state; applying a quantum gate or a series of gates; and finally, measuring the qubit's state. Each of these steps is prone to errors, so we need a way to isolate errors caused only by the quantum gates from the state-preparation-and-measurement (SPAM) errors.

The π-π and QPT methods are susceptible to SPAM errors. Randomized benchmarking is a standard method of characterizing the error in quantum gates to

---

2. By definition the projector $P = |i\rangle\langle i|$ acts on a quantum state and projects it onto the state $|i\rangle$.

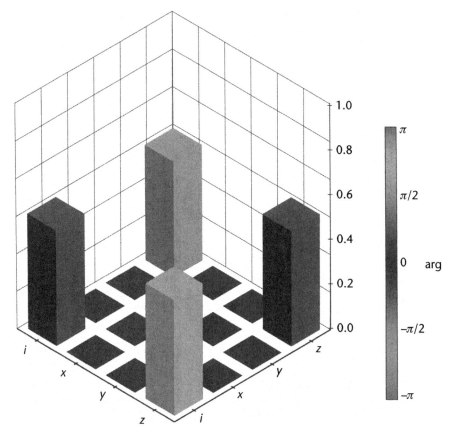

Figure 2.4 QPT for a $\pi/2$ phase gate.

estimate their average fidelity. The goal is to characterize a set of gates by applying them in a randomized sequence and track how much error increases with the length of the circuit. The sequence is chosen randomly from the Clifford group generators ($R_u = e^{\pm i\sigma_u \pi/4}$, where $u = x, y$). This method makes it possible to isolate the errors due to the quantum gate from those due to the SPAM. This is achieved by observing changes in error as we increase circuit length, effectively canceling out errors from measurement when looking at error differences. Another advantage of some versions of randomized benchmarking is their efficiency, as they can be efficiently applied to a large number of qubits. Figure 2.5(a) shows a typical plot of the sequence gate fidelity versus the circuit length, corresponding to the number of Clifford gates used. Figure 2.5(b) shows a single microwave control pulse, and Figure 2.5(c) shows a pulse sequence used for randomized benchmarking.

2.2.4 Quantum Circuits

We can construct quantum circuits by combining single-qubit and two-qubit gates. Figure 2.6 illustrates a quantum circuit comprised of four input qubits, q_0 to q_3, Hadamard, and CNOT gates. The meter symbol represents the measurement operation. Quantum circuits are typically written in the order of application from left

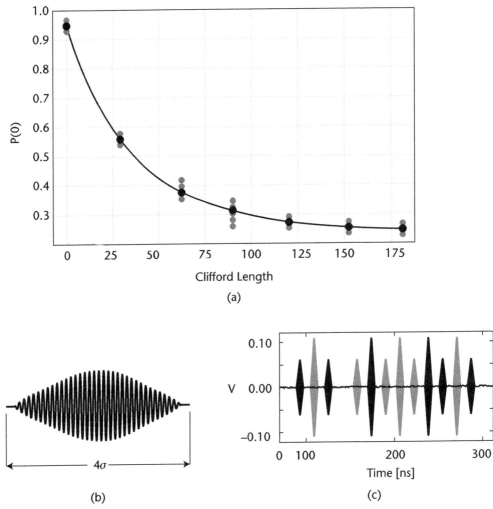

Figure 2.5 (a) Randomized benchmarking: sequence fidelity versus the number of Clifford gates; (b) a single microwave qubit control pulse; and (c) example of a randomized benchmarking pulse sequence. Black (gray) pulses correspond to rotations in the y(x)-axis.

to right. However, when calculating the result of a gate sequence, the calculations are performed from right to left. Therefore, if we apply $U_1 \cdots U_n$ gates to the input state $|\psi_{in}\rangle$ starting with U_1 on the left, the resulting output state $|\psi_{out}\rangle$ is given by $|\psi_{out}\rangle = U_n \cdots U_2 U_1 |\psi_{in}\rangle$.

2.2.5 Quantum Algorithm

The general quantum algorithm, as shown in Figure 2.7, involves the following steps:

- Using a single-qubit operation like the Hadamard gate to bring an initialized register of qubits into a superposition state.

2.2 Quantum Information Processing

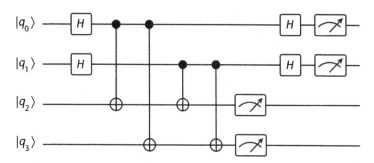

Figure 2.6 An example of a quantum circuit with Hadamard and CNOT gates. All the qubits are measured in the end.

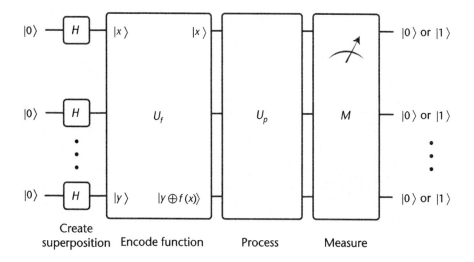

Figure 2.7 The structure of a quantum algorithm.

- Applying a multiqubit unitary operation to encode the function f using single-qubit and two-qubit gates.
- Implementing a processing step that involves single-qubit and two-qubit gates, which allows the interpretation of the qubit register on a computational basis.
- Measuring some or all of the qubits.

Let us look at how we can encode a classical function $f(x)$ [5]. As all calculation steps in quantum information must be reversible (as explained in Chapter 1), a specific version of $f(x)$ is required, which is defined through the following mapping U_f:

$$U_f : |x\rangle|y\rangle \rightarrow |x\rangle|y \oplus f(x)\rangle \tag{2.11}$$

It is possible to evaluate $f(x)$ and have the result in the output register by starting with the output register of qubits in $|0\rangle$ state

$$U_f(|x\rangle|0\rangle) = |x\rangle|f(x)\rangle \tag{2.12}$$

Consider two qubits with initial state $|00\rangle$ and apply the Hadamard gate on both.

$$\begin{aligned}|\alpha\rangle &= (H \otimes H)|00\rangle = \left(\frac{|0\rangle+|1\rangle}{\sqrt{2}} \otimes \frac{|0\rangle+|1\rangle}{\sqrt{2}}\right)|00\rangle \\ &= \frac{1}{2}(|00\rangle + |01\rangle + |10\rangle + |11\rangle)\end{aligned} \tag{2.13}$$

This gives a maximal superposition state involving all the computational states in the two-qubit input register. Now by combining (2.12) and (2.13), we obtain

$$\begin{aligned}U_f|\alpha\rangle|0\rangle &= \frac{1}{2}U_f(|00\rangle + |01\rangle + |10\rangle + |11\rangle) \otimes |0\rangle \\ &= \frac{1}{2}U_f(|00,0\rangle + |01,0\rangle + |10,0\rangle + |11,0\rangle) \\ &= \frac{1}{2}(|00, f(00)\rangle + |01, f(01)\rangle + |10, f(10)\rangle + |11, f(11)\rangle)\end{aligned} \tag{2.14}$$

Although we only operate $f(x)$ once by applying U_f, the output register contains evaluations of the function $f(x)$ for all possible values of x. The challenge is how to extract this information. Measuring the qubits in the output register produces $f(x)$ at a random value of x, which is not unique to quantum computing. What sets quantum computing apart is the ability to utilize the interference effect to enhance the probability of obtaining the correct answer upon output measurement. This involves applying additional unitary gates after encoding the function to establish relationships between multiple evaluations of $f(x)$ for different values of x. So, instead of evaluating individual values of $f(x)$, combinations of $f(x)$ values are used to determine the correct answer. This demonstrates how quantum computing utilizes quantum operations and interference effects to solve problems more efficiently.

Developing quantum algorithms is a challenging endeavor that demands a radically different mindset than developing classical algorithms. The following are a few examples of well-known quantum algorithms [5]:

- *Grover's algorithm:* A quantum search algorithm that can find a marked item in an unsorted database of size N with $O(\sqrt{N})$ queries to the oracle instead of $O(N)$ queries required by classical algorithms.
- *Shor's algorithm:* A quantum algorithm for integer factorization that is exponentially faster than the best-known classical algorithm. Shor's algorithm has important implications for cryptography as much of modern cryptography,

such as the widely used RSA encryption scheme, relies on the difficulty of factoring large numbers.
- *Deutsch-Jozsa algorithm:* One of the earliest examples of a quantum algorithm exponentially faster than any possible deterministic classical algorithm. The Deutsch-Jozsa algorithm currently has limited application in practice.
- *Quantum Fourier transform:* A quantum version of the classical discrete Fourier transform plays a central role in many quantum algorithms, such as Shor's algorithm.

2.2.5.1 Deutsch's Algorithm

The Deutsch-Jozsa problem involves a quantum computer called an oracle that performs a specific function, denoted as $f: \{0, 1\}^n \to \{0, 1\}$. This function takes n-bit binary inputs and produces either a 0 or 1 output for each input. The function is guaranteed to be either constant (outputting 0 for all inputs or 1 for all inputs) or balanced such that it outputs 1 for half of the inputs and 0 for the other half. The goal of the problem is to determine whether the function is constant or balanced using the oracle.

Deutsch's algorithm is a specific instance of the more general Deutsch-Jozsa algorithm, where the input size is $n = 1$ in $f: \{0, 1\}^n \to \{0, 1\}$. The condition to be checked is whether $f(0) = f(1)$. This can also be represented as $f(0) \oplus f(1)$, where $\oplus$ is addition modulo 2, which can also be thought of as a quantum XOR gate implemented as a CNOT gate. If the result is zero, the function f is constant; otherwise, it is not constant. The algorithm begins with a two-qubit state $|0\rangle|1\rangle$, which is then subjected to a Hadamard transform on each qubit, yielding $(|0\rangle + |1\rangle)(|0\rangle - |1\rangle)/2$. The algorithm is provided with a quantum implementation of the function f that maps $|x\rangle|y\rangle$ to $|x\rangle|f(x) \oplus y\rangle$. Applying this function to our current state, we obtain

$$\frac{1}{2}\Big(|0\rangle\big(|f(0) \oplus 0\rangle - |f(0) \oplus 1\rangle\big) + |1\rangle\big(|f(1) \oplus 0\rangle - |f(1) \oplus 1\rangle\big)\Big)$$
$$= (-1)^{f(0)} \frac{1}{2}\Big(|0\rangle + (-1)^{f(0) \oplus f(1)}|1\rangle\Big)\big(|0\rangle - |1\rangle\big) \quad (2.15)$$

Ignoring the last bit and the global phase results in $(|0\rangle + (-1)^{f(0) \oplus f(1)}|1\rangle)/\sqrt{2}$.

After applying a Hadamard transform to this state, we obtain

$$\frac{1}{2}\Big(|0\rangle + |1\rangle + (-1)^{f(0) \oplus f(1)}|0\rangle - (-1)^{f(0) \oplus f(1)}|1\rangle\Big)$$
$$= \frac{1}{2}\Big(\big(1 + (-1)^{f(0) \oplus f(1)}\big)|0\rangle + \big(1 - (-1)^{f(0) \oplus f(1)}\big)|1\rangle\Big) \quad (2.16)$$

$f(0) \oplus f(1) = 0$ if and only if we measure $|0\rangle$ and $f(0) \oplus f(1) = 1$ if and only if we measure $|1\rangle$. Therefore, we can determine with complete certainty whether the function $f(x)$ is balanced or constant.

2.2.6 Quantum Error Correction

As discussed in Sections 1.3 and 3.2, phenomena such as thermal noise or quasi-particle poisoning cause qubit decoherence and lead to the loss of quantum information. Consequently, quantum computers must be able to correct errors caused by decoherence and limited control accuracy to ensure fault-tolerant operation. Classical bits are only subject to the bit-flip error, while quantum bits can suffer from both bit-flip and phase-flip errors. The bit-flip error occurs when the state of a qubit flips from $|0\rangle$ to $|1\rangle$ or vice versa. The phase-flip error affects the phase of the qubit, resulting in a state such as $\alpha|0\rangle + \beta|1\rangle$ that is expected to become $\alpha|0\rangle + \beta e^{i\phi}|1\rangle$, but instead becomes $\alpha|0\rangle + \beta e^{i(\phi+\delta)}|1\rangle$. To address these errors, corresponding codes such as the bit-flip code and phase-flip code can be employed for correction.

The threshold theorem states the existence of a critical error rate, below which a quantum computer can suppress the logical error rate to arbitrarily low levels by applying quantum error–correction schemes [7]. The threshold value depends on factors such as the chosen quantum error–correction scheme. For example, the surface code, one of the most widely studied codes, requires gate fidelities of at least 99% for effective error correction in a noisy system. Achieving such high gate fidelities is a significant challenge in practice. However, gate fidelities of over 99% have been achieved for both single and two-qubit gates in various quantum computing platforms, including superconducting qubits [6].

Quantum error–correction schemes encode a single fault-tolerant logical qubit in multiple physical qubits. Like classical error-correction schemes, additional qubits encode quantum information to check for errors, correct them, and enable continued operation. The quantum Hamming bound indicates that at least five physical qubits are required to encode a single logical qubit to allow for arbitrary error correction. There are proposed quantum error correction codes with 5, 7, and 9 qubits [7]. Thus, significant overhead in quantum computers arises from quantum error–correction schemes that necessitate a large number of physical qubits. Recent experiments have demonstrated that fault-tolerant quantum computation is practically achievable by performing repeated, fast, and high-performance quantum error–correction cycles [11, 12].

Currently, the field of quantum computing is in the noisy intermediate-scale quantum (NISQ) era [13]. Noisy refers to the fact that no continuous error-correcting codes are presently applied to quantum algorithms, and intermediate refers to the limited number of qubits. These two obstacles prevent quantum computers from effectively reaching their true computational power to solve complex problems. However, some small-scale problems designed for the current quantum processors can be solved using NISQ algorithms. To pass beyond the NISQ era, a more sophisticated fault-tolerant quantum computer with error-correction capabilities and a significantly higher number of qubits is needed.

2.2.7 Quantum Supremacy

Quantum supremacy is a significant milestone that proves that a quantum computer can outperform classical computers by solving a particular computational task that the state-of-the-art classical computer would never be able to solve in a reasonable amount of time. Researchers at Google have created the first quantum processor to demonstrate quantum supremacy [14]. This processor, which consisted of 53 superconducting transmon qubits, was used to complete a particular computational task in a few hundred seconds, while for the state-of-the-art supercomputer, it would take around 10,000 years to complete this task.[3] Another experiment using a photonic quantum processor has shown evidence of solving a task that would take the best available algorithms and supercomputers more than 9,000 years in only 36 μs [16].

2.3 Quantum Computing Platforms

Recall that any two-level quantum system can function as a qubit, whether natural (like ions and atoms) or artificial (like superconducting circuits). However, these qubits are not useful unless their quantum states can be controlled and read out. Thus, the first step is establishing a trapping mechanism to keep the qubits in place using external or internal trap mechanisms. External traps, such as ion or neutral atom traps, can trap ions and neutral atoms, while in other cases, such as spin qubits in the AlGaAs heterostructure, the qubits are internally trapped through material or circuit engineering. After establishing a trap, control and measurement mechanisms must be developed. Depending on the qubit type, lasers, microwaves, and magnetic fields are used for control and read out of qubits.

Sections 2.3.1–2.3.3 briefly review several qubit platforms, including trapped atoms and ions, spin qubits, superconducting, and topological qubits.

2.3.1 Ions

Ions, such as the ytterbium ion, are naturally occurring quantum systems that can be utilized as qubits. In the trapping process, a laser is used to extract an electron, resulting in the atom having a positive electrical charge and a single-valence electron. Because ions are charged atoms, they can be controlled and trapped using an electric field. An engineered system of electrodes, with appropriate voltages, generates electric forces that trap the ions in place. The first ion trap was developed in the 1950s by Wolfgang Paul, and it has since evolved into modern planar traps, such as linear ion traps (see Figure 2.8). Lasers are used to perform all aspects of quantum-information processing, from initial preparation to final readout [17].

2.3.2 Neutral Atoms

Neutral atoms can also serve as qubits, with rubidium atoms being a common choice due to their well-established technological solutions, particularly in lasers [18]. Since neutral atoms have no charge, trapping them is more challenging. Various laser cooling and trapping techniques, such as magneto-optical traps (MOTs) and Doppler cooling, can trap an ensemble of neutral atoms. An optical tweezer isolates individual atoms from the ensemble to create an array of atoms

---

3. There are controversies surrounding this experiment [15].

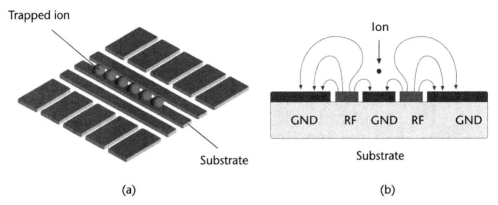

Figure 2.8 (a) A planar ion trap; using rapidly oscillating voltages allows for holding the ion into its position. (b) Cross-section of a planar ion trap.

in a small vacuum chamber. Figure 2.9 shows neutral atoms trapped in an optical potential. Qubits can be created by precisely controlling the spin of each atom's nucleus. Entanglement between neighbors is created by pumping the atoms into a Rydberg state using light pulses.

Naturally occurring qubits, such as ions and neutral atoms, do not require a huge dilution fridge and complicated cabling. On the other hand, semiconductor qubits (spin, superconducting, and topological) need a cryogenic environment and have challenging cabling, as described in Section 2.3.3. However, semiconductor qubits have benefits, such as utilizing the established semiconductor fabrication methods already used to fabricate electronic devices.

2.3.3 Semiconductor Qubits

Semiconductor qubits are synthetic qubits that are created by semiconductor fabrication technologies. Sections 2.3.3.1–2.3.3.3 discuss the three types of semiconductor qubits: spin, superconducting, and topological.

These quantum circuits are extremely sensitive to noise. As a significant portion of the noise is proportional to temperature, we place these circuits in a cold environment called a dilution fridge to minimize the noise that couples to these circuits.

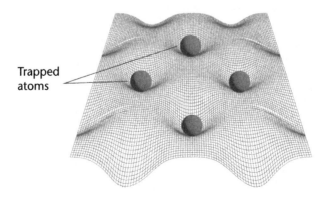

Figure 2.9 Trapped neutral atoms in an optical lattice.

2.3 Quantum Computing Platforms

2.3.3.1 Spin Qubit

Electron spins, which are naturally two-level systems, can serve as qubits. Creating spin qubits involves using a two-dimensional electron gas (2DEG). Shaping the band structure creates a potential well to trap individual electrons in an electrostatic potential defined by metallic surface gates (see Figure 2.10). The temperature must be decreased to below 4K to confine electrons within this well, as otherwise, unsuppressed thermal energy can add or remove electrons from the well. Consequently, spin-qubit processors are kept in a dilution fridge.

Typically, qubit manipulation can be performed via either electric or magnetic excitations. The qubit state is read out by performing a spin-to-charge conversion, which only allows spin-up electrons to tunnel out of the quantum dot. A charge sensor is used to measure the electron occupation of the well, indirectly determining the electron spin and qubit state [19].

2.3.3.2 Superconducting Qubit

Superconducting qubits are implemented using solid-state superconducting circuits, with charge and flux qubits being the two main types. Chapter 3 demonstrates how a superconducting circuit can create an artificial atom with adjustable parameters, including the transition frequency between quantum states.

Microwave pulses control the qubit's state, while the readout process uses the concept of cavity quantum electrodynamics (CQED). In CQED, the qubit's state is determined by examining the frequency shift in the resonator circuit coupled to the qubit [6]. Figure 2.11 illustrates a superconducting qubit coupled to a planar microwave resonator.

Let us explore the rationale behind utilizing microwave frequencies for superconducting qubits. Consider the scenario where the energy difference between the ground and excited states is relatively low, allowing the thermal energy present

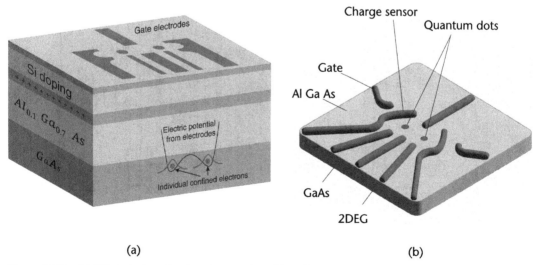

Figure 2.10 (a) AlGaAs heterostructure hosts spin qubits. The gate electrodes are used for the control and readout of the qubit. (b) The metallic gates are used for the control and readout of the qubit.

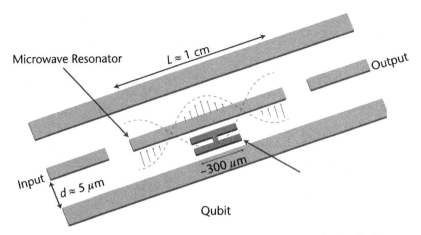

Figure 2.11 A superconducting qubit is coupled to a coplanar waveguide (CPW) microwave resonator.

inside the refrigerator to cause the qubit to transition from the ground state to the first excited state. This uncontrolled thermal transition is undesirable since we seek precise external control over the qubit's state.

Now, let us perform a rough calculation to determine the frequency associated with thermal energy in a standard dilution fridge. The thermal energy, denoted as E_{th}, can be expressed as Boltzmann's constant (k_B) multiplied by the temperature (T): temperature $E_{th} = k_B T$. The frequency (f_{th}) corresponding to this thermal energy is given by $f_{th} = k_B T/h$.

A typical dilution fridge can achieve a temperature of 20 mK, corresponding to a frequency of $f_{th} = 0.4$ GHz. To minimize the occurrence of thermal excitations from the ground to the excited state of the qubit, it is essential for the transition energy (E_{01}) associated with the external field to be significantly higher than the thermal energy (i.e., $E_{01} \gg E_{th}$). By "significantly higher," we mean at least 10 times greater, which leads to a transition frequency $f_{01} = 10 f_{th} = 4$ GHz.

Most charge qubits, such as the transmon, operate in the microwave frequency range of 4–8 GHz. Microwave frequencies are particularly appealing since they are sufficiently high to leverage standard cryogenic techniques and well-established microwave components and techniques commonly used in the telecommunications industry. While higher frequencies may offer certain advantages, they pose challenges regarding component cost, design complexity, and fabrication capabilities.

2.3.3.3 Topological Qubit

The topological qubit is also a synthetic qubit fabricated using lithographic techniques similar to those used for spin and superconducting qubits. The main advantage of a topological qubit is its inherent immunity to decoherence. Since active error correction is not required, far fewer physical qubits are needed, allowing the physical topological qubit to perform the same functions as a logical qubit.

The topological qubit operates based on emergent particles known as non-Abelian anyons and a braiding operation that ideally results in perfect quantum gates [see Figure 2.12(a)] [20]. One example of non-Abelian anyons is Majorana particles, which emerge in hybrid superconductor-semiconductor systems, such as superconductors on topological insulator substrates [see Figure 2.12(b)]. At the time of writing, no topological qubit has been created, although some works have demonstrated the existence of Majorana particles [21].

2.4 Challenges and Opportunities in Quantum Computing

Quantum computers are no longer science fiction; users can now program quantum computers online. However, most experts agree that quantum computers are at least a decade away from accomplishing the tasks for which they are ultimately anticipated. A quantum computer with thousands or millions of qubits must still be built to achieve the computational power required to address complex practical problems. All qubit technologies face the same major technical challenge: the scaling issue. Even if specific platforms appear easier to scale than others, it is still unclear which technology or technologies will ultimately be scaled up successfully. While quantum computing holds promise, setting realistic goals and expectations is crucial.

Sections 2.4.1–2.4.2 address the technical challenges related to scaling, and explore the necessary skill sets for quantum hardware engineers.

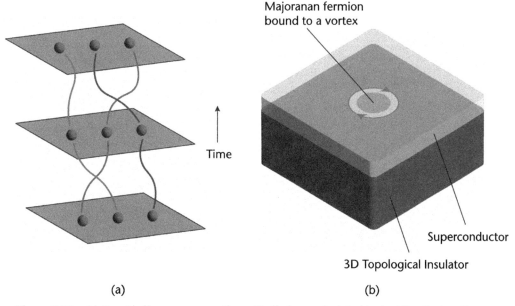

Figure 2.12 (a) Non-Abelian anyons can theoretically be manipulated by braiding their paths to change their states and perform topological quantum computation. Due to their robustness against errors and decoherence, the braids remain unchanged even under minor disturbances. (b) Majorana fermions can be hosted in a hybrid semiconductor-superconductor system in a two-dimensional setup. These fermions are formed in vortices created on the superconductor surface.

2.4.1 Technical Challenges of Scaling

The current state-of-the-art quantum computer can operate with up to a few hundred qubits. However, order-of-magnitude scaling is required to fully realize the potential of quantum computers. Recall that each quantum-computing platform has distinct features and operating requirements. Regardless of the qubit type, scalability issues currently affect all platforms. Sections 2.4.1.1–2.4.1.5 discuss the technical challenges of scaling semiconductor qubits [22].

2.4.1.1 Noise and Crosstalk

Like communication systems, qubit systems are susceptible to various impairments, including crosstalk and noise. Let us examine each of these effects individually. First, qubits are susceptible to environmental noise, such as stray electromagnetic fields, cosmic radiation, or thermal fluctuations. The noise can also be coupled to the qubit through the control and readout lines that enter the dilution fridge from the ambient temperature. As the number of qubits grows, more control and readout cables are needed, increasing the amount of noise coupled to the system by expanding the cross-section available for noise pickup.

Second, crosstalk in the form of inductive or capacitive coupling between the control and readout cables can also degrade qubit fidelity and cause correlated errors across multiple qubits. Although these errors are deterministic and can be nullified, they add significant complexity to the operation of the qubits.

2.4.1.2 Input/Output (IO) Management

While classical computers can have multiple inputs and outputs for a single logic gate, governed by Rent's rule dictating fan-in/fan-out numbers, quantum computers do not enjoy this luxury. In quantum computing, the number of distinct IO signals must be at least as large as the number of qubits. This is due to the "'no-cloning" theorem in quantum mechanics, which establishes the impossibility of creating an exact copy of a quantum state. Consequently, a single quantum gate cannot produce multiple outputs. Effectively managing the IO challenge requires the design of appropriate interconnects, cabling, and packaging and the development of strategies for time, space, and frequency multiplexing.

Another crucial consideration is the diameter of the magnet bore required for spin qubits and topological qubits, which demand strong magnetic fields. This imposes an additional constraint on the density of connectors.

2.4.1.3 Heat and Power Dissipation

As the number of qubits increases, so does the number of control cables that need to be connected to the fridge. Despite having a control interface inside the fridge at elevated temperatures (e.g., 4K), these cables must reach the qubits at millikelvin temperatures. As Section 7.4.3.1 discusses, attenuators help minimize the thermal noise coupled to the qubits. However, as signals pass through these attenuators, heat is generated and needs to be efficiently removed from the dilution refrigerator.

Managing this heat dissipation can become a significant challenge, especially when dealing with hundreds of thousands of cables.

Although a qubit, such as a superconducting qubit, is theoretically dissipation-less, the actual operation of the qubit generates heat. Therefore, in addition to the heat generated by the cables, the qubits release heat during operation, which must be removed from the fridge. This is challenging as the cooling power at millikelvin temperatures, where the qubits are located, is extremely low, typically in the range of a few milliwatts. To address this issue, researchers have developed hot spin qubit systems operating above 1K, which significantly facilitates the cooling process by providing higher cooling power in the watts and kilowatts range [23]. However, superconducting qubits still need to operate at millikelvin temperatures.

2.4.1.4 Size of the System

As the number of qubits increases, the size of the system will inevitably expand. Therefore, it is necessary to design an integrated architecture that minimizes the system size and improves signal integrity and power consumption. Let us investigate what occurs in a system with low integration.

There is a significant phase shift between the lines when using lengthy transmission lines with different path lengths. This makes the synchronization of control signals formidable, especially for qubits where a microwave signal's precise phase and amplitude are used to manipulate the qubit's state. Although this effect can be calibrated out, it becomes challenging for wideband signals, where the spectral content can be affected by parameters such as dispersion and mutual coupling between the lines, as discussed in Chapter 4.

The second variable impacted by system size is the propagation time of the control signals reaching the qubit in the lowest part of the fridge. Typical propagation times on the order of tens of nanoseconds are comparable to qubit gate times but much shorter than typical readout times in the range of hundreds of nanoseconds or microseconds.

2.4.1.5 Slew Rate, Rise Time, and Bandwidth

As discussed in Chapter 4, sufficient bandwidth is required to ensure that the shape of control pulses is not distorted. Chapter 7 describes the use of attenuators to reduce thermal noise, with the signal attenuated by 50–60 dB as it passes through the cable and arrives at the qubit. This condition requires a generator to create signals in the volt range to achieve millivolt amplitudes and picosecond rise times where the qubit is located. This, in turn, requires amplifiers that significantly increase the noise coupled to the qubit. Comparing this distributed solution to an on-chip solution, it becomes apparent that a standard integrated CMOS circuit can quickly achieve rise times shorter than ten picoseconds for millivolt pulses. Like op-amp circuits, the on-chip slew rate is determined by the RC time constant of the transistors and interconnects.

As discussed in Chapter 7, scientists have proposed a scalable control approach based on Cryo-CMOS to address these issues. This integrated solution uses a cryogenic chip inside the fridge to replace the room-temperature control and measurement system.

2.4.2 Skillsets for Quantum Hardware Engineers

Quantum workforce development is a challenging task that requires the collective efforts of industry and academia to train a new workforce or reskill existing ones, thus addressing the demand for skilled professionals within the quantum industry [24]. This section examines the skill sets required for quantum hardware engineers working on semiconductor qubits such as superconducting, spin, and topological qubits, allowing them to concentrate on the skills that will most impact their technical success in their careers. Quantum hardware engineers must be proficient in four areas while working on semiconductor quantum computers: nanofabrication, cryogenic techniques, microwave techniques, and data acquisition and measurement; Sections 2.4.2.1–2.4.2.4 discuss these topics.

2.4.2.1 Cryogenic Engineering

Semiconductor qubit experiments require operating at extremely low temperatures close to absolute zero, typically around 10–50 mK, to suppress the noise that could cause the qubit's fragile quantum states to decohere. Therefore, it is essential to understand the components and functions of a dilution fridge, which provides a cold and isolated environment for the qubit. Chapter 8 discusses the principles of dilution refrigeration.

One of the primary tasks of a quantum engineer working with semiconductor qubits is to identify and mitigate the paths through which noise can couple to the qubit. These paths can be conducted or radiated, requiring different noise and interference-blocking techniques using filters, circulators, attenuators, shielding, and lossy microwave cables. Chapters 6 and 7 discuss these techniques.

2.4.2.2 Microwave Engineering

Measurement and control of semiconductor qubits occur at RF and microwave frequencies. Thus, engineers working with them must have a solid understanding of cryogenic and room-temperature microwave components and systems. This includes amplifiers, mixers, filters, power combiners, attenuators, and up-and-down converters. The knowledge required in this area can be divided into the following three main categories:

- Noise-suppression techniques;
- Working with measurement instruments such as network analyzers and spectrum analyzers, which are used for component-level, system-level, and qubit measurements;
- Microwave signal-processing techniques, including amplification, down-conversion, up-conversion, and filtering. These techniques are essential for controlling and measuring the qubits by sending appropriate modulated

signals to the fridge and down-converting the signals from the fridge for digital postprocessing.

Chapters 4 and 5 cover the fundamentals of microwave systems and components.

2.4.2.3 Nanofabrication

Fabricating high-quality qubits requires engineers to become familiar with lithography techniques, such as electron-beam lithography, and metal-deposition techniques, such as sputtering or thermal evaporation. Chapter 3 briefly discusses the superconducting qubit fabrication process.

2.4.2.4 Data Acquisition

The final stage in the chain of semiconductor qubit experiments involves communication with various machines, their synchronization, and the acquisition of measured data. The acquired data is sent to computers for postprocessing. Therefore, quantum engineers must be familiar with various communication protocols such as serial communication, GPIB, TCP/IP, and USB. Additionally, they must be proficient in using software tools such as MATLAB, Python, QCoDeS, pyCQED, qKIT, and Labber for data acquisition and controlling the experiment.

References

[1] Feynman, R. P., "Simulating Physics with Computers," *International Journal of Theoretical Physics*, Vol. 21, 1982, pp. 467–488.

[2] Lloyd, S., "Universal Quantum Simulators," *Science*, Vol. 273, Issue 5278, 1996, pp. 1073–1078.

[3] Divincenzo, D. P., "The Physical Implementation of Quantum Computation," *Fortschritte der Physik*, Vol. 48, 2000, pp. 771–783.

[4] Shor, P., "Algorithms for Quantum Computation: Discrete Logarithms and Factoring," in *35th Annual Symposium on Foundations of Computer Science*, Santa Fe, NM ,1994.

[5] Nielsen, M. A., and I. L. Chuang, *Quantum Computation and Quantum Information*, Cambridge, England: Cambridge University Press, 2000.

[6] Krantz P., et al., "A Quantum Engineer's Guide to Superconducting Qubits," *Applied Physics Reviews*, Vol. 6, No. 2, arXiv:1904.06560, 2021.

[7] Gottesman, D., "An Introduction to Quantum Error Correction and Fault-Tolerant Quantum Computation," in *The Physics of Information Technology* (ed. by F. J. Dyson, D. E. Osheroff and R. Shankar), Cambridge, England: Cambridge University Press, 2009, pp. 221–246.

[8] Barends, R., et al., "Logic Gates at the Surface Code Threshold: Superconducting Qubits Poised for Fault-Tolerant Quantum Computing," *Nature*, Vol. 508, 2014, arXiv:1402.4848.

[9] Chow, et al., "Randomized Benchmarking and Process Tomography for Gate Errors in a Solid-State Qubit," *Phys. Rev. Lett.*, Vol. 102, 2008, arXiv:0811.4387.

[10] Aharonov, Dorit; Ben-Or, Michael (2008-01-01). "Fault-Tolerant Quantum Computation with Constant Error Rate," *SIAM Journal on Computing*, Vol. 38, No. 4, pp. 1207–1282. arXiv:quant-ph/9906129. doi:10.1137/S0097539799359385. ISSN 0097-5397. S2CID 8969800.

[11] Krinner, S., et al., "Realizing Repeated Quantum Error Correction in a Distance-Three Surface Code," *Nature*, Vol. 605, 2022, pp. 669–674.

[12] Google Quantum AI, "Suppressing Quantum Errors by Scaling a Surface Code Logical Qubit," *Nature*, Vol. 614, 2023, pp. 676–68, https://doi.org/10.1038/s41586-022-05434-1.

[13] Preskill, J., "Quantum computing in the NISQ era and beyond, Quantum," 2018. arXiv:1801.00862.

[14] Arute, F., et al., "Quantum supremacy using a programmable superconducting processor," Nature 574, 505–510 (2019).

[15] Pednault, E., et al. "Leveraging Secondary Storage to Simulate Deep 54-qubit Sycamore Circuits," 2019, arXiv:1910.09534.

[16] Madsen, L. S., "Quantum Computational Advantage with a Programmable Photonic Processor," *Nature*, Vol. 606, 2022, pp. 75–81.

[17] Xiang, Z., et al., "Hybrid Quantum Circuits: Superconducting Circuits Interacting with Other Quantum Systems," *Rev. Mod. Phys.*, Vol. 85, 2013, pp. 623–653.

[18] Henriet, L., et al., "Quantum Computing with Neutral Atoms," *Quantum 4*, 2020, arXiv:2006.12326v2, 2020.

[19] Botzem, T., et al., "Quadrupolar and Anisotropy Effects on Dephasing in Two-Electron Spin Qubits in GaAs," Nat Commun, Vol. 7, Article No. 11170, 2016, https://doi.org/10.1038/ncomms11170.

[20] Sarma, S., et al., "Majorana Zero Modes and Topological Quantum Computation," *npj Quantum Inf*, Vol. 1, Article No. 15001, 2015, https://doi.org/10.1038/npjqi.2015.1.

[21] Mourik, V., et al., "Signatures of Majorana Fermions in Hybrid Superconductor-Semiconductor Nanowire Devices," *Science*, Vol. 336, Issue 6084, April 12, 2012, pp. 1003–1007, DOI: 10.1126/science.1222360.

[22] Reilly, D. J., "Challenges in Scaling Up the Control Interface of a Quantum Computer," *2019 IEEE International Electron Devices Meeting* (IEDM), 2019, pp. 31.7.1–31.7.6, doi: 10.1109/IEDM19573.2019.8993497.

[23] Yang, C. H., et al., "Operation of a Silicon Quantum Processor Unit Cell Above One Kelvin," *Nature*, Vol. 580, 2020, pp. 350–354.

[24] Hughes C., et al., "Assessing the Needs of the Quantum Industry," in *IEEE Transactions on Education*, Vol. 65, No. 4, Nov. 2022, pp. 592–601, doi: 10.1109/TE.2022.3153841.

CHAPTER 3
Superconducting Qubits

> The task is not to see what has never been seen before, but to think what has never been thought before about what you see every day.
>
> —*Erwin Schrödinger*

This chapter explores superconducting qubits, with a specific emphasis on the transmon qubit. Section 3.1 provides a brief introduction to superconductivity. Section 3.2 covers the analysis, design, and figure of merits of the transmon qubit.

Sections 3.3 and 3.4 delve into single-qubit control and readout using microwave pulses and CQED and extend these concepts to two-qubit gates. Finally, Sections 3.5 and 3.6 conclude the chapter by examining the qubit calibration performance tests of quantum processors, such as randomized benchmarking.

3.1 Introduction to Superconductivity

In 1911, Heike Kamerlingh Onnes discovered superconductivity. He observed that the electrical resistance of mercury drops to a value indistinguishable from zero when cooled below a certain temperature known as the critical temperature T_c, as shown in Figure 3.1(a).

Superconductivity, also known as charged superfluidity, is a fascinating quantum mechanical phenomenon that exhibits the following macroscopic properties [1–3]:

- Zero electrical resistance: When cooled below a certain temperature called the critical temperature T_c [as shown in Figure 3.1(a)], the electrical resistance of a superconductor drops to a value indistinguishable from zero, resulting in dissipationless current flow. Section 3.1.1 explores how this supercurrent is carried by Cooper pairs, which are formed by the attractive interaction between electrons and the lattice.
- Meissner effect: A superconductor is not just a perfect conductor; it has a unique property called the Meissner effect. This effect causes a superconductor to expel magnetic fields from its interior, which can levitate magnetic objects and is utilized in ultra-high-speed trains [Figure 3.1(b)]. For a perfect conductor, the time rate of change of the magnetic field is zero (i.e., $dB/dt = 0$). This means the interior magnetic field must remain constant with time but can have a zero or nonzero value, as shown in Figure 3.2(a). With no initial magnetic field, the interior field remains zero if a magnetic field is

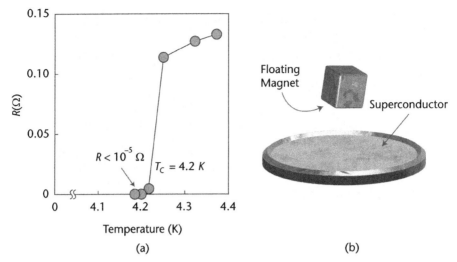

Figure 3.1 (a) Transition to zero resistance for superconducting mercury and (b) magnetic levitation due to the Meissner effect. The superconductor repels the magnet by producing screening currents on its surface which perfectly cancels the magnet's applied fields at the surface. The field produced is an image of the original magnet dipole, such that the magnet is repelled by an image field of itself produced by the superconductor.

applied to a perfect conductor. If the magnetic field is initially present, the field remains inside the specimen after transitioning to a perfect conductor.

As depicted in Figure 3.2(b), unlike a perfect conductor, field cooling of a superconductor results in expelling magnetic flux. This expulsion occurs due to Lenz's law, which results in the induction of circulating currents that oppose the development of the magnetic field within the conductor. This unique characteristic is known as diamagnetism, and a superconductor exhibits perfect diamagnetism, as the induced currents would exist at whatever magnitude necessary to cancel the external field.

Figure 3.2(c) illustrates a superconductor for $x > 0$, with a weak external magnetic field B_0 applied in the z-direction from vacuum $x < 0$. The magnetic field inside the superconductor is given by $B(x) = B_0 \exp(-x/\lambda_L)$, where λ_L is the London penetration depth. Due to the Meissner effect, the magnetic field B_0 penetrates into a superconductor over a small distance called London's penetration depth λ_L. Over the distance of λ_L the magnetic field becomes equal to $1/e$ times that of the magnetic field at the surface of the superconductor. The screening currents that flow at the surface generate magnetization M inside the superconductor, which perfectly compensates the applied field H, resulting in $B = 0$ inside the superconductor. The London penetration depth is given by

$$\lambda_L = \sqrt{\frac{m}{\mu_0 n e^2}} \tag{3.1}$$

for charge carriers of mass m, and density of superconducting electrons n and charge e. Typical values of λ_L range from 50 to 500 nm.

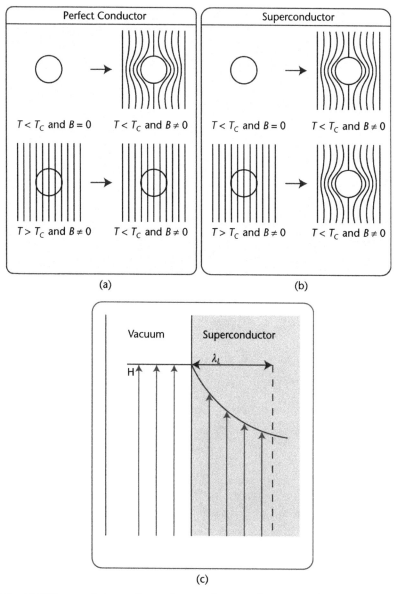

Figure 3.2 (a) Difference between the perfect conductor and a superconductor. Inside the specimen, for a perfect conductor, $dB/dt = 0$, whereas for (b) a superconductor, $B = 0$. (c) Visualization of London penetration depth.

- Flux quantization: Although a magnetic field cannot generally penetrate a superconductor, it can penetrate a hole created within it (see Figure 3.3). However, the flux passing through the hole must be quantized, meaning it can only take on discrete values.

The canonical momentum $\mathbf{p}$ in a superconductor in the presence of a magnetic field is given by $\mathbf{p} = \hbar \nabla \theta = e^* \Lambda \mathbf{J}_s + e^* \mathbf{A}$ where θ is the phase of the macroscopic wavefunction for the ensemble of Cooper pairs $\psi = |\psi(\mathbf{r})| e^{i\theta(\mathbf{r})}$ (see Section 3.1.1), J_s is the Cooper pair current density, e is the electron charge, $\mathbf{A}$ is the magnetic

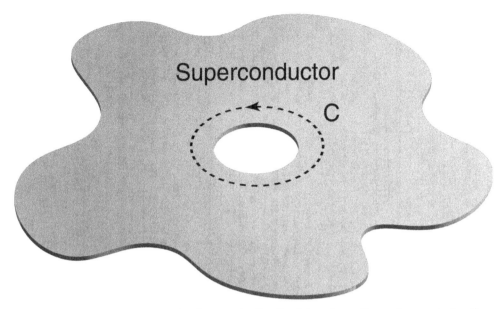

Figure 3.3 A hole in a superconducting sample. The line integral over the contour enclosing the hole is used to calculate the magnetic flux quantum.

vector potential, and Λ is given by $\Lambda = m^*/n^*_s(e^*)^2$, where, the effective mass and charge are $m^* = 2m_e$ and $e^* = -2e$, respectively, and n^*_s is the pair density [1, 2]. Taking the line integral of the canonical momentum over the closed path, as shown in Figure 3.3, results in

$$\oint \mathbf{p} \cdot \mathbf{dl} = \hbar \oint \nabla \theta \cdot \mathbf{dl} = e^* \oint \mathbf{A} \cdot \mathbf{dl} \tag{3.2}$$

The $\mathbf{J_S} = 0$ if the contour is sufficiently far from the surface (i.e., much larger than the London penetration depth). The line integral of the phase gradient is $\int_{\theta_1}^{\theta_2} \nabla \theta \cdot \mathbf{dl} = \theta_2 - \theta_1$. We know that the wave function must be single-valued. Therefore, the phase difference for a closed contour is a multiple of 2π

$$\hbar \oint \nabla \theta \cdot \mathbf{dl} = 2\pi n \hbar = e^* \oint \mathbf{A} \cdot \mathbf{dl} \tag{3.3}$$

Use of Stokes's theorem makes it possible to express the line integral of $\mathbf{A}$ as the surface integral of $\nabla \times \mathbf{A}$. Moreover, we know from Maxwell's equations that $\nabla \times \mathbf{A} = \mathbf{B}$. Thus, we have

$$e^* \oint \mathbf{A} \cdot \mathbf{dl} = e^* \int_S (\nabla \times \mathbf{A}) \cdot \mathbf{ds} = e^* \int_S \mathbf{B} \cdot \mathbf{ds} = e^* \Phi_S = 2\pi n \hbar \tag{3.4}$$

where Φ_S is the total flux enclosed by contour C. As the flux density in the bulk of the superconductor is zero, Φ_S represents the flux through the hole. Equation (3.4) results in

3.1 Introduction to Superconductivity

$$\Phi_s = \frac{2n\pi\hbar}{e^*} = \frac{nh}{e^*} \qquad (3.5)$$

As a result, the flux quantum is defined as $\Phi_0 = h/2e = 2.07 \times 10^{-15}$ Wb.

So applying a magnetic flux to a ring modifies the phase of the macroscopic wavefunction around the ring. However, only certain values of the magnetic flux can make the phase return to its initial value after a complete turn. These values are an integer multiple of the flux quantum Φ_0. The superconducting quantum interference device (SQUID) is an ultrasensitive magnetometer with a loop of superconducting wire that can detect changes in quantum flux. SQUIDs are also used in superconducting qubits, as discussed in Section 3.1.3.

3.1.1 Cooper Pairs

In 1957, Bardeen, Cooper, and Schrieffer developed the BCS theory, a theory that describes the microscopic mechanism of superconductivity [4]. Ginzburg-Landau and Bogolyubov also contributed to successful theories of superconductivity.

The mechanism behind superconductivity can be described in terms of Cooper pairs, which are pairs of electrons formed by attractive forces due to the electron-lattice interaction. The supercurrent is carried by pairs of electrons rather than individual electrons. Let us now examine how Cooper pairs are formed.

Figure 3.4(a) shows an illustration of atomic sites of a superconducting material, where each circle corresponds to a positively charged atom. Each atom has already contributed one electron that may freely move between atomic sites through normal electron conduction. This free electron moving in the lattice attracts the positively

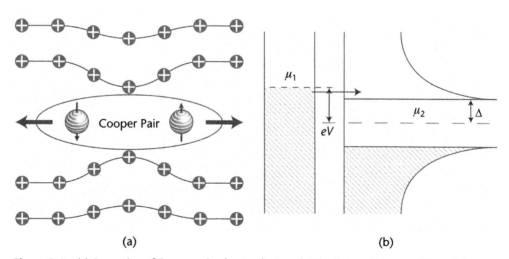

Figure 3.4 (a) Formation of Cooper pairs due to electron-lattice interactions and (b) model of electron tunneling. [Density of states is plotted horizontally and energy vertically. Shading denotes states occupied by electrons. Normal-superconductor tunneling at $T = 0$, with bias voltage just above the conduction threshold (i.e., eV slightly exceeds the energy gap Δ). Horizontal arrow depicts electrons from the left tunneling into empty states on the right.]

charged atoms. Due to the local increase in positive charge density, another free electron with opposite spin is attracted to this region, forming a Cooper pair. Therefore, a Cooper pair consists of two electrons with opposite spin and momentum. The binding energy of a single Cooper pair is extremely weak.

Now, let us examine the behavior of multiple Cooper pairs and how they act collectively. While a single electron is a fermion subject to the Pauli exclusion principle, which means that each energy level can only accommodate two electrons with opposite spins, a Cooper pair is a boson and does not follow the exclusion principle. The pairing of electrons near the fermi surface lowers the energy of the superconductor and forms a gap to excitations. Moreover, because of this gap, all pairs (and also electrons) act as a single macroscopic wavefunction that represents the collective fluid of Cooper pairs with a well-defined amplitude and phase, as shown by $\psi = |\psi|e^{i\theta}$. The density of Cooper pairs is given by $|\psi|^2$, and θ is the phase of the condensate. As can be seen, all the pairs have the same amplitude and phase (i.e., the pairs lock phase and move in lock steps). Therefore, there is no scattering by other electrons resulting in zero resistance.

Superconductivity occurs below a specific temperature known as the critical temperature, which is explained by the fact that cooling down the sample minimizes lattice vibrations and reduces the probability of electrons scattering off atoms and breaking the Cooper pairs.

Figure 3.4(b) depicts the band diagram of normal and superconducting metals. In the normal metal, all energy states are filled up to the Fermi level, which is shown by μ_1. On the other hand, in the superconducting part, Cooper pairs are condensed inside the gap at the Fermi level, which is shown by μ_2. Below the Fermi level and the gap, all single electron states are occupied, while above the gap, there are empty single electron states. The density of single electron states peaks at the so-called BCS ears on both sides of the gap and then decays away from the gap. No energy states are allowed for single electrons within the gap, which protects the Cooper pairs with an energy gap equal to the energy needed to break a pair.

Let us examine the tunneling effect from a normal metal to a superconducting electrode. For single electrons to tunnel to the superconducting electrode, the Fermi energy of the normal metal must be brought above the superconducting gap, enabling the existence of empty states for the single electrons to tunnel. In this case, a current flows from the normal conductor to the superconductor. The theoretical I-V of a normal-superconductor (N-S) junction is shown in Figure 3.5(a). The N-S junction in Figure 3.5(b) demonstrates that the derivative of the I-V curve at each point (i.e., dI/dV) provides the density of states [1–3]. By changing the gate voltage, the superconductor's accessible states are changed, and the band diagram is reconstructed.

The BCS theory predicts that the value of the energy gap Δ at temperature T depends on the critical temperature T_C. The ratio of the value of the energy gap at zero temperature to the value of the superconducting transition temperature T_c (expressed in energy units) takes the following universal value, which is independent of material, for materials with no strong electron-phonon interaction $\Delta(T=0) = 1.76 k_B T_c$.

This energy gap is about 100 μV/K. So, aluminum with a $T_C = 1.2$K has an energy gap of 120 μV and niobium with $T_c = 9.7$K has an energy gap of about 1 meV.

3.1 Introduction to Superconductivity

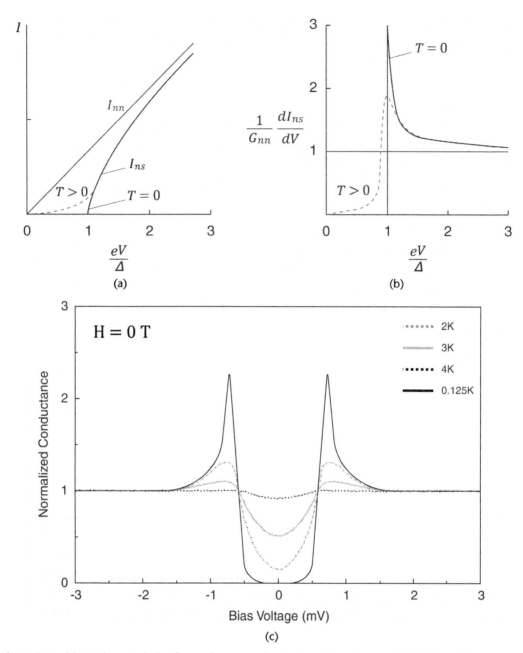

Figure 3.5 (a) I-V characteristic of normal-superconductor tunnel junctions and (b) differential conductance of normal-superconductor tunnel junctions. (Solid curves refer to $T = 0$; dashed curves refer to a finite temperature.) (c) The superconducting gap shrinks with increasing temperature.

Near the critical temperature and according to Ginzburg-Landau theory, the energy gap is equal to [1–3]

$$\Delta(T \rightarrow T_c) \approx 3.06 k_B T_c \sqrt{1 - \left(\frac{T}{T_c}\right)} \qquad (3.6)$$

So, the gap shrinks by increasing the temperature and completely disappears for temperatures above the critical temperature, as shown in Figure 3.5(c).

The coherence length of a superconductor refers to the distance over which superconductivity persists and measures the size of a Cooper pair, which is typically around 1 μm. In other words, it is the length over which the superconducting state decays into a normal state, and the density of Cooper pairs, or equivalently, $|\psi|^2$, gradually decreases. This transition from superconducting to normal happens over a thickness related to the coherence length, as illustrated in Figure 3.6. The coherence length ξ_{BCS} given by

$$\xi_{BCS} = \frac{\hbar v_f}{\pi \Delta} \tag{3.7}$$

determines the characteristic size of Cooper pairs, where v_f is the Fermi velocity, and Δ is the superconducting gap parameter.

The ratio $\kappa = \lambda/\xi$, where λ is the London penetration depth, is known as the Ginzburg-Landau parameter. Type-I superconductors are those with $0 < \kappa < 1/\sqrt{2}$, and type-II superconductors are those with $\kappa > 1/\sqrt{2}$. Figure 3.6 shows the type-I and type-II superconductors based on the Ginzburg-Landau parameter. Section 3.1.2 examines type-I and type-II superconductors.

3.1.2 Types of Superconductors

A superconductor can expel external magnetic fields up to a certain field intensity, beyond which its superconductivity is destroyed. Hence, in addition to the critical temperature, a critical magnetic field B_c is defined for superconductors. Superconductors are classified based on how they react to the external magnetic fields above the critical magnetic field B_c. The critical magnetic field is strongly correlated with temperature and is given by the following equation for temperatures below the critical temperature T_c according to Ginzburg-Landau theory [1–3]

$$B_c \approx B_c(0)\left[1 - \left(\frac{T}{T_c}\right)^2\right] \tag{3.8}$$

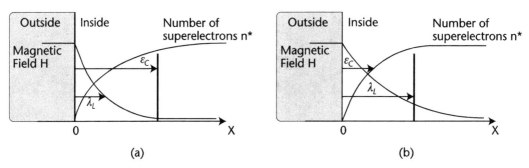

Figure 3.6 Coherence length and London penetration depth comparison for (a) a type-I superconductor and (b) a type-II superconductor.

3.1 Introduction to Superconductivity

where $B_c(0)$ is the critical magnetic field at zero Kelvin. The critical magnetic field reduces with increasing temperature, according to (3.8). In other words, the superconductor can withstand smaller magnetic fields as its temperature approaches its normal state.

Type-I superconductors exhibit a sharp transition from the superconducting state to the normal state when the applied magnetic field exceeds the critical field B_c, as shown in Figure 3.7(a). This is known as a first-order phase transition. These materials are typically elemental metals such as mercury, lead, and aluminum or alloys like niobium-germanium and niobium-titanium. These materials have relatively low critical temperatures, typically below 10K, and exhibit weak-coupling behavior, which means that the interaction between Cooper pairs and the lattice

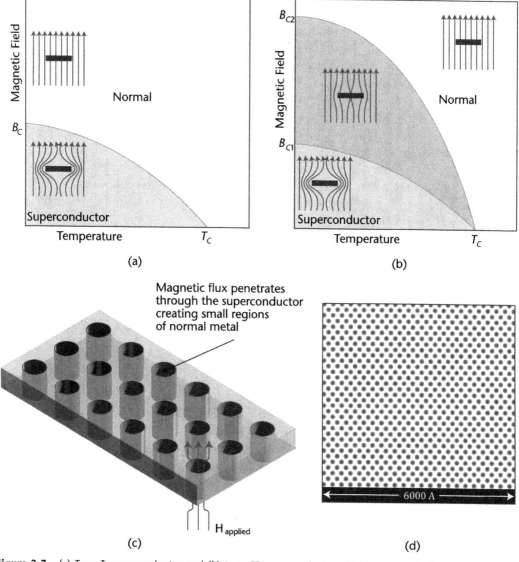

Figure 3.7 (a) Type-I superconductor and (b) type-II superconductor, (c) the magnetic flux penetration in the superconductor, and (d) STM image of Abrikosov vortices in a superconducting material.

is relatively weak. As a result, type-I superconductors have relatively low critical magnetic fields, and external magnetic fields or thermal fluctuations can quickly suppress their superconducting properties. Nevertheless, type-I superconductors are used in various applications, including constructing superconducting magnets for medical imaging, particle accelerators, and superconducting devices such as SQUIDs and qubits.

Type-II superconductors have higher critical temperatures in bulk and exhibit strong coupling behavior due to the stronger interaction between the Cooper pairs and the lattice. As shown in Figure 3.7(b), a type-II superconductor has two critical magnetic fields, B_{c1} and B_{c2}.

As the applied magnetic field is increased beyond B_{c1}, a mixed state is formed in which the superconducting and normal phases coexist. Above the B_{c1}, the applied magnetic field penetrates the material in the form of vortices, where each vortex carries a quantized magnetic flux. The screening supercurrents circulate in the λ_L-vicinity of the vortex core to screen the rest of the superconductor from the vortex magnetic field. The diameter of the vortex core is on the order of the superconducting coherence length ξ. Figure 3.7(c, d) show the vortices, where the areas of normal material are surrounded by superconducting material. As the temperature increases, the normal cores become more closely packed and eventually overlap, leading to the loss of the superconducting state.

Passing above the second critical field strength, B_{c2}, results in a full breakdown of superconductivity, as shown in Figure 3.7(b).

Almost all impure and compound superconductors are type-II. Note that due to the existence of a critical magnetic field, there is a maximum current that can be carried in a superconducting wire, as the current generates a magnetic field in accordance with Ampere's law.

3.1.3 Josephson Junction

Brian Josephson first proposed the Josephson junction in 1962. It is one of the most critical circuit components with applications in high-speed digital circuits, microwave amplifiers (Josephson parametric amplifier), and quantum computing (qubits).

The Josephson junction is a superconducting circuit element that consists of two superconducting electrodes separated by a thin insulating layer, as shown in Figure 3.8(a). The equivalent circuit of a Josephson junction is shown in Figure 3.8(a). Aluminum electrodes are widely used for quantum computing applications. After depositing the first aluminum electrode through thermal evaporation, we allow an insulating oxide layer to grow on top of it. Then, the second aluminum layer is deposited, as shown in Figure 3.8(b).

In a Josephson junction, Cooper pairs can tunnel coherently across the insulating barrier, producing a supercurrent I that is given by $I = I_c \sin(\phi(t))$, where I_c is the critical current (i.e., the maximum supercurrent that the junction can support before entering the normal state), and $\phi(t)$ is the time-dependent phase difference across the junction [5]. Once the critical current is exceeded, the normal single electron tunneling becomes dominant.

3.1 Introduction to Superconductivity

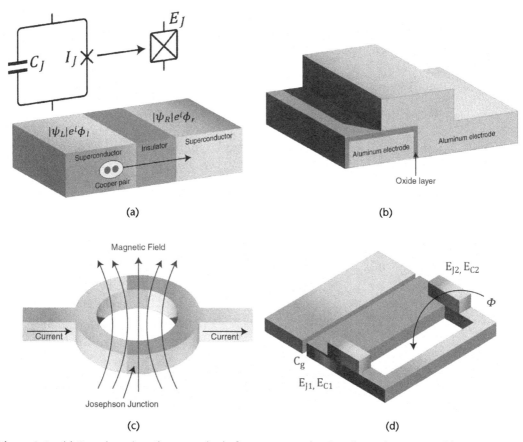

Figure 3.8 (a) Josephson junction comprised of two superconducting electrodes separated by an extremely thin insulating layer. Cooper pairs can coherently tunnel across the insulating barrier. The equivalent circuit of a Josephson junction is also shown. (b) Implementation of a Josephson junction using aluminum electrodes and an oxide layer (Al/AlOx/Al). (c) Two parallel Josephson junctions form a SQUID. The Josephson energy or the total inductance can be controlled using an external flux. (d) Implementation of a SQUID with Josephson and charging energy of E_{Ji} and E_{Ci}. C_g is the gate capacitance to control the charge.

The phase difference across the junction evolves in time in the presence of a potential V (i.e., $\hbar(d\phi/dt) = 2eV$). Now by taking the time-derivative of Josephson current, which is given earlier, we obtain

$$\frac{dI}{dt} = \left(I_c \cos\phi\right)\dot\phi = \frac{2e\dot V I_c}{\hbar}\cos\phi \tag{3.9}$$

The Josephson inductance L_J can be found using Faraday's law $V = -L_J \dot I$ as $L_J = \Phi_0/2\pi I_c \cos\phi = L(0)/\cos\phi$, where $\Phi_0 = h/e$ is the magnetic flux quantum, and $L(0) = \Phi_0/2\pi I_c$ the inductance at zero bias [5]. This nonlinear inductance combined with a parallel capacitor constitutes an anharmonic oscillator that serves as the basis for various superconducting qubit topologies [56].

The Josephson energy can be calculated using Faraday's law. Assuming that the Josephson phase is φ_1 at time t_1, and φ_2 at time t_2, the change in energy can be calculated as

$$\Delta E = \int_1^2 IV\,dt = \int_1^2 I\,d\Phi = \int_{\varphi_1}^{\varphi_2} I_c \sin\varphi\, d\left(\Phi_0 \frac{\varphi}{2\pi}\right) = -\frac{\Phi_0 I_c}{2\pi}\Delta\cos\varphi \qquad (3.10)$$

So, the energy stored in a Josephson junction can be considered a state function that depends only on the initial and final state of the junction. This energy is defined as

$$E(\varphi) = -\frac{\Phi_0 I_c}{2\pi}\cos\varphi = -E_{Jmax}\cos\varphi \qquad (3.11)$$

The $E_{Jmax} = \Phi_0 I_c/2\pi$ is called Josephson energy, and it is related to the Josephson inductance by $E_{Jmax} = L(0)I_c^2$.

Note that while a coil stores energy in the magnetic field generated by the current passing through it, a Josephson junction does not generate a magnetic field from the supercurrent. Instead, it stores energy in the kinetic energy of the charge carriers. The parameters of a Josephson junction can be determined by measuring the junction resistance R_n at room temperature (e.g., by using a four-terminal measurement). The critical current shows temperature dependence and can be determined using [6]

$$I_c = \frac{\pi\Delta(T)}{2eR_n}\tanh\left(\frac{\Delta(T)}{2k_B T}\right) \qquad (3.12)$$

where $\Delta(T)$ is the superconducting energy gap given in (3.6). For $T \to 0$ the relation can be simplified as $I_c = \pi\Delta(0)/2eR_n$, where $\Delta(0) = 1.76\,k_B T_c$. So having the critical temperature of the material and normal state junction resistance allows for calculating the critical current.

The Josephson energy can also be calculated using the normal state resistance. We define the resistance quantum as $R_q = h/4e^2 \approx 6{,}453.20\,\Omega$ and the Josephson energy is also given by $E_{Jmax} = (R_q/R_n)(\Delta(0)/2)$.

A SQUID uses two Josephson junctions in parallel, as shown in Figure 3.8(c, d). It is possible to tune the Josephson energy of a SQUID by applying an external magnetic field that generates a flux of ϕ. The Josephson energy of the SQUID can then be expressed as:

$$E_J(\phi) = E_{Jmax}\left|\cos\frac{\pi\phi}{\phi_0}\right| \qquad (3.13)$$

The nonlinearity of the Josephson junction and the tunability of Josephson energy are essential features that enables the use of Josephson junctions as a building block for qubits, which are artificial atoms. This tunability makes it possible to adjust the transition frequency of the qubit, enabling the performance of specific qubit gates through the utilization of an external flux bias line. Section 3.2 explains how a Josephson junction is employed to construct an artificial atom.

3.2 Superconducting Qubit

This section reviews the concepts related to the Cooper pair box and transmon qubit. Next, in this section we will examine how the qubit-environment interaction causes decoherence and explore the corresponding time scales and sources of decoherence.

3.2.1 Artificial Atom

Quantization is a fundamental consequence of quantum mechanics, whereby a quantum system can only respond to specific energies E or, equivalently, certain frequencies ($E = hf$), where the frequency f is associated with the transition energy E between two levels of the system. Atoms themselves are quantum systems and can be conceptualized as finely tuned receivers that exclusively respond to specific frequencies [see Figure 3.9(a)]. These unique frequencies act as fingerprints for each atom and can be used for identification purposes. The transition frequencies of a hydrogen atom are depicted in Figure 3.9(b).

As seen in Figure 3.9(b), each transition in a hydrogen atom corresponds to the photon emission at a specific frequency or wavelength. This is due to the nonequidistant energy levels of the atom. As discussed in Chapter 1, an anharmonic quantum oscillator can be used to model this behavior, which in contrast to a quantum harmonic oscillator, has nonequidistant energy levels. Anharmonicity is crucial for controlling the transition between energy levels, since without it a single-frequency pulse with a broad spectrum could cause excitations to many states, and thus the circuit would not exhibit the gate operations of a qubit. Isolating an anharmonic oscillator's first two energy levels allows them to be used as a qubit.

Natural atoms are utilized in various quantum computing platforms, including atom and ion qubits, as discussed in Chapter 2. However, the properties of natural atoms, such as the transition frequency, cannot be tuned. To emulate an artificial atom, we require a system that mimics an anharmonic oscillator with nonequidistant energy levels. As observed in Chapter 1, a parallel LC circuit can be quantized and employed as a quantum harmonic oscillator. By introducing nonlinearity, we can transform the LC circuit into an anharmonic oscillator. This is where the Josephson junction, with its nonlinear inductance becomes relevant. Figure 3.9(c, d) provides a comparison between the harmonic and anharmonic oscillators constructed from electrical elements.

Since Josephson junctions are superconducting components, we must use a refrigerator to cool these circuits below their critical temperature. Moreover, as discussed in Chapter 2, these quantum circuits are extremely sensitive to noise. Since a significant portion of the noise is temperature-dependent, these circuits are placed in a refrigerator to minimize the noise coupled to them. Chapter 8 discusses the operation of a dilution fridge.

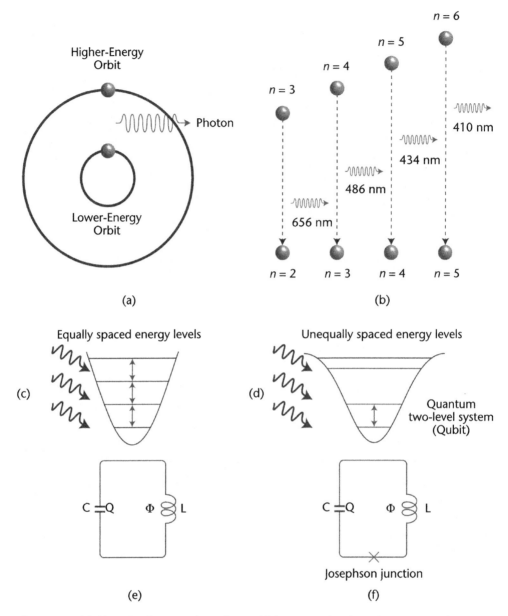

Figure 3.9 (a) When the electron relaxes from a higher energy level to a lower one, a photon with a certain frequency corresponding to the difference in energy levels is emitted. (b) Balmer series transtions of hydrogen atom; each transition has a unique frequency. (c) A quantum harmonic oscillator has equidistant energy levels, so there is no unique transition between energy levels. (d) An anharmonic quantum oscillator has nonequidistant energy levels. This results in unique transition energies between energy levels. (e) An LC circuit without a Josephson junction is a harmonic oscillator with equidistant energy levels. (f) An LC circuit with Josephson junction is an anharmonic quantum oscillator with nonequidistant energy levels. This allows one to solely address two energy levels, as required for a qubit.

By designing superconducting circuits or qubits, we can adjust their parameters to set the transition frequency. However, if the energy gap between the ground state (corresponding to |0⟩) and the first excited state (corresponding to |1⟩) is too small, thermal energy within the dilution fridge can cause unwanted transitions between these states. To ensure external control over the qubit, we must design the qubit to have a sufficiently high transition energy.

Let us roughly calculate the frequency associated with thermal energy in a standard dilution fridge. The thermal energy E_{th} is equal to Boltzmann's constant times the temperature, and the frequency associated with this energy is equal to Planck's constant times the frequency: $E_{th} = k_B T = h f_{th}$. A standard dilution fridge can reach a temperature of 20 mK, corresponding to a frequency of $f_{th} = 0.4$ GHz. To ensure that thermal excitations from the ground to the excited state of the qubit rarely occur, the transition energy corresponding to the external field needs to be much higher than the thermal energy (i.e., $E_{01} \gg E_{th}$). This results in a transition frequency $f_{01} = 10 f_{th} = 4$ GHz. Many charge qubits, such as the transmon, operate within the microwave frequency range of 4–8 GHz. The following qualities make microwave frequencies advantageous:

- As we have shown, these frequencies are high enough for known cryogenic techniques to be used.
- One can use commercially available microwave components and well-known microwave techniques.

Although going higher in frequency can have some advantages, it causes severe issues related to the cost of the components, design complexity, and fabrication capabilities.

3.2.2 Cooper Pair Box

This section examines the Cooper pair box (CPB) charge qubit. Recall that a Josephson junction can be modeled as an inductor in parallel with the junction's capacitance, C_J. As shown in Figure 3.10(a), the electrodes of the Josephson junction, which are called the superconducting island and superconducting reservoir, allow for the coherent tunneling of Cooper pairs due to the Josephson effect. To manipulate the number of charges on the island, a voltage source is capacitively coupled through C_g, utilizing the Coulomb blockade tunneling effect (see Figure 3.10). The gate voltage can adjust the energy levels on the island, permitting a Cooper pair to hop to or from the island, causing Cooper pairs to tunnel. The basis states of the CPB qubit are largely determined by the number of Cooper pairs residing on the island.

Next, let us examine the eigenenergies and eigenstates of the CPB. The Hamiltonian of a CPB consists of the charging energy required to add a Cooper pair and the junction's coupling energy or the Josephson energy. The Hamiltonian is given by [7]

$$H = \frac{\hat{Q}^2}{2C} + \frac{\hat{\Phi}^2}{2L} = \frac{\hat{Q}^2}{2C} - E_J \cos\left(\frac{2\pi\hat{\Phi}}{\Phi_0}\right) = \frac{(2e)^2}{2C} \hat{n}^2 - E_J \cos\hat{\phi}$$
$$= 4E_C \hat{n}^2 - E_J \cos\hat{\phi}$$

(3.14)

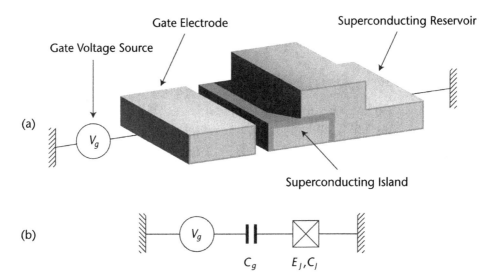

Figure 3.10 (a) Cooper pair box and (b) its equivalent circuit.

where the charge operator $\hat{Q}$ is related to the Cooper pair number operator $\hat{n}$ as $\hat{Q} = -(2e)\hat{n}$, and $\hat{\phi}$ is related to the flux operator $\hat{\Phi}$, and $E_J = I_c\Phi_0/2\pi$. The total capacitance C is the sum of the gate capacitance C_g and the junction capacitance C_J (i.e., $C = C_g + C_J$). Including the charge number n_g from the gate results in $H = 4E_C(\hat{n} - n_g)^2 - E_J\cos\hat{\phi}$.

By comparing the electrical and mechanical quantum harmonic oscillators (see Chapter 1), we observe that the charge and phase operators are analogous to the position and momentum operators. Specifically, the phase operator can be described as the position operator, while the charge number operator can be described as momentum operator, as illustrated in (3.15):

$$\hat{n} \rightarrow -i\frac{\partial}{\partial\phi}$$
$$\left(\hat{p} \rightarrow -i\hbar\frac{\partial}{\partial x}\right)$$
(3.15)

The commutation relation for these conjugate variables is $[\hat{\phi},\hat{n}] = -i$.

To find the wavefunction and the energies corresponding to this Hamiltonian, we need to solve the eigenvalue problem with energy E_k and wavefunction $|\psi_k\rangle$ satisfying $\hat{H}|\psi_k\rangle = E_k|\psi_k\rangle$. Working in the phase basis $\hat{\phi}|\phi\rangle = \phi|\phi\rangle$, we need to solve

$$\langle\phi|\hat{H}|\psi_k\rangle = E_k\langle\phi|\psi_k\rangle = E_k\psi_k(\phi)$$
(3.16)

After replacing the operators in (3.26), we obtain

$$\left[4E_C\left(-i\frac{\partial}{\partial\phi} - n_g\right)^2 - E_J\cos\phi\right]\psi_k(\phi) = E_k\psi_k(\phi)$$
(3.17)

3.2 Superconducting Qubit

which is equivalent to Mathieu's equation, and its solution is written in terms of functions of the same name. In Figure 3.11(a), the first three energy levels of the CPB are shown as a function of charge offset n_g for $E_J/E_C = 1$. Each energy level varies with charge n_g. Figure 3.11(b) illustrates a two-level approximation of the first two energy levels. Notably, the transition frequency between the two energy levels is not constant; rather, it depends on the gate charge n_g. This feature makes the qubit highly sensitive to charge fluctuations, such as those resulting from tunneling charges in the imperfect dielectrics and interfaces which are named two level system. As a result, dephasing can occur (see Section 1.6.6.2) because the energy levels are not sharply defined and are modulated by the charge fluctuations. One method of reducing the effect of charge fluctuations in a CPB is to operate the qubit at so-called sweet spots, as depicted in Figure 3.11(a), where the dE/dQ is minimal. Section 3.2.3 demonstrates that a better way to minimize the sensitivity of the energy levels to charge fluctuations is to control the E_J/E_C ratio.

3.2.3 Transmon Qubit

Another way to reduce the sensitivity of the CPB's energy levels to charge fluctuations is to increase the E_J/E_C ratio to operate in the transmon regime. As depicted in Figure 3.12(a), the energy levels become less sensitive to changes in charge as the value of E_J/E_C rises. Notably, when $E_J/E_C = 50$, the energy levels become nearly flat. Note that to operate in the transmon regime, the E_J/E_C ratio must be between 20 and 100, with 50 being the typical value. The transmon qubit was invented by a group of researchers at Yale University [7]. The term "transmon" is a shortened version of "transmission line shunted plasma oscillation qubit."

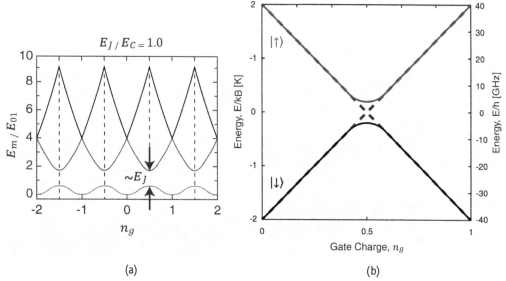

(a)   (b)

Figure 3.11 (a) Eigenenergies E_m (first three levels, $m = 0,1,2$) of the CPB as a function of the effective offset charge n_g for $E_J/E_C = 1$. The bottom of the $m = 0$ level is chosen as the zero point of energy. Energies are given in units of the transition energy E_{01}, evaluated at the degeneracy point $n_g = 1/2$. The vertical dashed lines in (a) are placed at the charge sweet spots at half-integer n_g. (b) Two-state approximation near a charge degeneracy point.

The E_J/E_C ratio can be increased by decreasing the charging energy $E_c = q^2/2C$ using a large capacitor. Typically, an interdigital capacitor is shunted with the Josephson junction, as illustrated in Figure 3.12(b). To enable the transition frequency to be tuned, the Josephson junction is replaced with a SQUID. An external coil or a flux bias line is then utilized to adjust the transition frequency, as shown in Figure 3.12(c).

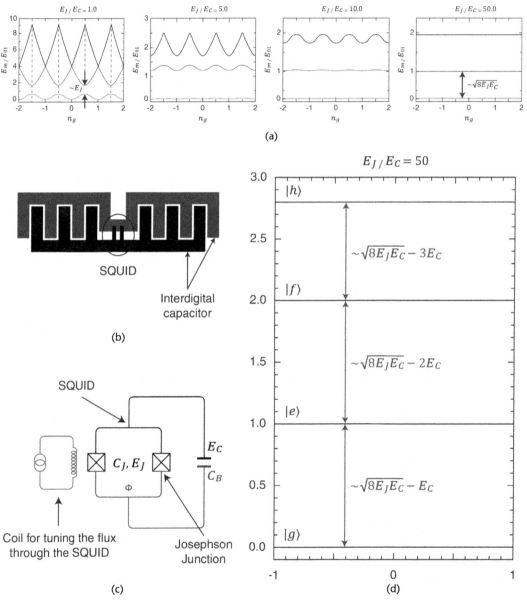

Figure 3.12 (a) Eigenenergies E_m (first three levels, $m = 0,1,2$) of the CPB as a function of the effective offset charge n_g for various values of E_J/E_C; (b) lithography implementation of the equivalent circuit; and (c) equivalent circuit of the transmon qubit, where the two parallel Josephson junctions form a SQUID and allow tuning the qubit frequency using an external flux. (d) The transition energy is proportional to the geometric mean of the inductor and capacitor energies. (e) The flux bias can adjust the qubit's transition frequency (f) relative anharmonicity of the qubit as a function of E_J/E_C.

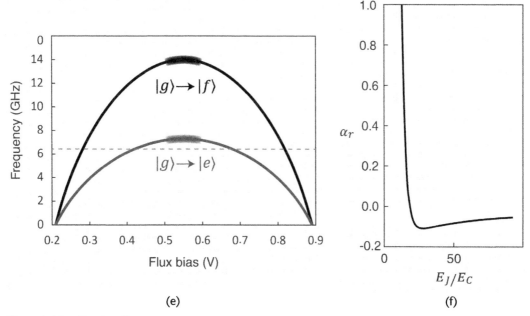

Figure 3.12 *(Continued)*

The qubit's transition frequency ω_{01} is directly proportional to the geometric mean of the Josephson junctions' and capacitor's characteristic energies, E_J and E_C, as shown in Figure 3.12(d) [7]

$$\omega_{01} \approx \frac{\sqrt{8E_J E_C} - E_C}{\hbar} \quad (3.18)$$

The desired transition frequency can be achieved by engineering the geometry and parameters of the capacitor and the Josephson junction, which usually lies within the 5–7-GHz range. Using SQUID allows tuning the qubit frequency ω_q by an external magnetic field and adjusting the Φ flux

$$\omega_q \approx \omega_{01} \sqrt{\left|\cos\left(\frac{\pi\Phi}{\Phi_0}\right)\right|} \quad (3.19)$$

In Figure 3.12(e), it can be observed how the frequency changes with the flux bias. A sweet spot for the flux bias can be identified, where the impact of changes in flux on the qubit's frequency is minimized.

Operating in the transmon regime significantly reduces the qubit's sensitivity to charge fluctuations that could otherwise cause decoherence. However, there is a trade-off between sensitivity to charge fluctuations quantified by charge dispersion and the qubit's anharmonicity, as we will see shortly. Specifically, as the E_J/E_C ratio increases to reduce sensitivity to charge fluctuations, the anharmonicity of the qubit decreases. This results in energy levels that become almost equally spaced, which affects the qubit levels' addressability.

Now, let us look at the effect of E_J/E_C on charge dispersion and anharmonicity. The charge dispersion for the mth energy band ϵ_m is defined as

$$\epsilon_m = E_m(n_g = 0) - E_m(n_g = 1) \tag{3.20}$$

The charge dispersion is $\epsilon_m \approx 4E_C$ for CPB with $E_J < E_C$. For $E_J/E_C \gg 1$, charge dispersion is reduced exponentially, leading to exponential decrease in sensitivity to charge fluctuations, as given by [7]

$$\epsilon_m \simeq (-1)^m E_C \frac{2^{4m+5}}{m!} \sqrt{\frac{2}{\pi}} \left(\frac{E_J}{2E_C}\right)^{\frac{m}{2}+\frac{3}{4}} e^{-\sqrt{\frac{8E_J}{E_C}}} \tag{3.21}$$

The absolute anharmonicity α is defined as the energy difference between transitions from the ground to the first excited state E_{01} and from the first to the second excited state E_{12}

$$\alpha \equiv E_{12} - E_{01} \cong -E_C \tag{3.22}$$

Anharmonicity of zero refers to a harmonic oscillator with equal energy levels. Recall that zero anharmonicity creates issues for the two-level addressability. The relative anharmonicity α_r is defined as

$$\alpha_r \equiv \frac{\alpha}{E_{01}} = \frac{\alpha}{\hbar\omega_{01}} \cong -\left(\frac{8E_J}{E_C}\right)^{-\frac{1}{2}} \tag{3.23}$$

As seen from (3.23), the E_J/E_C ratio also affects the relative anharmonicity, as shown in Figure 3.12(f).

Compared to charge dispersion, anharmonicity given in (3.23) reflects a weaker algebraic dependence on E_J/E_C. A sufficiently large anharmonicity is needed to prevent qubit operations from exciting other transitions in the system. As there is a trade-off between the charge dispersion and anharmonicity, it is essential to choose a value of E_J/E_C that is large enough to suppress the charge dispersion and yet small enough to have enough anharmonicity.

According to (3.23), the absolute anharmonicity is given by $\alpha = \hbar\omega_{01}\alpha_r$. The frequency spread of a bandwidth-limited pulse allows estimating the corresponding minimum pulse duration to be $\tau_p \sim |\omega_{01}\alpha_r|^{-1}$ [7]. So, the anharmonicity limits the pulse duration and therefore the speed of qubit operations. For the coherent control of the qubit, the pulse duration must remain small compared to T_1 and T_2. Large anharmonicity results in a short pulse duration; however, the length of microwave pulses cannot be significantly shorter than 10 ns. So, for a typical pulse length of 10 ns, a minimum anharmonicity of $|\alpha_r^{min}| \sim (\tau_p\omega_{01})^{-1} \sim (10 \text{ ns} \times 2\pi \times 10 \text{ GHz})^{-1} = 1/200\pi$ corresponding to $20 \lesssim E_J/E_C \ll 5 \times 10^4$ is required. This allows a large range of E_J/E_C with exponentially decreased sensitivity to charge noise and yet sufficiently large anharmonicity for qubit operations [7].

Moreover, by increasing the E_J/E_C ratio, the sensitivity to charge noise is reduced without sacrificing too much anharmonicity, which results in significantly improved dephasing time.

Now, let's examine some numerical values for a real-world qubit. Table 3.1 shows that the junction's resistance at room temperature R_n, measured at a probe station, determines the Josephson energy E_J, which ranges from 10 to 60 GHz. Parameters such as oxidation time, deposition thickness, and junction area determine the normal resistance R_n. So, several tests and feedback from fabrication are necessary to adjust these parameters to reach the desired R_n and, consequently, the desired Josephson energy.

The interdigital capacitor shown in Figure 3.12(b) is used to build the transmon's capacitance, where a typical design has a finger length of 70 μm, a finger width of 10 μm, and a gap of 10 μm. The number of fingers is chosen such that the length of the qubit is around 300–400 μm. Simulation software is used to determine the total capacitance consisting of the capacitance of the interdigital capacitor, coupling capacitance to the resonator, and other parasitic capacitances. Using the simulated capacitance makes it possible to estimate the charging energy E_c. The transition energy of the qubit is then proportional to the geometric mean of the inductor and capacitor characteristic energies (i.e., E_J and E_C). When using SQUID, the transition frequency of the qubit can be tuned using an external flux, as given in Table 3.1.

Other actively investigated superconducting qubit types include the RF-SQUID (prototype of a flux qubit) and the current-biased junction (prototypal phase qubit); for a review, see [8, 9].

Table 3.1 Summary of Transmon Qubit Design Formulas

	Formula	Range
Transmon regime	$E_J/E_C \gg 1$	$20 < E_J/E_C < 100$
Josephson junction	$\Delta(0) = 1.76\, k_B T_c$	Josephson energy 10–60 GHz
	Resistance quantum	R_n: 2–10 Kohm
	$R_q = h/4e^2 \approx 6{,}453.20\,\Omega$	
	$\Phi_0 = h/2e = 2.07 \cdot 10^{-15}\,\text{Tm}^2$	
	R_n is the normal state resistance	
	$E_{Jmax} = (R_q/R_n)(\Delta(0)/2)$	
	$I_c = \pi \Delta(0)/2_e R_n$	
	$L_J = \Phi_0/2\pi I_c \cos\Delta \approx \hbar^2/4e^2 E_J$	
Charging energy	$E_C = e^2/2C_t$	300–400 MHz
	$C_t = C_J + C_B + C_g$	$C_J = 3$–10 fF
Qubit frequency	$\omega_{01} \approx 1/\sqrt{L_J C_t} \approx \sqrt{8E_J E_C}/\hbar$	5–7 GHz
	Using SQUID	
	$\omega_q \approx \omega_{01}\sqrt{|\cos(\pi\Phi/\Phi_0)|}$	
Anharmonicity	$\alpha = E_{12} - E_{01} \approx -E_C$	
	$\alpha_r = \alpha/E_{01} = -\sqrt{E_C/8E_J}$	

Lithography techniques, such as electron beam lithography (EBL), are utilized to fabricate nanometer-sized structures like the Josephson junction. Once the fabrication is complete, the chip is placed in a sample holder, and a bonding machine is employed to connect the micron-sized pads on the chip to the PCB traces on the sample holder, as depicted in Figure 3.13. The sample holder, containing the chip, is then placed in a dilution refrigerator and cooled to temperatures between 10 and 20 mK [10–12]. Once the desired temperature is reached, qubit operations can be conducted.

3.2.4 Qubit Coherence Time Scales

As discussed in Chapter 1, the interaction between a qubit and its surrounding environment can be understood as the environment effectively measuring the qubit, resulting in the loss of its coherent superposition state and ultimately leading to the collapse of its wave function. The coherence of a single qubit is characterized by "coherence times" referred to as T_1, the longitudinal relaxation time, and T_2, the transverse relaxation time. These notions of T_1 and T_2 are borrowed from the field of nuclear magnetic resonance (NMR).

Section 3.2.4.3 discusses various sources of decoherence and how to minimize their effects. Chapter 7 further demonstrates the extensive use of noise-suppression techniques to reduce the interaction between the qubit and its environment. However, it is impossible to completely isolate the qubit due to the need for control and readout access. While the error rate in a transistor is $1/10^{20}$, for a qubit, it is much higher. Therefore, as discussed in Section 2.2.6 quantum error correction is unavoidable for quantum computations.

3.2.4.1 Longitudinal Relaxation

The qubit undergoes interactions with its environment, resulting in both qubit excitation and relaxation processes. Thermal excitations are significantly suppressed at the temperature of 20 mK in a dilution refrigerator, allowing us to primarily

Figure 3.13 A quantum processor chip is connected through the pads to the microwave lines by bonding wires. The chip communicates with the outside world through these SMP connectors, allowing it to control and read out the state of the qubits on the chip.

3.2 Superconducting Qubit

focus on the relaxation process. On the Bloch sphere representation, relaxation is depicted by the transition of the Bloch vector from the $|1\rangle$ state to the $|0\rangle$ state.

One of the fundamental distinctions between quantum mechanics and classical mechanics lies in the role of the vacuum. In classical mechanics, the vacuum is typically perceived as an empty space with no influence on physical systems. However, in quantum mechanics, atoms and other quantum systems interact with the vacuum due to vacuum quantum fluctuations, and this interaction holds a central significance in quantum electrodynamics. For example, atoms in a high energy state eventually decay to their ground state. This behavior arises from the coupling of atoms to the continuum of radiation modes present in the vacuum, where resonant vacuum modes interact with the atom's excited state, leading to the phenomenon of spontaneous emission. Refer to Figure 3.14 for a visual depiction of this process.

So, the interaction with resonant modes of the vacuum results in relaxation. The interaction of the atom with nonresonant modes of the continuum results in Lamb shift, which causes a tiny correction of the atomic energy levels. Section 3.3.2.3 covers the same phenomena in a qubit, where the energy levels of the qubit also undergo Lamb shift due to interaction with the vacuum.

The relaxation time T_1 is the average time it takes for a qubit in the excited state $|1\rangle$ to relax to the ground state $|0\rangle$ spontaneously. The relaxation is also called depolarization due to its origin in NMR, where an external magnetic field polarizes the sample and aligns the spins. The polarization decays due to the spin interaction with the lattice.

As shown in Figure 3.15, to measure T_1, we bring the qubit to the excited state $|1\rangle$ using a π pulse or, equivalently, the X-gate. Then we wait for time Δt and measure the population of the excited state. We want to measure a temporal range of 800 μs with a time step of 80 μs. Then we excite the qubit, wait for 80 μs and report the percentage of measurements that return $|1\rangle$.

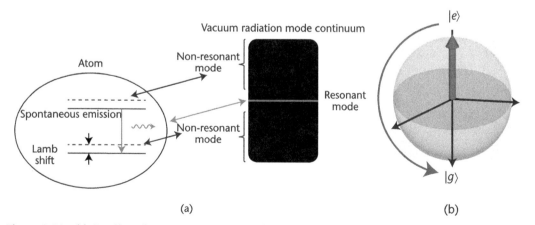

Figure 3.14 (a) Coupling of an atom to vacuum modes. The vacuum resonant mode results in spontaneous emission, while interaction with the nonresonant modes of the vacuum continuum results in Lamb shift. (b) Bloch sphere representation of a qubit relaxation.

Next, we excite once more, wait for 160 μs, report the percentage of measurements that return |1⟩, and repeat it for the next steps. For each point, it is better to repeat the measurement and take the average to increase the accuracy. T_1 can be extracted by interpolating the measured points using the following fit function, where T_1 represents the time the excited state population reaches $1/e$ of its initial value and C and D are constants.

$$x(t) = Ce^{-\frac{t}{T_1}} + D \tag{3.24}$$

3.2.4.2 Transverse Relaxation

Transverse relaxation time T_2 is a combination of longitudinal relaxation with Γ_1 rate and pure dephasing with Γ_ϕ rate.

First, let us look at pure dephasing. As discussed in Chapter 1, processes that randomly modify the effective transition frequency of the qubit result in dephasing, which causes the qubit to accumulate a random phase. Dephasing is frequently brought on by low-frequency noise much below the transition frequency of the qubit. As discussed in Chapter 1, the qubit state precesses about the Z-axis at the qubit frequency, f_{10}; that is, the qubit phase ϕ advances in time according to $\phi(t) = 2\pi f_{10} t$. A frequency offset of δf for a period t creates a phase offset $\delta\phi = \delta f t$. Thus, frequency noise ($\delta f(t)$) over time produces phase noise ($\delta\phi$). Of course, if we had a constant frequency offset, we could change or remeasure our bare frequency and eliminate the phase offset. Unlike energy relaxation, pure dephasing can be caused by noise at a large range of frequencies. Moreover, during pure dephasing,

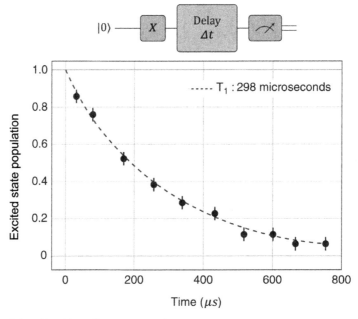

Figure 3.15 Relaxation time, T_1 measurement.

no energy is exchanged with the environment; therefore, it can be reversed by applying unitary operations (e.g., dynamical decoupling pulses) [10].

Any energy relaxation process will also dephase the qubit at a rate $\Gamma_1/2$, as the Bloch vector of a superposition state, is halfway between the $|0\rangle$ and $|1\rangle$ states. The phase information gets lost in such a relaxation event, as the Bloch vector points to the $|0\rangle$ pole after relaxation. Therefore, it is no longer known which way the Bloch vector had been pointing along the equator; therefore, the relative phase of the superposition state is lost.

Fluctuations on shorter time scales occurring during a single experiment (during one decay lifetime) cause dephasing at a rate Γ_ϕ called the pure dephasing rate. One can define a dephasing quality factor $Q_\phi = \omega_a/\Gamma_\phi$, which is the number of coherent oscillations before the qubit accumulates a random π phase shift.

Fluctuations occurring on longer timescales create an ensemble dephasing denoted by the rate Γ^*_ϕ, called the inhomogeneous dephasing rate. This last type is analogous to spatial magnetic field inhomogeneities in NMR and is often called inhomogeneous broadening. Like its NMR counterpart, it can theoretically be reduced using spin-echo techniques because it is not intrinsic to the system. Though it is possible to compensate for inhomogeneous broadening, doing so uses up time in operations.

The transverse relaxation rate is the sum of all the abovementioned rates:

$$\Gamma_2 = \frac{\Gamma_1}{2} + \Gamma_\phi + \Gamma^*_\phi \tag{3.25}$$

Two measurements are performed to determine the transverse relaxation time. Figure 3.16 shows the Hahn echo experiment for measuring the pure dephasing time and the Ramsey experiment for measuring the inhomogeneous dephasing time.

As shown in Figure 3.16(a), the Hahn echo experiment consists of three pulses. State $|+\rangle = (|0\rangle + |1\rangle)/\sqrt{2}$ is obtained by rotating the initial state $|0\rangle$ by $R_y(\pi/2)$ around the y-axis. We apply a delay for time $\Delta t/2$, followed by a π rotation around the y-axis. This would take the state to $|-\rangle = (|0\rangle - |1\rangle)/\sqrt{2}$ in the absence of dephasing noise. After one more delay of $\Delta t/2$, we apply another $R_y(\pi/2)$ rotation, which would bring the state back to $|0\rangle$ in the absence of dephasing.

As in (3.24), an exponential curve is fitted to the experimental data, where T_2 is the appropriate time constant:

$$f(t) = Ae^{-\frac{t}{T_2}} + B \tag{3.26}$$

As shown in Figure 3.16(a), the curve decays to 50% and not zero. The reason is that during the delay, the superposition state is also subject to T_1 relaxation. If the state decays to $|0\rangle$, the second $\pi/2$ rotation takes the state to $|+\rangle$, resulting in an equal chance of measuring $|0\rangle$ and $|1\rangle$. In contrast, the states $|0\rangle$ and $|1\rangle$ are

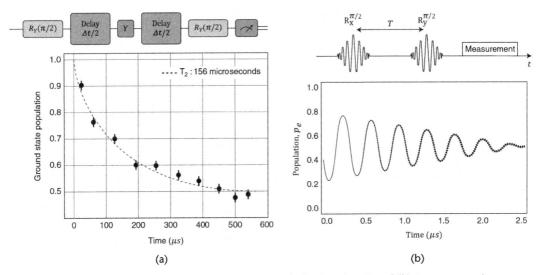

Figure 3.16 (a) Hahn echo experiment to measure pure dephasing time T_2 and (b) Ramsey experiment to measure the inhomogeneous dephasing time T_2^*.

not sensitive to a phase change (they are located at poles and cannot have phase change), so dephasing does not impact the results of the T_1 experiment.

The Ramsey experiment is shown in Figure 3.16(b). An $X_{\pi/2}$-pulse, intentionally detuned from the qubit frequency by $\delta\omega$, is applied to bring the qubit to the equator. This causes the Bloch vector to precess in the rotating frame at a rate $\delta\omega$ around the z-axis. After a delay time T, a second $X_{\pi/2}$ pulse is used to project the Bloch vector on the equator to the z axis. This is shown in Figure 3.16(b). The oscillations decay with an approximately exponential decay function, which gives us the T_2^*. Figure 3.17 shows Bloch sphere diagram of a Ramsey experiment.

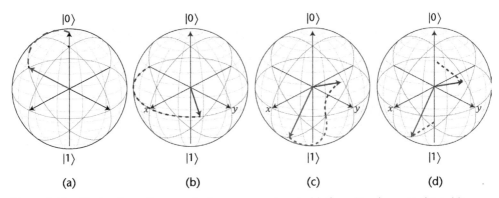

Figure 3.17 Bloch sphere diagram of a Ramsey experiment: (a) the $\pi/2$ pulse puts the qubit on the equator; (b) The qubit evolves for time t; (c) the recovery $\pi/2$ pulse rotates the state about the x-axis or the y-axis; and (d) the measurement projects the y or x (coordinate) to the probability of $|1\rangle$.

3.2.4.3 Sources of Qubit Decoherence

A qubit's coherence times, T_1 and T_2, are influenced by both internal and external sources. Internal sources encompass trapped vortices, quasiparticle tunneling, and material-based two-level systems. On the other hand, external sources consist of parasitic electric and magnetic fields, voltage fluctuations on the gate, fluctuations in magnetic flux, and interactions with photons, phonons, and the environmental circuit modes. Additionally, ionizing radiation stemming from environmental radioactive materials and cosmic rays can increase the quasiparticle density, thereby limiting the coherence times of superconducting qubits [13]. Various decoherence channels in a superconducting qubit and their impacts on coherence times are outlined in Table 3.2.

When the temperature is above zero and the bias voltage is below the conduction threshold at zero temperature (indicating that the bias voltage is smaller

Table 3.2 Sources of Qubit Decoherence and Their Effect on Coherence Times

Coherence Time		Description
T1	Spontaneous emission	Relaxation from the excited state to the ground state.
	Purcell effect	Modification of the spontaneous emission rate when placing the qubit in a cavity.
	Dielectric loss	Dielectric losses of the substrates cause relaxation and dephasing of the qubit. Amorphous SiO_2 has loss tangents as large as 5×10^{-3} [13]. Crystalline Si and sapphire offer favorably low-loss tangents ($\tan\delta$) in the order of 10^{-6} and 10^{-8} at cryogenic temperature.
	Quasi-particle tunneling	Single electron tunneling causing an overall odd number of electrons or thermal breaking of Cooper pairs leads to both relaxation and dephasing in qubits.
	Flux coupling	An external magnetic flux bias allows for tuning the qubit frequency. It also opens up additional channels for energy relaxation.
T2	Charge noise	Charge fluctuations cause qubit dephasing.
	Flux noise	Noise in the externally applied flux translates into fluctuations of the effective Josephson coupling energy E_J.
	Critical current noise	Fluctuations in Josephson energy due to the noise in the critical current generated by trapping and detrapping of charges associated with spatial reconfigurations of ions inside the tunneling junction.
	Quasi-particle tunneling	In addition to relaxation, it also causes dephasing. Adding or removing a single charge from the island alters the transition frequency of the qubit drastically if not operating in the transmon regime.
	E_c noise	Noise in the charging energy or fluctuations in the effective capacitances of the circuit. No evidence to be a limiting factor for transmon.
	Interaction with photons and phonons	Cosmic rays. Trapped vortices.

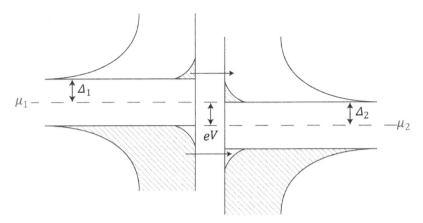

Figure 3.18 SIS junction quasiparticle tunneling at temperatures $T > 0$. The bias voltage is below the threshold for conduction at $T = 0$ (i.e., $eV = \Delta_1 + \Delta_2$). (Horizontal arrows depict tunneling involving thermally excited electrons or holes.)

than the combined superconducting energy gaps), tunneling of nonequilibrium quasi-particles can take place between the superconductors. This tunneling process entails the transfer of electron-like or hole-like quasi-particles, as depicted by horizontal arrows in Figure 3.18.

Previously, it was demonstrated that, in theory, single electrons cannot tunnel into the superconducting energy gap. However, single-particle excitations known as quasiparticles can also tunnel across the Josephson junction because they are above the energy gap. Figure 3.18(b) illustrates the band structure of an SIS junction or a Josephson junction. At temperatures above zero ($T > 0$), certain quasiparticle states may exist in the unoccupied band above the gap on the left electrode, as well as some empty states below the right electrode. Both scenarios result in a flow of single-electron current across the junction. The undesired tunneling of single electrons can introduce errors when stable quantum numbers, such as charge number in a charge qubit or parity in a topological qubit, are required. The interaction between quasiparticles and the phase degree of freedom creates an undesired pathway for qubit energy relaxation [14].

The population of quasi-particles decreases exponentially as the temperature decreases. This is another reason to reach millikelvin temperatures, which are significantly below the critical temperature of the superconducting metal. However, even at these low temperatures, nonequilibrium quasiparticles still exist, leading to relaxation effects. Keeping quasi-particles away from the active components of the qubit presents a viable approach to enhance the device's performance. Methods such as gap engineering or vortex traps have been employed but proved ineffective for transmon qubits. However, a normal metal trap has proven effective for transmon qubits. The concept involves using a normal metal, such as copper, on top of the qubit with an oxide layer acting as a barrier between them, as depicted in Figure 3.19(a). As shown in Figure 3.19(b, c), quasi-particles in the superconducting region can tunnel into the normal metal and become trapped there. They dissipate their energy through phonon emission or inelastic electron-electron scattering [15].

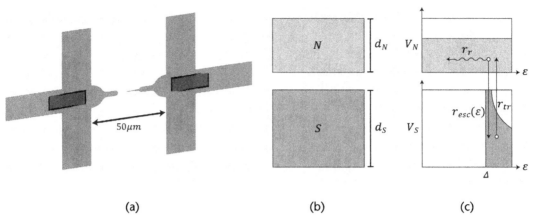

Figure 3.19 (a) Implementation of a quasi-particle trap in a transmon qubit; (b) a thin insulating layer separating a normal metal N of thickness, d_N, from a tiny superconductor S of thickness, d_S; and (c) trapping process: tunneling from S to N with a rate r_{tr} and from N to S with a rate $r_{esc}(\varepsilon)$, and relaxation in N with a rate r_r. (*After:* [15].)

3.2.4.4 Strategies for Qubit Coherence Improvement

The coherence time of superconducting qubits can be enhanced through various strategies, including the following:

- Improving the qubit itself through better materials and fabrication techniques;
- Improving the environment through filtering and shielding techniques (discussed in Chapter 7);
- Improving the control by utilizing high-performance equipment such as low-phase noise microwave sources (see Chapter 7).

Some strategies for improving qubits and minimizing the effects of decoherence are listed as follows.

- Enhance the characteristics of the junctions and materials to eliminate excess sources of 1/f noise.
- Use shielding and filtering to block external sources of decoherence, such as thermal noise and ionizing radiation (see Chapter 8).
- Design qubits with reduced sensitivity to specific sources of decoherence (e.g., transmon design) or operate qubits at sweet spots with minimum sensitivity to charge or flux fluctuations.
- Take advantage of symmetries in design and operation to cancel out unwanted effects.
- Use dynamic methods, such as spin echo and geometric manipulations, to counteract specific sources of decoherence.

The improvement of qubit decoherence time over the years shows an improvement of about three orders of magnitude for each decade. The number of operations per error has also improved by four orders of magnitude in a decade, reaching 10^4, the threshold for implementing error correction.

3.3 Qubit Control and Readout

We have observed the process of creating an artificial atom, or a qubit, through lithography. However, simply possessing a qubit is insufficient. It is crucial to control the state of the qubit on a Bloch sphere and ultimately determine the qubit's state using a readout technique. Sections 3.3.1 and 3.3.2 delve into how these two operations—control and readout of a superconducting transmon qubit—can be accomplished using cavity quantum electrodynamics (CQED), also referred to as circuit QED (cQED) when implemented on-chip.

To achieve this, comprehending the interaction between a qubit and light is essential, as this interaction is crucial for controlling and reading the qubit's state. Many principles from quantum optics, which explore light-matter interactions, can be directly applied to qubits. Sections 3.3.1 and 3.3.2 explore the interaction between a qubit and a free-space electromagnetic (EM) field, an EM field in a cavity, as well as phenomena such as the Rabi oscillations, ac-Stark shift, Lamb shift, and Kerr effect.

3.3.1 Qubit Control

As discussed in Chapter 2, quantum algorithms are implemented using single-qubit and two-qubit gates. Furthermore, Chapter 2 shows that a single-qubit gate can be represented as a rotation on a Bloch sphere. The positioning on the Bloch sphere can be manipulated through pulses of light, specifically microwave radiation in the case of a superconducting qubit. Section 3.3.1.1 explores the interplay between radiation and a two-level system or qubit to comprehend how we can exercise control over the state of a single qubit on the Bloch sphere.

3.3.1.1 Rabi Oscillations

The most basic scenario for the interaction between light and matter involves a two-level system, or a qubit interacting with a single-mode EM field. This interaction leads to oscillations of the qubit state between the ground and excited states at a frequency referred to as the Rabi frequency. Figure 3.20(a) demonstrates that the Rabi oscillation corresponds to the rotation of the state vector on the Bloch sphere. As we will soon discover, the speed of this rotation is determined by the Rabi frequency, which is influenced by the field's amplitude and the qubit's dipole moment.

Let us now explore two models for light-matter interaction: the semi-classical Rabi model and the Jaynes-Cummings model. The semiclassical model utilizes a quantized atom model combined with a classical description of the electromagnetic field. On the other hand, the Jaynes-Cummings model quantizes both the atom and the field.

The Rabi model is based on the assumption of a classical representation of the field $E(t)$, which is given as follows $E(t) = E_x(t) = E_0\cos(\omega t)$ [16]. The atom is considered to be quantized into two states: the ground state $|g\rangle$ and the excited state $|e\rangle$. The interaction between the field and the atom occurs through the atom's dipole moment, denoted as $\boldsymbol{d} = -e\boldsymbol{r}$, where e represents the charge of the electron, and $\boldsymbol{r}$ corresponds to the size of the atom. The Hamiltonian describing the interaction

3.3 Qubit Control and Readout

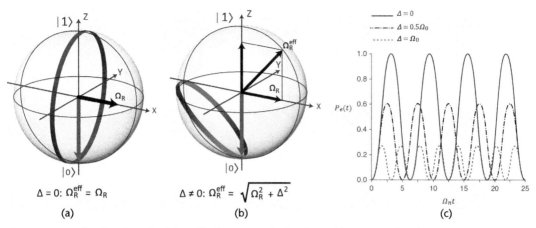

Figure 3.20 Visualization of Rabi oscillations on a Bloch sphere: (a) resonant (no detuning), (b) nonresonant (finite detuning), and (c) Rabi oscillations for various detuning values. The vertical axis is the probability of being in the excited state.

between the atom and the field is given by $\hat{H}^{(I)}(t) = \hat{d} \cdot E(t) = -\hat{d}E_0\cos(\omega t)$, where $\hat{d}$ is the dipole moment operator.

Recall from Section 1.5.2.4 that the Hamiltonian of an isolated noninteracting atom with the ground state energy of zero is given by $\hat{H}_{atom} = \hbar\omega_{eg}|e\rangle\langle e|$. The total Hamiltonian consists of the atom and the interaction Hamiltonian given by

$$\hat{H} = \hat{H}_{atom} + \hat{H}^{(I)} = \hbar\omega_{eg}|e\rangle\langle e| - \hat{d}E_0\cos(\omega t) \quad (3.27)$$

We assume that the probability of finding the atom in the ground state $|C_g(t)|^2$ and excited state $|C_e(t)|^2$ are time-dependent. Therefore, the following Ansatz is used to solve the Schrödinger equation.

$$|\psi(t)\rangle = C_g(t)|g\rangle + C_e(t)e^{\frac{-i(E_e-E_g)t}{\hbar}}|e\rangle = C_g(t)|g\rangle + C_e(t)e^{-i\omega_{eg}t}|e\rangle \quad (3.28)$$

Substituting this Ansatz in the time-dependent Schrödinger equation $i\hbar\partial|\psi(t)\rangle/\partial t = \hat{H}|\psi(t)\rangle$ results in a system of coupled first-order differential equations for the probability amplitudes

$$\begin{aligned}\dot{C}_g &= -\frac{i}{\hbar}E_0\cos(\omega t)d_{eg}e^{i\omega_{eg}t}C_e \\ \dot{C}_e &= -\frac{i}{\hbar}E_0\cos(\omega t)d_{eg}^*e^{-i\omega_{eg}t}C_g\end{aligned} \quad (3.29)$$

The initial conditions for solving this system of differential equations at t close to zero are $C_g(0) \approx 1$ and $C_e(0) \approx 0$, meaning that all the population is in the ground state at $t \approx 0$. By substituting these initial conditions in (3.29), we obtain

$$\dot{C}_g = 0$$

$$\dot{C}_e = -\frac{i}{\hbar} E_0 \frac{e^{+i\omega t} + e^{-i\omega t}}{2} d_{eg}^* e^{-i\omega_{eg} t} = -\frac{i}{2\hbar} E_0 d_{eg}^* \left(e^{+i(\omega - \omega_{eg})t} + e^{-i(\omega_{eg} + \omega)t} \right) \quad (3.30)$$

Using the rotating-wave approximation (RWA), the $e^{-i(\omega_{eg}+\omega)t}$ term can be neglected, which results in

$$\dot{C}_e = -\frac{i}{2\hbar} E_0 d_{eg}^* e^{-i(\omega_{eg} - \omega)t} = -\frac{i}{2\hbar} E_0 d_{eg}^* e^{i\Delta t} \quad (3.31)$$

where the difference between the frequency of the electromagnetic field ω and the transition frequency of the atom ω_{eg} is defined as detuning $\Delta = \omega - \omega_{eg}$.

Let us now discuss the RWA. Consider a radiation frequency ω that is close to the atom's transition frequency ω_{eg}, such that $|\omega_{eg} - \omega| \ll \omega_{eg} + \omega$. For the near-resonant term with frequency $(\omega_{eg} - \omega)$, it takes $\tau = 2\pi/\Delta$ to complete one full oscillation, while the nonresonant term with frequency $(\omega_{eg} + \omega)$ will complete many cycles. As a result, the nonresonant term will average to zero over the full oscillation timescale of $\tau = 2\pi/\Delta$. This allows us to neglect the nonresonant term since its effect is negligible. This approximation is known as the RWA.

Integrating the differential equation in (3.31) results in

$$C_e(t) = -i \frac{d_{eg}^* E_0}{\Omega_R \hbar} e^{\frac{-i\Delta t}{2}} \sin\left(\frac{\Omega_R t}{2}\right) \quad (3.32)$$

where Ω_R is the Rabi frequency. The generalized Rabi frequency, denoted as Ω_R^{eff}, represents the frequency at which the state oscillates between the ground and the excited states. The generalized Rabi frequency is given by

$$\Omega_R^{\text{eff}} = \sqrt{\Delta^2 + \Omega_R^2} = \sqrt{\Delta^2 + \frac{(d_{eg}^* E_0)^2}{\hbar^2}} \quad (3.33)$$

For zero detuning $\Delta = 0$, the generalized Rabi frequency becomes equal to the Rabi frequency (i.e., $\Omega_R^{\text{eff}} = \Omega_R = d_{eg}^* E_0/\hbar$). This means that the rate at which the state vector rotates on the Bloch sphere depends on the amplitude of the radiation E_0 and the atom's dipole moment d_{eg}^*. The resonant and nonresonant Rabi oscillations on a Bloch sphere are depicted in Figure 3.20(a, b).

The probability that the atom is in the excited state is given by

$$P_e(t) = |C_e(t)|^2 = \frac{(d_{eg}^* E_0)^2}{(\Omega_R^{\text{eff}} \hbar)^2} \sin^2\left(\frac{\Omega_R^{\text{eff}} t}{2}\right) \quad (3.34)$$

The probability that the atom is in the ground state can be calculated using the fact that $|C_g(t)|^2 + |C_e(t)|^2 = 1$. Equation (3.34) is plotted in Figure 3.20(c) for various values of the detuning Δ. The smaller the detuning, the larger the probability of finding the atom in its excited state. At resonance, $\Delta = 0$, the time it takes for all the atomic population to be transferred to the excited state is given by $t = \pi\hbar/(d^*_{eg}E_0)$. Hence, larger excitation fields and a greater dipole moment lead to quicker oscillations on the Bloch sphere. Section 3.3.1.2 shows how these same parameters influence the speed of quantum gates, since Rabi oscillations serve as the fundamental principle for controlling qubit states and performing qubit gates.

3.3.1.2 Qubit Rotations

To control the rotation angle of a qubit on the Bloch sphere, pulsed radiation can be employed instead of continuous wave (CW) radiation, as depicted in Figure 3.21. By employing the optical Bloch equations, one can determine the complete dynamics of a state vector on the Bloch sphere.

In the case of resonance with zero detuning, the rotation angle φ of a rectangular pulse with width T is given by $\varphi = \Omega_R T$. For a pulse with an arbitrary shape, the rotation angle is given by $\varphi = \int \Omega_R(t) dt$. It is important to note that any rotation on the Bloch sphere can be achieved using no more than three microwave pulses. It can be demonstrated that a simple microwave drive with a controllable phase can be utilized for x and y rotations. Direct z rotations can be accomplished through either fast flux tuning or an off-resonant drive that induces an ac-Stark shift, as we will see shortly.

Gate speed refers to the rate at which the state of a qubit can be manipulated on the Bloch sphere. A higher gate speed enables the execution of a greater number of gate operations within a specific time frame. As illustrated in (3.51), a larger dipole moment leads to a higher Rabi frequency, which necessitates a shorter pulse

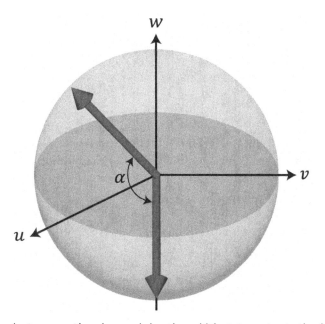

Figure 3.21 An electromagnetic pulse can bring the qubit's state vector to the desired position on the Bloch sphere.

duration to achieve a specific angle on the Bloch sphere. In comparison to natural atoms, artificial atoms like the transmon qubit are considerably larger, spanning a length of a few hundred micrometers. Consequently, they possess a substantial dipole moment and a correspondingly large Ω_R. This implies that gate implementation is faster for artificial atoms compared to natural atoms, owing to the larger Rabi frequency.

Alongside gate speed, gate fidelity plays a crucial role in qubit operations, as having gates with high fidelity is essential for quantum-error correction. Gate fidelities of 99.9% (equivalent to one error per 1,000 operations) have already been achieved. For more comprehensive information on gate fidelity and quantum-error correction, please refer to Chapter 2.

3.3.1.3 Rotation around X- and Y-Axes

The qubit drive can be represented by [17]

$$\mathcal{E}(t) = \begin{cases} \Omega^x(t)\cos(\omega_d t) + \Omega^y(t)\sin(\omega_d t), & 0 < t < t_g \\ 0, & \text{otherwise} \end{cases} \quad (3.35)$$

which shows a single-frequency carrier ω_d and two independent quadrature controls, $\Omega^x(t)$ and $\Omega^y(t)$; t_g is the time taken for one gate operation.

With zero detuning between the drive and the qubit frequency, rotations either around the x or y axes will follow the relative weight of $\Omega^x(t)$ and $\Omega^y(t)$. For example, a π-pulse or bit-flip gate σ_x can be generated by choosing a drive $\Omega^x = \Omega^\pi$ and $\Omega^y = 0$, which is on for a time t_g, with $\int_0^{t_g} \Omega^\pi dt = \pi$. This takes the qubit population from the ground state to the excited state and vice versa. The π-pulse can similarly be performed as a y-rotation just by switching, $\Omega^y = \Omega^\pi$ and $\Omega^x = 0$. To make superpositions of the qubit, $\pi/2$ pulses around x and y can also be performed to generate $(|0\rangle \pm |1\rangle)/\sqrt{2}$ and $(|0\rangle \pm i|1\rangle)/\sqrt{2}$, respectively. Arbitrary rotations about any axis can be performed using combinations of x and y rotations.

3.3.1.4 Rotation Around the Z-Axis

Recall that a qubit state in the x-y plane initiates precession around the z-axis at the qubit transition frequency. Therefore, manipulating the frequency of the qubit enables adjustments to the rate and direction of its precession. This can be achieved through two methods: external flux or ac-Stark shift.

3.3.1.4.1 Flux Bias Gate

Using a flux bias line makes it possible to change the Josephson energy of the SQUID and therefore change the qubit's frequency. For a transmon qubit with $E_J \gg E_C$ the qubit Hamiltonian reads $H_q = \hbar\sqrt{8E_J^{\max}E_C|\cos(\pi\Phi/\Phi_0)|}\sigma_z$. As a result, the amount of z-phase θ_z can be controlled by controlling Φ over a gate period t_g such that $\theta_z = \int_0^{t_g} dt\sqrt{8E_J^{\max}E_C|\cos(\pi\Phi(t)/\Phi_0)|}$.

3.3.1.4.2 ac-Stark Shift

The ac-Stark shift can also be used to control rotations around the z-axis. As discussed in Section 3.3.2.3, the qubit frequency can be shifted depending on the number of photons in the cavity.

A direct rotation around the Z-axis to shift the phase of a single qubit can be implemented by employing the off-resonant ac-Stark shift effect (see Section 3.3.2.3 for ac-Stark shift), where the qubit transition frequency, and subsequently its phase, is changed due to virtual photon transitions. Note that the drive must be sufficiently detuned from the qubit transition frequency to avoid inducing direct transitions via the σ_x term.

3.3.1.5 Qubit Drive Pulse Shaping

Rectangular pulses are not employed in practical applications for controlling the qubit state; instead, Gaussian or other pulse shapes are preferred. Shortly, we will understand why this is the case. Furthermore, due to the weak anharmonicity of the transmon qubit, it becomes necessary to employ optimal pulse-control techniques to mitigate phase and population errors caused by leakage into the third energy level.

Pulse shaping is widely used in various fields of physics and engineering to tailor the pulse shape for specific purposes, such as reducing occupied bandwidth. Different pulse-shaping filters, including raised cosine, sinc, and Gaussian, are employed. The choice of the filter involves trade-offs, considering factors such as sidelobes, spectral width, and ease of implementation.

In qubit experiments, it is desirable to have a band-limited pulse to avoid exciting higher-order transitions, such as the second excited state $|f\rangle$. (Recall that $|g\rangle$ represents the ground state and that $|e\rangle$ is the first excited state.) Let's explore how the pulse shape can impact higher-order qubit transitions.

Figure 3.22(a) illustrates that a rectangular pulse exhibits a sinc shape in the frequency domain, occupying a relatively broad bandwidth. The presence of non-zero frequency components at sufficiently high frequencies can inadvertently excite undesired higher-order qubit transitions. By smoothing out the sharp edges of the rectangular pulse, we can eliminate the high-frequency components and restrict the pulse's bandwidth. Now, let's examine the Fourier transform of a Gaussian pulse. The Gaussian pulse is unique in that its Fourier transform is also a Gaussian function. Figure 3.22(b) shows that the frequency content of the Gaussian pulse approaches zero as frequency increases, and its bandwidth is limited, thereby preventing higher-order qubit excitations.

Figure 3.22(c) illustrates the Fourier transform of Gaussian pulses with different widths in the time domain. According to the scaling property of the Fourier transform, reducing the pulse width in the time domain leads to an increase in the bandwidth in the frequency domain, and vice versa. A Gaussian pulse with a standard deviation σ in the time domain corresponds to a Gaussian standard deviation of $\sigma_f = (2\pi\sigma)^{-1}$ in the frequency domain.

When dealing with longer pulses in the time domain, they result in shorter pulses in the frequency domain. By adjusting the pulse width, we can ensure that the frequency bandwidth remains significantly smaller than the anharmonicity of the third level, which is typically in the frequency range of approximately 300–450 MHz. It is important to control the pulse width to avoid significant effects from the third level, especially when using Gaussian pulses with a range of $\sigma = 1$–2 ns. However, employing longer pulses is generally not preferred since shorter pulses allow for shorter gate operations, enabling the implementation of more qubit gates before dephasing and relaxation occur. Even with longer pulses, the width in the frequency domain may still encompass the qubit $|e\rangle \leftrightarrow |f\rangle$ transition, particularly in designs with relatively low anharmonicity.

To further improve the performance of the Gaussian pulse, we can utilize a technique known as Derivative Removal by Adiabatic Gate (DRAG) pulse [18]. The DRAG pulse effectively eliminates the frequency component at the $|e\rangle \leftrightarrow |f\rangle$ transition frequency, as shown in Figure 3.22(d). By optimizing the qubit drive pulse using DRAG, we can successfully suppress leakage outside the computational subspace and minimize phase errors resulting from coupling to higher energy levels.

In upcoming chapters, we will explore in more detail the generation of pulses using techniques such as arbitrary waveform generators (AWGs).

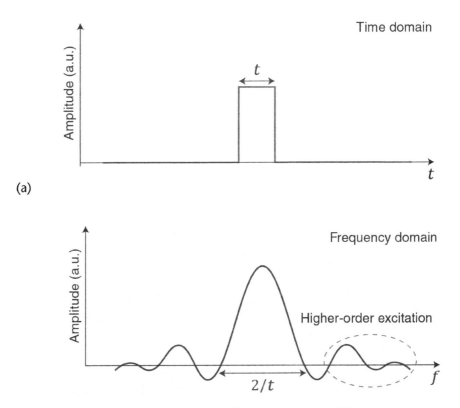

Figure 3.22 (a) A rectangular pulse occupies a relatively large bandwidth in the frequency domain; high-frequency components can cause higher-order qubit transitions. (b) Using a Gaussian pulse limits the bandwidth and prevents higher-order qubit transition. The Gaussian pulse bandwidth for various pulse widths is compared. (c) Comparison between a regular Gaussian qubit drive pulse and a derivative removal by Adiabatic gate (DRAG) qubit drive pulse. The DRAG pulse suppresses the unintended excitation of the qubit $|f\rangle$ state.

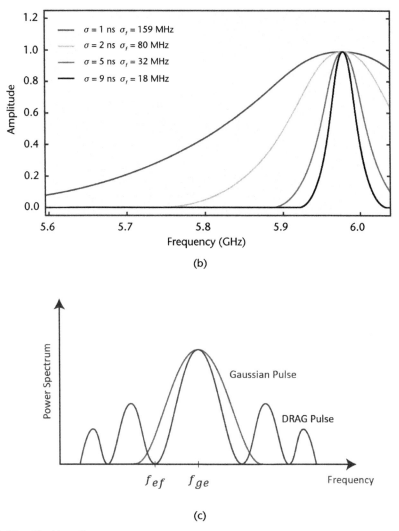

Figure 3.22 *(Continued)*

3.3.2 Qubit Readout

In order to read out the state of a qubit, it is necessary to determine whether it is in the ground state or the excited state. This process must be carried out with minimal interaction to avoid strong disturbance to the system, which can lead to changes in the qubit's state and the collapse of the wavefunction. One approach to achieve this is through a technique known as quantum nondemolition (QND) measurement, wherein the qubit is coupled to a cavity, and the state of the qubit is determined by observing the shift in the resonant frequency of the cavity. This technique, referred to as CQED, has been widely studied in the field of quantum optics. It offers a means to investigate the interaction between light and matter, encompassing the physics of a spin and an oscillator in mutual interaction. This methodology can be directly applied to superconducting qubits in the microwave range, as we will soon explore.

3.3.2.1 Purcell Effect

As shown in Section 3.2 coupling the atom to resonant modes of the vacuum results in spontaneous emission or qubit relaxation. The atom's interaction with vacuum modes can be engineered by modifying the atom's environment. By putting atoms in a cavity (for now, think of a cavity as a box that confines light), the density of vacuum modes and its spatial distribution are modified, as shown in Figure 3.23(a).

Let's consider a scenario where an atom is placed inside a resonant cavity (i.e., a cavity whose fundamental mode aligns with the atom's transition frequency). In this setup, the atom's dipole is aligned with the polarization of the cavity mode, and it is positioned at the maximum field location, as depicted in Figure 3.23(b). When the emitter's linewidth is narrower than the cavity linewidth, the probability of spontaneous emission is increased beyond its bulk value. This enhancement of the spontaneous emission rate is determined by the Purcell factor, denoted as $F_P = (3\lambda_c^3/4\pi^2)(Q/V)$, where V is the volume of the resonant mode, Q is the cavity's quality factor, and λ_c the wavelength in the material [16]. The volume of the resonant mode is a measure of the spatial confinement of EM fields inside the cavity. A smaller mode volume enhances the light-matter interaction. This enhancement of spontaneous emission has to do with the increased density of vacuum modes due to the presence of the cavity. Decreasing vacuum modes' density can inhibit the spontaneous emission rate (see Purcell filter in Section 3.3.2.2). This shows that a classical macroscopic system (i.e., a cavity) can be coupled coherently to an atom and control its radiative properties. This technique will be used for the qubit readout, as we will see shortly.

3.3.2.2 Cavity QED

A cavity serves as a resonator that confines EM energy through reflection. The choice of reflecting mechanisms depends on the frequency of the EM field. Mirrors are used for reflection in optical cavities, open-ended microwave transmission lines

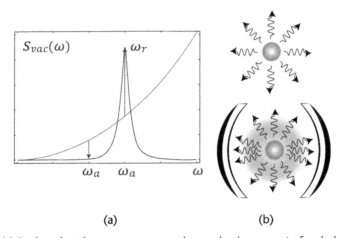

Figure 3.23 (a) Engineering the vacuum, suppression, and enhancement of emission ($\gamma \sim \Omega^2 S_{vac}(\omega_a)$). The spontaneous emission rate can be enhanced by weak coupling to a cavity and also due to the way that the photon excitation may not be recovered due to the cavity's damping. This is due to the increase in the local density of modes compared with their density in free space. (b) The rate of spontaneous emission increases when the atom is placed in a cavity.

(2D) or metallic enclosures (3D) form microwave cavities, and acoustic cavities can be created using two acoustic mirrors (Bragg reflectors). Due to inherent losses, cavities can only store photons for a limited period. The rate at which a photon exits the cavity, denoted as κ, is inversely proportional to the photon's lifetime τ inside the cavity, which represents the duration a photon remains in the cavity before being lost. The photon loss rate κ is also related to the quality factor Q of the cavity, expressed as $\kappa = \omega/Q$. The spectrum of a lossy resonator exhibits a Lorentzian shape with a finite linewidth determined by κ. Chapter 5 covers resonators in more detail.

Now, let us consider the combination of an atom and a cavity. Figure 3.24(a) illustrates a generic CQED system, consisting of two reflecting surfaces with an atom positioned at the cavity's field maximum. Standing waves are formed within the cavity through the superposition of incoming and reflecting waves from the mirrors. The standing wave interacts with the atom, resulting in alternating cycles of atom as observed from the excitation (due to photon absorption) and relaxation (due to photon emission). The speed at which the atom interacts with the cavity field is determined by the coupling factor g.

Table 3.3 compares the length scales for various cavity QED systems. A larger atom size corresponds to a larger dipole moment, which results in a larger coupling factor. With optical light, the wavelength is approximately $\lambda \approx 10^{-7} - 10^{-6}$m, and the radius of natural atoms is of the order of $r \approx 10^{-10}$m. Hence, the size of the atom is small compared to the light wavelength, and the light field can be considered constant over the atom. However, for a transmon coupled to surface-acoustic waves (SAWs), the wavelength is comparable to the size of the transmon, making it a giant atom. In this case, multiple coupling points cause strong interference effects, resulting in frequency-dependent relaxation rate and Lamb shift. Exploiting this interference effect in giant atoms allows for the design of relaxation rate, Lamb shift, and anharmonicity [19].

Figure 3.24(b) compares the elements of optical and microwave CQED systems. Figure 3.24(c, d) shows the planar (2D) and 3D CQED systems for transmon qubits. Later, we will discuss the details of each structure.

Jaynes-Cummings (JC) Model

We previously analyzed the interaction of free-space EM field with qubit using the semiclassical Rabi model. In this model, we treated the EM field classically without quantization. We next use the well-known Jaynes-Cummings (JC) Hamiltonian to model the interaction between the qubit and cavity, which employs the quantization of the EM field as well as the atom.

Quantization of EM Field

In EM field quantization, one finds a way to express the Hamiltonian of an EM field in terms of a quantum harmonic oscillator. Consider a one-dimensional cavity with length L along the z-axis with E-field polarized in $\hat{x}$ direction such that $\mathbf{E}(\mathbf{r}, t) = E_x(z, t)\hat{x}$. It can be shown that the EM field Hamiltonian can be written in terms of raising and lowering operators $\hat{a}$ and $\hat{a}^\dagger$ as follows

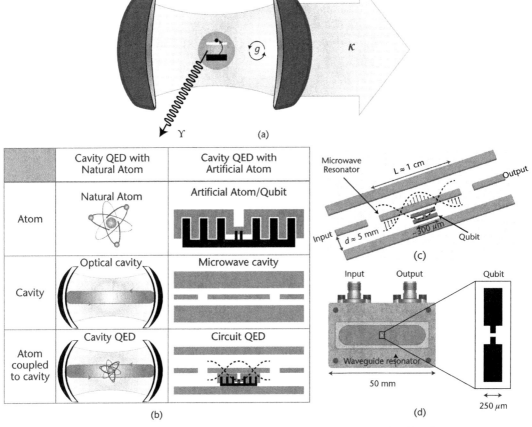

Figure 3.24 (a) A two-level atom in a resonant cavity. The cavity is characterized by three parameters: g, κ, and γ, which, respectively, quantify the atom-cavity coupling, the photon decay rate in the cavity, and the nonresonant decay rate. (b) Comparison of the optical and microwave cavity QED systems with natural and artificial atoms. (c) A superconducting qubit coupled to a CPW cavity. The electric fields are formed between the two sides of the central conductor and the grounds. The qubit is placed in the middle of the cavity where the electric field is maximum. (See Chapter 5 for details on microwave cavities.) (d) A superconducting qubit coupled to a 3D microwave cavity.

Table 3.3 Atom Sizes in Different Cavity QED Systems Discussed in [43]

Type of atom interacting field	Length scale		
	r = radius l = length	λ	r/λ or (l/λ)
Natural atom/optical light	$r \approx 10^{-10}$ m	$10^{-7} - 10^{-6}$ m	$10^{-4} - 10^{-3}$
Rydberg atom/microwaves	$r \approx 10^{-8} - 10^{-7}$ m	$10^{-3} - 10^{-1}$ m	$10^{-7} - 10^{-4}$
Transmon/microwaves	$l \approx 10^{-5} - 10^{-3}$ m	$10^{-3} - 10^{-1}$ m	$10^{-4} - 1$
Transmon/SAW	$l \approx 10^{-5} - 10^{-4}$ m	$10^{-6} - 10^{-5}$ m	$1-100$

$$\hat{H} = \hat{H}_{\text{field}} = \hbar\omega_c\left[\hat{a}(t)\hat{a}^\dagger(t) + \frac{1}{2}\right] \approx \hbar\omega_c\hat{a}(t)\hat{a}^\dagger(t) = \hbar\omega_c\hat{a}\hat{a}^\dagger \tag{3.36}$$

where the zero-point energy is at $\hbar\omega_c/2$. The electric and magnetic fields can be expressed in terms of raising (creation) and lowering (annihilation) operators

$$\begin{aligned}\hat{E}_x(z,t) &= E_0\left[\hat{a}(t) + \hat{a}^\dagger(t)\right]\sin(kz) \\ \hat{B}_y(z,t) &= B_0\left[\hat{a}(t) - \hat{a}^\dagger(t)\right]\cos(kz)\end{aligned} \tag{3.37}$$

JC Hamiltonian

Like the semiclassical Rabi model, the JC Hamiltonian consists of three terms, the isolated atom, the EM field, and their interaction: $\hat{H} = \hat{H}_{\text{atom}} + \hat{H}_{\text{field}} + \hat{H}_I$. The free-field Hamiltonian is given by $\hat{H}_{\text{field}} = \hbar\omega_c\hat{a}\hat{a}^\dagger$, as obtained in (3.37).

Two-level atom ground and excited states can be written as $|g\rangle = (0 \ \ 1)^T$, $|e\rangle = (1 \ \ 0)^T$. The Hamiltonian can then be written as

$$\begin{aligned}\hat{H} &= E_g|g\rangle\langle g| + E_e|e\rangle\langle e| = \begin{pmatrix} E_e & 0 \\ 0 & E_g \end{pmatrix} = \frac{1}{2}\begin{pmatrix} E_g + E_e & 0 \\ 0 & E_g + E_e \end{pmatrix} \\ &+ \frac{1}{2}\begin{pmatrix} E_e - E_g & 0 \\ 0 & -(E_e - E_g) \end{pmatrix} = \frac{1}{2}(E_g + E_e)\hat{1} + \frac{1}{2}(E_e - E_g)\hat{\sigma}_z\end{aligned} \tag{3.38}$$

with ω_a the atomic transition frequency, where the energy difference can be written as $\hbar\omega_a = E_e - E_g$. As we only care about energy differences, we can shift the zero-point energy to $E_g + E_e$. Using the $\hat{\sigma}_z$ Pauli operator, the free-atom Hamiltonian becomes

$$\hat{H}_{\text{atom}} = \frac{1}{2}\hbar\omega_a(|e\rangle\langle e| - |g\rangle\langle g|) = \frac{1}{2}\hbar\omega_a\hat{\sigma}_z \tag{3.39}$$

The field interacts with the atom through its dipole moment $\hat{d}$. We also assume that we put the atom where the electric field is a maximum so $\sin(kz) = 1$. The interaction Hamiltonian then reads

$$\hat{H}_I = -\hat{d}\cdot\hat{E} = -\hat{d}\left(\frac{\hbar\omega}{\varepsilon_0 V}\right)^{\frac{1}{2}}(\hat{a} + \hat{a}^\dagger) \tag{3.40}$$

where $\hat{d} = \hat{d}\cdot e$, and e is the unit vector in the field direction. We now introduce the atomic transition operators $\hat{\sigma}_+ = |e\rangle\langle g|$ and $\hat{\sigma}_- = |g\rangle\langle e|$. Because of parity consideration, one can say $\langle e|\hat{d}|e\rangle = 0 = \langle g|\hat{d}|g\rangle$. We assume that the matrix element $d = \langle g|\hat{d}|e\rangle$ is real. Therefore, we have $\hat{d} = d|g\rangle\langle e| + d^*|e\rangle\langle g| = d(\hat{\sigma}_- + \hat{\sigma}_+)$. The total Hamiltonian of the system becomes.

$$\hat{H} = \frac{1}{2}\hbar\omega_a\hat{\sigma}_z + \hbar\omega_c\hat{a}^\dagger\hat{a} + \hbar g\left(\hat{\sigma}_+ + \hat{\sigma}_-\right)\left(\hat{a} + \hat{a}^\dagger\right) \qquad (3.41)$$

Where the single photon dipole coupling strength is defined as

$$g = -\frac{d}{\hbar}\left(\frac{\hbar\omega_c}{\varepsilon_0 V}\right)^{\frac{1}{2}} \qquad (3.42)$$

Two of the four atom-field coupling terms $(\hat{\sigma}_+ + \hat{\sigma}_-)(\hat{a} + \hat{a}^\dagger)$ can be neglected by applying the rotating wave approximation, as explained next.

We use Heisenberg's equation of motion (See Section 1.6.5) for the annihilation operator of a harmonic oscillator

$$\begin{aligned} \frac{d\hat{a}}{dt} &= \frac{i}{\hbar}\left[\hat{H},\hat{a}\right] = \frac{i}{\hbar}\left[\hbar\omega\left(\hat{a}^\dagger\hat{a} + \frac{1}{2}\right),\hat{a}\right] \\ &= i\omega\left[\left(\hat{a}^\dagger\hat{a}\hat{a} + \frac{1}{2}\hat{a}\right) - \left(\hat{a}\hat{a}^\dagger\hat{a} + \frac{1}{2}\hat{a}\right)\right] = -i\omega\left[\hat{a},\hat{a}^\dagger\right]\hat{a} = -i\omega\hat{a} \end{aligned} \qquad (3.43)$$

Using (3.43), we find the time-dependent solution $\hat{a}(t) = \hat{a}(0)e^{-i\omega t}$ for the annihilation operator. In the same way, it can be shown that $\hat{a}^\dagger(t) = \hat{a}^\dagger(0)e^{i\omega t}$, $\hat{\sigma}_+(t) = \hat{\sigma}_+(0)e^{i\omega_0 t}$, and $\hat{\sigma}_-(t) = \hat{\sigma}_-(0)e^{-i\omega_0 t}$. The operator products look like $\hat{\sigma}_+\hat{a} \sim e^{i(\omega_0-\omega)t}$ (atom goes to the excited state due to photon absorption), $\hat{\sigma}_+\hat{a}^\dagger \sim e^{i(\omega_0+\omega)t}$ (atom goes to the excited state, and a photon is created), $\hat{\sigma}_-\hat{a} \sim e^{-i(\omega_0+\omega)t}$ (atoms go to the ground state, and a photon is absorbed), $\hat{\sigma}_-\hat{a}^\dagger \sim e^{-i(\omega_0-\omega)t}$ (atom goes to the ground state, and a photon is created). Only the $\hat{\sigma}_+\hat{a}$ and $\hat{\sigma}_-\hat{a}^\dagger$ conserve energy, and the other two terms are not energy-conserving as they express a simultaneous excitation or relaxation of the atom and field mode. Because the latter terms have a small probability at moderate ac fields, we only keep the two energy-conserving terms. This is called RWA. This approximation holds if the energy of adding a photon or adding a qubit excitation is much larger than the coupling or the energy difference between them $(\omega + \omega_a) \gg g, |\omega_r - \omega_a|$. In particular, when the coupling strength $g \ll \omega, \omega_a$ and the two systems are close to degeneracy $\sim \omega, \omega_a$ the nonconserving energy terms can be dropped. Therefore, the Hamiltonian becomes

$$\hat{H} = \frac{1}{2}\hbar\omega_a\hat{\sigma}_z + \hbar\omega_c\hat{a}^\dagger\hat{a} + \hbar g\left(\hat{\sigma}_+\hat{a} + \hat{\sigma}_-\hat{a}^\dagger\right) \qquad (3.44)$$

The following section examines the solutions of the JC Hamiltonian.

Solutions of JC Hamiltonian

The JC Hamiltonian in (3.41) can be solved analytically. The uncoupled eigenstates of the atom (ground and excited states $|g\rangle$, $|e\rangle$) and the field (photon number state

$|n\rangle)$ are called bare states. The bare states are no longer the eigenstates of the JC Hamiltonian. The atom-field coupling lifts their degeneracy and results in the entanglement of bare states, creating dressed states $|n, e\rangle$ and $|n + 1, g\rangle$. This means that the only way to describe the atom-cavity system is by describing all of it. In other words, the system cannot be described by describing the state of the atom or the cavity separately (see Section 2.2). The Hamiltonian can be expressed as [16]

$$H_n = \begin{pmatrix} \hbar\omega_c\left(n + \frac{1}{2}\right) + \frac{\hbar\Delta}{2} & \hbar g\sqrt{n+1} \\ \hbar g\sqrt{n+1} & \hbar\omega_c\left(n + \frac{1}{2}\right) - \frac{\hbar\Delta}{2} \end{pmatrix} \quad (3.45)$$

where the detuning is $\Delta = \omega_a - \omega_c$. The eigenenergies of this Hamiltonian are

$$E_{1n} = \left(n + \frac{1}{2}\right)\hbar\omega_c + \frac{1}{2}\hbar\Omega_n \quad (3.46)$$

$$E_{2n} = \left(n + \frac{1}{2}\right)\hbar\omega_c - \frac{1}{2}\hbar\Omega_n \quad (3.47)$$

with the generalized Rabi frequency Ω_n

$$\Omega_n = \sqrt{\Delta^2 + \Omega_0^2(n+1)} \quad (3.48)$$

$\Omega_0 = 2g$ is called the vacuum Rabi frequency. The eigenstates are

$$|1n\rangle = \sin\vartheta_n|e,n\rangle + \cos\vartheta_n|g,n+1\rangle \quad (3.49)$$

$$|2n\rangle = \cos\vartheta_n|e,n\rangle - \sin\vartheta_n|g,n+1\rangle \quad (3.50)$$

with coefficients

$$\cos\vartheta_n = \frac{\Omega_n - \Delta}{\sqrt{(\Omega_n - \Delta)^2 + 4g^2(n+1)}} \quad (3.51)$$

$$\sin\vartheta_n = \frac{2g\sqrt{(n+1)}}{\sqrt{(\Omega_n - \Delta)^2 + 4g^2(n+1)}} \quad (3.52)$$

At resonance $\Delta = 0$, the two maximally entangled symmetric and antisymmetric superposition states are split by $E_{1n} - E_{2n} = \hbar\Omega_0\sqrt{n+1}$, as shown in the level diagram in Figure 3.25. The energy-level splitting depends on the coupling

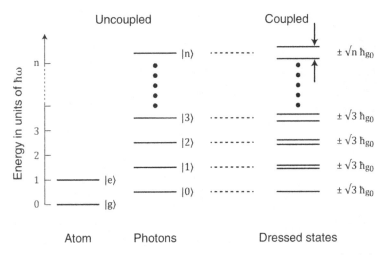

Figure 3.25 Dressed states and the JC ladder. Note that the $\{|g, n\rangle, |e, n-1\rangle\}$ basis is used here so that $\Omega_n = \sqrt{n}\hbar g_0$.

factor g and the number of photons in the cavity n. So, if we start with the atom in its ground state and one photon in the cavity, the atom absorbs the photon and goes to the excited state $|g, 1\rangle \rightarrow |e, 0\rangle$, and vice versa. Moreover, this interaction happens at the rate of vacuum Rabi frequency $\Omega_0 = 2g$. The next section discusses vacuum Rabi oscillations.

Vacuum Rabi Oscillations

The difference between the fully quantized and semiclassical formulations of the light-matter interaction is that the semiclassical expression for the generalized Rabi frequency (Rabi frequency with detuning) is replaced with $\Omega_0\sqrt{n+1}$. This striking difference shows that even the vacuum field with zero field amplitude or no photons ($n = 0$) can cause coherent oscillation and couple the two dressed states. This is why Ω_0 is called vacuum Rabi-frequency.

The coefficients $c_{1n}(t)$, $c_{2n}(t)$ of an arbitrary state $|\psi(t)\rangle = c_{1n}(t)|1n\rangle + c_{2n}(t)|2n\rangle$ can be expressed as

$$\begin{pmatrix} c_{2n}(t) \\ c_{1n}(t) \end{pmatrix} = \begin{pmatrix} \exp(i\Omega_n t) & 0 \\ 0 & \exp(-i\Omega_n t) \end{pmatrix} \begin{pmatrix} c_{2n}(0) \\ c_{1n}(0) \end{pmatrix} \quad (3.53)$$

In resonance with zero detuning $\Delta = 0$, this gives for a state initially in the excited state

$$\left|c_{e,n}(t)\right|^2 = \cos^2\left(g\sqrt{n+1}\,t\right) \quad (3.54)$$

$$\left|c_{g,n+1}(t)\right|^2 = \sin^2\left(g\sqrt{n+1}\,t\right) \quad (3.55)$$

Even with no photon in the cavity $n = 0$, there is a coherent exchange of one energy quantum between the atom and the field mode called vacuum Rabi oscillations

$$|c_{e0}(t)|^2 = \cos^2(gt) = \frac{1}{2}\left(1 + \cos(\Omega_0 t)\right) \quad (3.56)$$

The vacuum Rabi oscillation is very different than the irreversible exponential decay of an excited atom. The periodic energy exchange has an analogy with two coupled pendulums.

Strong Coupling

Strong coupling occurs when the rate of interaction between the atom and the cavity is significantly faster than both the spontaneous emission rate and the photon loss rate, denoted as $g \gg (\kappa, \gamma)$, where (κ, γ) denotes the larger of κ and γ. In such a scenario, spontaneous emission can be reversed, allowing an initially excited atom in a cavity to emit a photon(s) into a single-mode electromagnetic cavity and then subsequently reabsorb it. Over time, these oscillations will dampen due to cavity losses, eventually causing the photon to leave the cavity. The relaxation time of the atom, T_1, can be determined by analyzing the envelope of the damped Rabi oscillations. Strongly coupled systems are those in which the Rabi dynamics can persist despite dissipation [7–20].

It is possible to show that many characteristic features of strongly coupled quantum systems can be derived from classical analysis of a simple coupled harmonic oscillator, such as two coupled spring-mass systems. Both the classical and quantum coupled oscillators share an intriguing property in the strong coupling regime known as anticrossing. The eigenenergies as a function of detuning for two uncoupled oscillators are shown in Figure 3.26(a). As coupling is introduced, anticrossing occurs (i.e., a splitting at zero detuning emerges), known as the vacuum Rabi oscillation, a hallmark of strong coupling. This is shown in Figure 3.26(b). The interaction is considered a weak coupling if $g \ll (\kappa, \gamma)$. When two systems are far detuned (i.e., $\Delta = \omega_a - \omega_c \gg 1$), the cavity's resonance frequency is equal to its bare frequency. However, when $\Delta \approx 0$ (i.e., when the atom and cavity are in resonance), a splitting emerges, where its size is proportional to the coupling

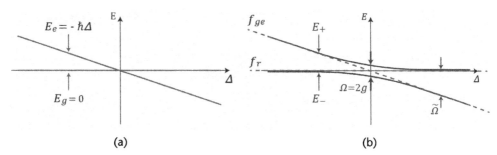

Figure 3.26 Eigenenergies of two coupled oscillators as a function of the detuning Δ: (a) Zero coupling, where one eigenenergy was kept at a fixed reference level; and (b) avoided crossing due to strong coupling. Avoided crossing in a qubit-cavity system can be observed by using a flux bias line to sweep the qubit frequency to bring it in and out of resonance with the microwave cavity.

factor (i.e., 2g). As we will see in the following section, having a large coupling factor g is essential for distinguishing the qubit's ground and excited states when performing the readout.

To generate the plot displayed in Figure 3.26(b), one of the systems must vary the resonant frequency to sweep through different detuning values. As the qubit has a tunable resonant frequency, this can be achieved by controlling the flux passing through the SQUID. Then, the transmission (S_{12} parameter) of the resonator can be measured and plotted versus detuning.

Strong Dispersive Regime

Recall that the wavefunction collapses when the qubit state is measured at resonance. Therefore, it is necessary to use a QND measurement, based on the dispersive shift, for probing the qubit state in a nondestructive way. In the dispersive limit, where the qubit and cavity are far detuned with respect to each other (i.e., $\Delta = |\omega_a - \omega_c| \gg g$), only virtual photon exchange is allowed, and no atomic transitions occur, nor is there any energy exchange. In this case, the coupled system experiences level shifts proportional to g^2/Δ, as shown in Figure 3.27(a, b). These level shifts arise due to dispersive interactions mediated by virtual photons.

Using the second-order time-dependent perturbation theory, the JC Hamiltonian can be approximated in the dispersive regime. Expanding the Hamiltonian terms into powers of g/Δ yields [20, 21]

$$H = \underbrace{\hbar\omega_r(a^\dagger a + 1/2)}_{\text{Single photon mode harmonic oscillator}} + \underbrace{\hbar\omega_a \sigma_z/2}_{\text{Two-level system}} + \underbrace{\hbar\chi(a^\dagger a + 1/2)\sigma_z}_{\text{Dispersive interaction}} \quad (3.57)$$

where $\chi = g^2/\Delta$ is called the dispersive shift. We can rewrite the Hamiltonian as:

$$H_{JC(\text{disp})} \approx \hbar\left(\omega_r + \frac{g^2}{\Delta}\sigma_z\right)\left(a^\dagger a + \frac{1}{2}\right) + \frac{\hbar\omega_a}{2}\sigma_z \quad (3.58)$$

As can be seen from (3.58), the qubit-cavity coupling is QND concerning both photon number and qubit polarization since it commutes with both. The qubit state-dependent shift of the cavity with the new resonance frequency $\tilde{\omega}_r = \omega_r \pm g^2/\Delta$ is observed in (3.58). This shift in the cavity frequency is called the dispersive shift.

As can be seen in Figure 3.27(a), for a large enough g and small enough detuning and photon loss rate κ, and qubit relaxation rate γ, the ground and excited states can be distinguished due to the state-dependent frequency shift of the cavity. The strong dispersive regime refers to the situation when the dispersive shift $\chi = g^2/\Delta$ is greater than the linewidths of the qubit, γ, and of the cavity, κ. In this regime, the transmission through the cavity becomes a nonlinear function of the qubit's state, which permits a projective QND readout. The cavity transmission is a Lorentzian, which can be written as $T = (\kappa/2)/(\omega - (\omega_C \pm \chi\sigma_z) + i\kappa/2)$.

It is now clear why a large coupling factor, g, is desired to increase the dispersive shift χ resulting in sufficiently large distinguishability between the qubit states. A large g also means strong coupling to the environment, which results in

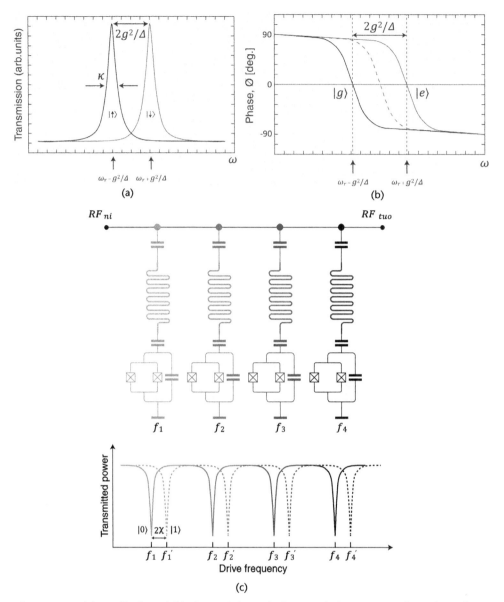

Figure 3.27 (a) Amplitude and (b) phase of the cavity's transmission spectrum depends on the qubit's state. Here Ω represents the transition frequency, ω_r the resonance frequency of the cavity, $\Delta = \Omega - \omega_r$ the detuning, g the coupling strength, and κ is the decay rate of the cavity. In the dispersive regime, qubit and resonator are far detuned from each other with $g/\Delta \ll 1$. (c) Multiplexed qubit readout scheme; each qubit has its own dispersive shift.

qubit relaxation. It can be shown that the Purcell relaxation rate of the qubit due to the resonator-environment coupling is given by $\gamma_{\text{res-env}}^{\text{Purcell}} = (g/\Delta)^2 \kappa$ [10]. This shows that the relaxation rate increases by reducing the detuning Δ between the qubit and the cavity, although increasing Δ will also increase the dispersive shift and the qubit state's distinguishability. Moreover, having a larger bandwidth κ for the resonator allows for fast readout, as we will see shortly however, it also increases the relaxation rate. Therefore, it is essential to engineer g and κ to allow fast readout without compromising the qubit coherence.

Let us look at real-world numerical values for a transmon qubit coupled to a CPW cavity. The qubit-cavity coupling strength is $g_0/\pi = 100$ MHz, the cavity frequency is $\omega_r/2\pi = 7$ GHz, photon decay rate is $\kappa/2\pi = 300$ kHz, and the qubit charging energy is $E_C/2\pi = 340$ MHz. The qubit is detuned from its flux sweet spot by ~1.5 GHz with a resonant frequency of $\omega_{01}/2\pi = 6$ GHz.

The same concept of a dispersive shift can be applied for the multiplexed readout of multiple qubits, as illustrated in Figure 3.27(c). Each resonator is coupled to a qubit and a shared RF readout line. By examining the dispersive shift in the power spectrum of each resonator, depicted in Figure 3.27(c), the state of each qubit can be determined.

Qubit Readout Requirements

Achieving high-fidelity qubit readout requires satisfying two requirements. First, it is necessary to have a sufficiently large signal-to-noise ratio (SNR) in order to clearly distinguish the peaks of the dispersive shift. This can be achieved by using quantum-limited amplifiers, such as a Josephson parametric or SLUG amplifier (see Chapter 5). Second, the readout process must be fast enough to complete it within a time significantly shorter than the qubit coherence time. A lengthy readout time increases the likelihood of qubit relaxation, thereby reducing the readout fidelity. It will be demonstrated that Purcell filters can enable fast readout. By combining quantum-limited amplifiers and Purcell filters, it becomes possible to achieve a readout fidelity of 99% or higher within approximately 100 ns, which is much shorter than the current qubits relaxation time [10].

High SNR

We saw that the dispersive shift of the cavity frequency could be used for the readout of qubits states. In order to resolve the separation between the |0⟩ and (|1⟩ peaks, the SNR needs to be sufficiently large. We can define a parameter called readout fidelity F as follows, which strongly depends on the SNR

$$F = \frac{1}{2}\left(1 + \text{erf}\left(\sqrt{\frac{SNR}{2}}\right)\right) \qquad (3.59)$$

where *erf* is the Gaussian error function [22]. A readout fidelity of 99% corresponds to one erroneous readout in 100 measurements. To achieve $F = 99.9\%$, an $SNR = 10$ is needed.

One way to achieve a sufficiently large SNR is to repeat qubit measurements and use time averaging, which helps to improve the SNR by a factor of $\sqrt{N}$, where N is the number of signals used in averaging. So, improving the SNR by a factor of 10 reduces the measurement time by a factor of 100. Figure 3.28(a, b) shows that by increasing the sampling time, the separation between the peaks shown in solid lines increases linearly in time, whereas the peak widths only increase as $\sqrt{t}$ [10].

Another way to improve SNR is to use quantum-limited amplifiers with an extremely low noise figure in the readout chain, which allows single-shot readout to reduce the measurement time significantly.

3.3 Qubit Control and Readout

Let us analyze how much signal power is typically necessary for the readout to achieve SNR = 10. As discussed in Chapter 4, SNR is defined as the ratio of the signal power to the noise power $SNR = P/N$. The vacuum noise power can be calculated as the noise power density [i.e., noise power per unit of bandwidth in the vacuum $PSD_{vac} = \hbar\omega/2 = -207$ dBm/Hz], times the bandwidth of the readout resonator, which is usually around 1 MHZ, so $\kappa = 2\pi \times 1$ MHz. Therefore, the $SNR = P/(\kappa \times PSD_{vac})$. So, for an $SNR = 10$, the readout power is $P \approx -130$ dBm at a frequency of 6 GHz.

Now, one might say that we could easily improve the SNR by increasing the readout power or equivalently increasing the average number of photons in the cavity n, where the power for an overcoupled resonator is related to the average photon number as $P \approx n\hbar\omega\kappa$. The issue is that for a cavity populated with a large number of photons, the frequency shift induced by the qubit will be dependent on both the number of photons in the cavity due to the ac-Stark shift, as well as the state of the qubit, as shown in Section 3.3.2.3

In practice, the PSD of the measurement instruments is typically in the range of −150 dBm/Hz, which is much higher than the vacuum noise floor. To reach an $SNR = 1$, a measurement time of 25 ms is needed, which is much larger than the qubit's relaxation time. Therefore, amplification is necessary to improve the SNR and reduce the measurement time. We examine quantum-limited amplifiers, such as the parametric amplification, in Chapter 5 to see how they can improve the overall SNR of the readout chain by the noise-squeezing technique.

Recently, researchers could perform fast, high-fidelity qubit readout even without a quantum-limited amplifier using shelving techniques, two-tone readout excitation of the readout resonator, and machine learning algorithms [24].

Fast Readout

According to DiVincenzo's criteria, a qubit needs to be sufficiently isolated from the environment to prevent decoherence yet still be accessible for the state read out, which represents two contradictory criteria. We have observed that coupling the qubit to the environment results in qubit relaxation due to enhanced spontaneous emission. The coupling to the environment occurs during measurement, requiring a coupling to the readout port. The readout fidelity F is affected by the readout time τ_{ro}, as given by $F(\tau_{ro}) = 1 - e^{-(\tau_{ro}/T_1)}$, where $\tau_{ro} = \tau_{rd} + \tau_s/2$ denotes the total time for the readout, τ_{rd} is the readout delay due to the resonator transient, and τ_s is the sampling time [10]. Therefore, it is necessary to perform a fast single-shot readout such that $\tau_{ro} \ll T_1$, which requires decreasing the two-time factors τ_{rd} and τ_s. The τ_s can be reduced by keeping the integration time τ_s as short as possible.

To reduce the τ_{rd}, the resonator's bandwidth needs to be increased as a larger bandwidth results in faster damping of the transients. The coupling regime of the resonator also plays an essential role in reducing the response time. We investigate various coupling regimes for resonators in Chapter 5. Strongly coupled (overcoupled) resonators with low-quality factors (high bandwidth) are appropriate for performing fast measurements. Moreover, a large bandwidth is beneficial for resetting a qubit to its ground state, which can be done by bringing its frequency near the cavity resonance and using the Purcell-enhanced decay rate. Undercoupled resonators with large quality factors adversely affect the qubit readout fidelity since fewer signal photons are collected during a qubit lifetime. Large-quality factor resonators can store photons in the cavity on a long-time scale, with potential use as a quantum memory [25].

While operating in the dispersive regime, the qubit and resonator are far detuned from each other, allowing for separate engineering of their environment through carefully designing the impedance seen by the resonator and the qubit using a so-called Purcell filter. The filter is designed to provide sufficiently large bandwidth with a strong coupling to the readout port at the resonator frequency while isolating the qubit from its environment at the qubit frequency, as shown in Figure 3.29(a). This technique allows for fast readout while protecting the qubit from relaxing into its environment.

Figure 3.29(b) shows the coupling between the qubit (ω_q), resonator (ω_r), Purcell filter, and the environment. The qubit is capacitively coupled to the resonator with a coupling strength shown by g. The resonator is coupled to the Purcell filter with a photon loss rate of κ. The Purcell filter is then coupled to the environment. The transmission spectrum of the Purcell filter shown in Figure 3.29(a) protects the qubit while allowing the resonator field's fast decay to the environment. The decay rate of the qubit-resonator-filter-environment chain is given by [10]

$$\gamma_{\text{res-filter-env}}^{\text{Purcell}} = \kappa \left(\frac{g}{\Delta}\right)^2 \left(\frac{\omega_q}{\omega_r}\right)\left(\frac{\omega_r}{2Q_F \Delta}\right) \qquad (3.60)$$

where Q_F denotes the quality factor of the Purcell filter.

The Purcell filter can be implemented depending on the required bandwidth, insertion loss (see Chapter 5), and qubit-resonator detuning. The Purcell filter can be as simple as a quarter-wavelength impedance transformer [26]. It can also be implemented as a wideband stepped-impedance filter [26] or the low-Q bandpass filter, which also acts as a quantum bus that makes it possible to connect several frequency-multiplexed readout resonators sharing the same amplifier chain.

3.3.2.3 ac-Stark Shift and Lamb Shift

The Hamiltonian in (3.58) can be interpreted either as a shift in the cavity frequency which depends on the state of the qubit, or as the ac-Stark shift plus the Lamb shift of the qubit frequency proportional to the number of photons in the cavity [27]. We can rewrite the Hamiltonian in (3.58) as follows:

$$H = \hbar\omega_r\left(a^\dagger a + 1/2\right) + 1/2(\hbar\omega_a + \overbrace{\hbar\chi}^{\text{Lamb shift}} + \overbrace{\hbar 2\chi a^\dagger a}^{\text{ac-Stark shift}})\sigma_z \qquad (3.61)$$

$$\underbrace{\hphantom{\hbar\chi + \hbar 2\chi a^\dagger a}}_{\text{photon number-dependent light shift}}$$

The dispersive interaction can be viewed as a constant Lamb shift[1] of g^2/Δ induced by the vacuum fluctuations in the cavity and a photon-number–dependent

---

1. Natural atoms such as hydrogen atom also undergo the Lamb shift due to the coupling of the atom to the vacuum field that should otherwise have the same energy in classical empty space.

3.3 Qubit Control and Readout

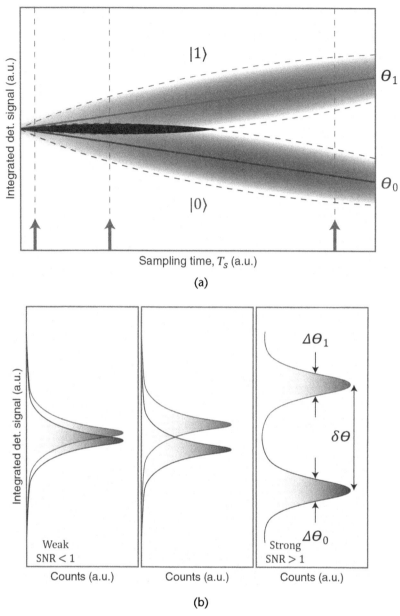

Figure 3.28 (a) The distribution of qubit states during the readout signal sampling in the presence of noise. As time progresses, the distance between the peaks (represented by solid lines) increases linearly, while the peak widths only increase as the square root of time $\sqrt{t}$. Three black arrows indicate the points where line cuts were taken for three different sampling times shown in (b), (c), and (d). (b) The states are not separated for a short sampling time, resulting in a weak measurement (SNR < 1). (c) After a longer sampling time, the peaks begin to separate, and in (d) finally become completely distinct, resulting in a strong measurement (SNR > 1). (*After:* [10].)

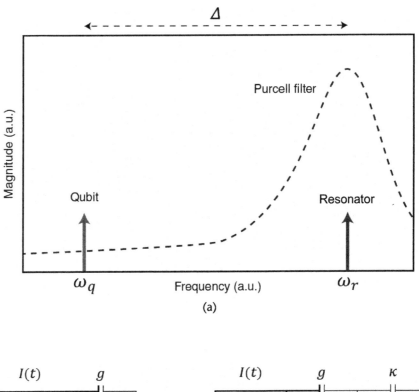

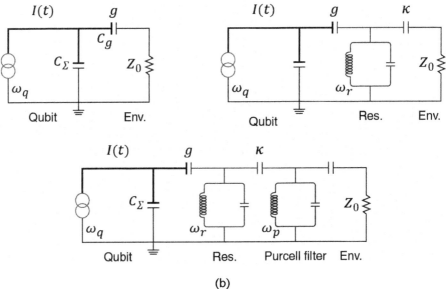

Figure 3.29 (a) The added Purcell filter protects the qubit while allowing the resonator field to decay fast to the environment. The transmission spectrum of a Purcell filter shows that the peak is centered around the resonator frequency, and the qubit frequency is far detuned. (b) The equivalent circuit of the qubit, the resonator, and the Purcell filter.

shift, called an ac-Stark shift, of $(2g^2/\Delta)n$ where $n = \langle a^\dagger a \rangle$ is the average number of photons present in the cavity. The shift in the qubit frequency due to the Lamb shift and ac-Stark shift is given by

$$\tilde{\omega}_q = \omega_q + \frac{g^2}{\Delta} + \frac{2g^2}{\Delta} n \qquad (3.62)$$

So, in this case, the atom and cavity remain uncoupled, and the energy levels of the atom undergo a constant Lamb shift of g^2/Δ and a photon-dependent ac-Stark shift of $2ng^2/\Delta$. This property makes it possible to tune the qubit transition frequency with microwave pulses to produce controlled rotations about the z-axis. Moreover, this shows that the measurement power, proportional to the number of photons in the cavity, can affect the qubit's frequency.

Photon Number Calibration Using ac-Stark Shift

The QND aspect of the readout depends on the photon population of the resonator, as the qubit's relaxation and dephasing rates are influenced by the number of photons in the resonator [23]. Moreover, we will discover that nonlinear effects arise as the number of photons in the cavity increases. As discussed in Chapter 4, some inherently nonlinear systems can be approximated as linear under small signal conditions; that is where the system is driven with sufficiently small amplitude. However, in the large-signal regime with sufficiently large excitations, the system's nonlinearity must be considered. The small-signal and large-signal regimes of the JC Hamiltonian are shown in Figure 3.30(a).

The dispersive JC Hamiltonian is inherently nonlinear; however, it can be approximated linearly if the number of photons is much smaller than the critical photon number (i.e., $n \ll n_{crit} = \Delta^2/4g^2$), so the n_{crit} determines the boundary between the linear and nonlinear approximation. For a $g = 150$ MHz and a detuning of $\Delta = 1.5$ GHz the $n_{crit} = 25$ photons. Recall that number of photons in a cavity is given by $n \approx P/\hbar\omega\kappa$. Therefore, to remain in the linear regime, the average population of the cavity must be in the range of a few photons corresponding to power levels as low as -150 dBm. So, when performing measurements, the amplitude of the tone must be adjusted to maintain the average population of the cavity well below the critical photon number. In the presence of insertion loss, the average number of photons n inside the cavity on resonance is given by $n = 2P_{in}\sqrt{IL}\,Q_L/\hbar\omega^2$, where IL is the insertion loss of the resonator, and Q_L is the cavity's quality factor. Figure 3.30(b) shows the ac-Stark shifted qubit transition frequency with a linear dependence on the applied measurement power.

As we increase the number of photons and reach the $n \approx n_{crit}$, the linear approximation starts to break down entirely, and the resonator's photons significantly populate excited levels. More terms in the JC Hamiltonian must be considered to take the nonlinear effects into account, where these terms are of the Kerr-type. Also, increasing the number of photons in the cavity comes at the cost of reduced coherence time [26].

Figure 3.31(a, b) shows the line broadening and the shift of the qubit frequency due to the increased number of photons. As shown in Figure 3.31(c), the spectrum

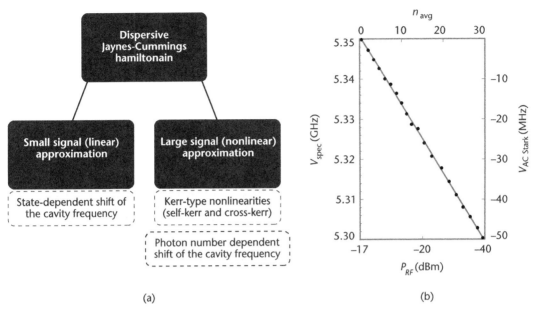

Figure 3.30 (a) Small- and large-signal regimes of the dispersive JC Hamiltonian. (b) ac-Stark shifted qubit transition frequency with a linear dependence $v_{ge}(n_{avg}) = v_{ge}(0) + 2\chi n_{avg}$ on the applied measurement power (dots). The detuning-dependent qubit shift per photon of 2χ. The relevant parameters are $\Delta/2\pi = -1.088$ GHz, $g/2\pi = 54$ MHz and the calculated ac-Stark shift per photon of $2\chi/2\pi = -1.69$ MHz. From the linear fit, a coherent measurement power of $P_{n=1} \approx -32$ dBm populates the cavity with, on average, a single photon.

of the qubit can be resolved into individual photon number peaks with $\chi > \gamma$, where this photon number–splitting pattern allows for the determination of the mean photon number of the cavity. As shown in Figure 3.31(c), the qubit frequency peaks will be located at $\omega_n = \omega_q + 2n\chi$. Note that there is always a nonzero number of thermal photons in the cavity ($n = 0.1$) due to leakage of infrared photons or insufficient cooling of an attenuator stage on the input or output microwave lines.

The weighting of the amplitudes for various photon numbers in Figure 3.31(d) can be used to find the average photon number in the cavity. A Poisson distribution $P(n) = e^{-\underline{n}}\underline{n}^n/n!$ with a mean value of $\underline{n}$ gives the average occupation number of the nth state.

The Kerr-type nonlinear terms result in the nonlinear photon number (n) dependence of the cavity's resonant frequency ω_r. Kerr-type terms represent the lowest-order approximation of the full nonlinearity arising from the JC interaction. The total qubit state-dependent dispersive shift of the cavity frequency is given by $\chi \cong \chi_{ge} - \chi_{ef}/2$, where $\chi_{ge} = g_{ge}^2/\Delta_{ge}$ dispersive shift for $|g\rangle \rightarrow |e\rangle$ transition, and $\chi_{ef} = g_{ef}^2/\Delta_{ef}$ dispersive shift for $|e\rangle \rightarrow |f\rangle$ transition. The photon number-dependent resonance frequency of the resonator is given as

$$\omega_r(n) = \tilde{\omega}_r + \chi\sigma_z + \zeta n\sigma_z + \zeta' n \qquad (3.63)$$

where the self-Kerr ζ' and cross-Kerr ζ terms are given by

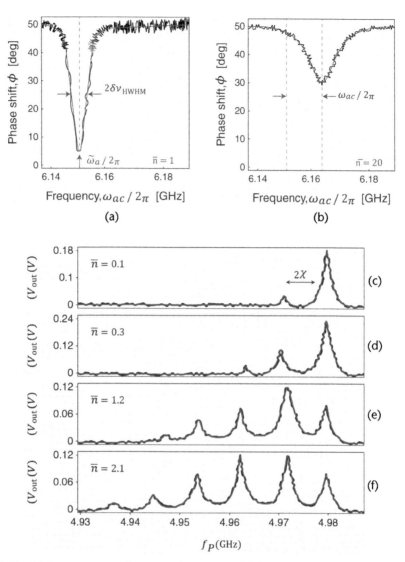

Figure 3.31 (a) Intraresonator photon number $\bar{n} \approx 1$ with fit to Lorentzian line shape. The qubit transition frequency $\tilde{\omega}_a$ at low measurement power and the half width at half maximum $\delta\nu_{HWHM}$ are shown. (b) Measurement at high power with average photon number $\bar{n} \approx 40$ with fit to Gaussian line shape. The ac-Stark shift ω_{ac} of the lines are indicated. (c) Photon number-splitting in the transmon spectrum. The primary qubit transition is seen at $\tilde{\omega}_{ge}/2\pi = 4.982$ GHz. The peak at 4.973 GHz is due to a residual population of thermal photons $n_{th} = 0.1$ in the resonator. (d) A cavity tone is applied at $\omega_c/2\pi = 5.474$ GHz and power $P_{rf} = 2.5$ aW, producing a population of $\bar{n} = 0.3$ photons in the resonator. (e) Cavity tone of power $P_{rf} = 20$ aW, $\bar{n} = 1.2$ photons in the resonator. (f) Cavity tone of power $P_{rf} = 160$ aW, $\bar{n} = 2.1$ photons in the resonator. Multiple peaks corresponding to different cavity-photon numbers. The weighting of these peaks is given by a simple Poisson distribution as a function of the mean number of photons in the cavity $\bar{n} = 2$.

$$\zeta' \approx \left(\chi_{ge} - \chi_{ef}\right)\left(\lambda_{ge}^2 + \lambda_{ef}^2\right) \tag{3.64}$$

$$\zeta \approx \chi_{ef}\lambda_{ef}^2 - 2\chi_{ge}\lambda_{ge}^2 + \frac{7\chi_{ef}}{4}\lambda_{ge}^2 - \frac{5\chi_{ge}}{4}\lambda_{ef}^2 \tag{3.65}$$

and where $\lambda_{j,j+1} \equiv g_{j,j+1}/\Delta_{j,j+1}$.

Although high-power measurements have their challenges, Reed et al. have shown that a simple and robust measurement protocol can be used for single-shot measurement of a transmon qubit when working at very large measurement power [27].

3.3.3 Spectroscopic Measurement Methods

Spectroscopic qubit measurements play a crucial role in circuit QED. They provide essential information about the qubit's energy levels such as the ground-to-first and first-to-second excited state transitions energies, which unveil the charging and Josephson energies and enable us to characterize the qubit spectrum. Figure 3.32(a) illustrates the commonly employed qubit spectroscopy techniques used to determine the qubit frequencies.

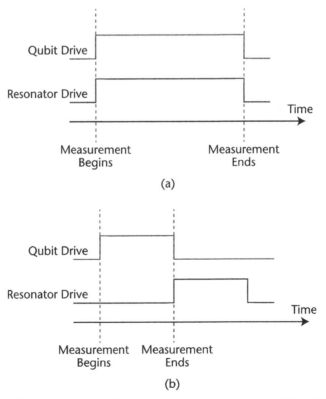

Figure 3.32 (a) Pulse sequences used for continuous wave spectroscopy; the qubit and the readout resonator are driven continuously and simultaneously. (b) Pulsed qubit spectroscopy; the readout signal is turned on only after the qubit drive has been switched off.

In CW spectroscopy, both the qubit and the readout resonator are simultaneously driven. However, a disadvantage of CW spectroscopy is that the photons within the readout resonator can cause ac-Stark shifts in the qubit frequencies during the qubit drive. Therefore, the measurement needs to be conducted at low resonator drive power to mitigate this effect.

Pulsed spectroscopy, as depicted in Figure 3.32(b), addresses the ac-Stark shift problem by activating the resonator drive only after turning off the qubit drive. This ensures that no photons are present in the resonator when the qubit is driven. To avoid power broadening of the linewidth, it is recommended to drive the qubit at low power, as higher power can reduce the resolution of the qubit frequency. However, to observe both the $|g\rangle \leftrightarrow |e\rangle$ and $|e\rangle \leftrightarrow |f\rangle$ transition frequencies, the measurement must be conducted at a higher qubit drive power compared to regular spectroscopy. This is necessary to induce a two-photon transition at $\omega_{ef} = 2\omega_{gf/2} - \omega_{ge}$.

3.3.4 Equivalent Circuit of Qubit-Cavity Coupling

The qubit, which is coupled to a single cavity mode, can be represented using an equivalent linear circuit. This circuit is comprised of LC resonators that correspond to the qubit and the cavity, interconnected by a coupling capacitor, as depicted in Figure 3.33. The utilization of an equivalent circuit approach facilitates a deeper comprehension of the parameters that influence the coupling dynamics and is valid for the lowest two qubit states. This understanding is valuable for the purpose of designing and optimizing the qubit-cavity system.

As mentioned in Chapter 1, an electrical oscillator comprised of an inductor and a capacitor can be likened to a mechanical oscillator consisting of a mass and spring. In this analogy, the charge on the capacitor represents the position coordinate of the mass, while the flux through the inductor emulates the role of momentum [48]. The process of circuit quantization follows a straightforward approach: We obtain a Hamiltonian for the circuit and reformulate it in a manner akin to the Hamiltonian of a quantum harmonic oscillator. Subsequently, we identify the variables that correspond to position and momentum, enabling us to treat the system as a quantum harmonic oscillator by defining commutation relations and utilizing lowering and raising operators. This section delves into the derivation of the Hamiltonian from the Lagrangian.

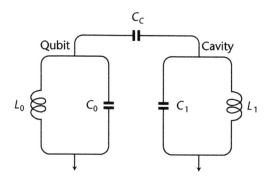

Figure 3.33 A pair of LC oscillators connected by the coupling capacitor C_c.

The circuit's topology in Figure 3.33 suggests that the nodal circuit analysis is appropriate. The unknown variable is the voltage of each node, which is equal to the time derivative of the inductor's flux connected to that node. We choose the flux of inductors Φ_i as the coordinate. The Lagrangian of the equivalent circuit can be written as [28]

$$\mathcal{L} = T - V = \frac{1}{2}C_1\dot{\Phi}_1^2 + \frac{1}{2}C_2\dot{\Phi}_2^2 + \frac{1}{2}C_0\left[\dot{\Phi}_1 - \dot{\Phi}_2\right]^2 - \frac{1}{2L_1}\Phi_1^2 - \frac{1}{2L_2}\Phi_2^2 \quad (3.66)$$

The kinetic energy (T) and potential energy (V) can be obtained by comparing the charge and flux to position and momentum. Matrix notation is often employed for convenience, where the Lagrangian can be written in matrix form as given by $\mathcal{L} = (1/2)\dot{\Phi}C\dot{\Phi} - (1/2)\Phi L^{-1}\Phi$, where the capacitance matrix is $C \equiv \begin{pmatrix} C_1 + C_0 & -C_0 \\ -C_0 & C_2 + C_0 \end{pmatrix}$ and the inverse inductance matrix is $L^{-1} \equiv \begin{pmatrix} 1/L_1 & 0 \\ 0 & 1/L_2 \end{pmatrix}$.

This equivalent circuit has the nice property that the inductance and capacitance matrices can be found easily by inspection. The ith diagonal element of the capacitance matrix is the sum of the capacitances connected to the ith nodes. For nondiagonal elements, the entry is minus the capacitance between the nodes. Now, we are going to find the Hamiltonian. The canonical momenta are given by $Q_i = \partial \mathcal{L}/\partial \dot{\phi}_i$. In terms of the inverse capacitance matrix, we obtain $\dot{\Phi} = C^{-1}Q$. The Hamiltonian can be written as $H = Q_i\dot{\Phi}_i - \mathcal{L} = 1/2(QC^{-1}Q + \Phi L^{-1}\Phi)$. Rewriting the above Hamiltonian gives

$$H = \frac{1}{2}QC^{-1}Q + \frac{1}{2}\Phi L^{-1}\Phi \quad (3.67)$$

with $Q = -iQ_{ZPF}(a_i - a_i^\dagger)$, where $Q_{ZPF} = \sqrt{C\hbar\Omega/2}|$ the resonant frequency of each resonant mode is $\Omega_i = 1/\sqrt{L_iC_i}$. The coupling factor or the g-factor can be found by looking at the coefficients of Q_iQ_j terms in the Hamiltonian given in (3.67).

$$C_{ij}^{-1}Q_iQ_j = -C_{ij}^{-1}Q_{ZPF,i}Q_{ZPF,j}\left(a_i - a_i^\dagger\right)\left(a_j - a_j^\dagger\right)$$
$$= -\hbar\frac{C_{ij}^{-1}}{2}\sqrt{\frac{\Omega_i\Omega_j}{C_i^{-1}C_j^{-1}}}\left(a_i - a_i^\dagger\right)\left(a_j - a_j^\dagger\right) \quad (3.68)$$

where the g-factor with the unit of angular frequency is given by

$$g_{ij} = \frac{C_{ij}^{-1}}{2}\sqrt{\frac{\Omega_i\Omega_j}{C_i^{-1}C_j^{-1}}} = \frac{C_0^{-1}}{2}\sqrt{\frac{\Omega_1\Omega_2}{C_1^{-1}C_2^{-1}}} \quad (3.69)$$

This demonstrates how the equivalent circuit components contribute to the g-factor and how they can be adjusted to control it.

Recall that the single mode of a resonator could be modeled using a parallel LC circuit with the resonant frequency of $\omega_r = \sqrt{1/L_r C_r}$. The photon loss rate or the resonator linewidth is given by $\kappa = \omega_r/Q$. The voltage across the resonator can be calculated as $\hat{V}_r = \hat{Q}_r/C_r$, where its average is zero $\langle 0|\hat{V}_r|0\rangle = 0$. However, the variance of the voltage is not zero $\sqrt{\langle 0|\hat{V}_r^2|0\rangle} = \sqrt{\hbar\omega_r/2C_r}$.

This is called the root mean square (RMS) vacuum voltage of the LC oscillator and is about ~1 μV for typical values of the resonance frequency. The qubit is coupled to this zero-point motion fluctuation. In a coplanar waveguide, a 5-μm distance between the central line and the ground results in an electric field of about $E = V/d = 1\mu V/5\mu m = 0.2$ V/m, which is large due to the nature of EM circuits. Vacancies and defects have dipole moments, and this large field can couple to them, so we can easily lose quantum information by transferring it to vacancies and defects. So, one way to solve this problem is to reduce the strength of the electric field (e.g., by using a 3D cavity), as shown in Figure 3.24(d). In order to compensate for the weak electric field, we need to make the qubit huge, such that d.E remains unchanged. This is why the qubit in a 3D cavity is so large, as shown in Figure 3.24(d). Using a qubit in a 3D cavity results in a large increase in coherence times T_1 and T_2.

Let us look at the Hamiltonian parameters for a transmon qubit coupled to a microwave cavity. We saw that RMS vacuum voltage of the LC oscillator is given by $V = \sqrt{\hbar\omega_r/2C_r}$. The quantum gate voltage $\hat{V}_g = V(a + a^\dagger)$ is related to the gate charge as $\hat{n}_g = C_g\hat{V}_g/(2C_r)$. By replacing this value in the electrostatic component of the charge qubit Hamiltonian described in (3.14) and then expanding the square, we derive a term that represents the coupling between the qubit and the resonator $H \propto -4E_c C_g \hat{V}_g \hat{n}$. This contains the charge qubit state $\hat{n}$ as well as the quantum field oscillator state $\hat{V}_g$ and can be simplified as $\hat{H} = 2\hbar g(a + a^\dagger)\hat{n}$ with the single qubit cavity coupling strength g = $(C_g/C_\Sigma)(eV/\hbar) = \beta(eV/\hbar)$. The ratio $\beta = C_g/C_\Sigma \in \{0, 1\}$ is the coupling capacitance divided by the total capacitance of the qubit.

It is noteworthy that the maximum dimensionless coupling strength between the qubit and cavity g/ω_r depends solely on the resonator's geometric and dielectric properties without considering the transition matrix element. This implies a correlation between the coupling strength of a single superconducting qubit and the fine structure constant α given approximately by $g/\omega_r \sim 4\beta\sqrt{\alpha/\epsilon_r}$ with $\alpha = e^2/(4\pi\epsilon_0\hbar c)$. For practical values of the dielectric constant, the greatest possible coupling strength of a single qubit is approximately $g/\omega_r \sim 0.1$. However, this limit only applies to a half-wave transmission line resonator. Moreover, an inductively coupled superconducting qubit can easily surpass this limit, as shown in [8]. The same applies when multiple qubits are coupled together to the resonator field.

3.3.5 Qubit Control and Readout in Practice

Figure 3.34(a) illustrates a transmon qubit coupled to a resonator, along with the associated drive and readout lines. The qubit control involves two lines: the qubit drive line (X, Y control) and the flux bias line (Z control). Figure 3.34(a) provides a zoomed-in view of the qubit, showcasing the capacitive couplings to the drive and readout lines. Figure 3.34(b) displays the equivalent circuit of the structure, showcasing the microwave readout and control pulses. Additionally, Figure 3.34(b) demonstrates how the current I_{bias} of the flux bias line tunes the qubit's frequency by applying a magnetic flux to the SQUID.

Figure 3.34(c) depicts the room-temperature and cryogenic control and readout hardware. Chapter 7 presents a comprehensive discussion of the qubit's hardware setup. In Figure 3.34(c), an AWG is responsible for generating flux bias pulses, while another AWG and a microwave signal generator are utilized to generate microwave pulses to control the qubit's state. The readout pulse, generated by the AWG, undergoes modulation by the local oscillator (LO), which is a CW microwave signal, and is then applied to the input of the readout resonator. After amplification and downconversion, the output of the readout resonator is directed to a digitizer. The digitized signal is then processed on the host PC.

3.4 Two-Qubit System

As discussed in Chapter 2, single-qubit and two-qubit gates are necessary for implementing quantum algorithms. This section focuses on two-qubit gates. The two-qubit gates in the transmon-like superconducting qubit architecture can be roughly divided into two main families. The first group relies on local magnetic fields to adjust the transition frequency of qubits, while the second group involves all-microwave control. Additionally, there are hybrid approaches that combine elements from both categories. Notably, the concepts of tunable coupling and

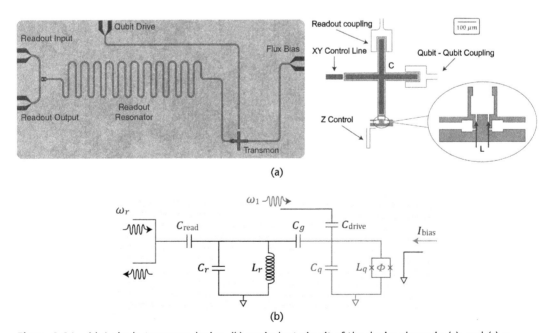

Figure 3.34 (a) A single transmon device, (b) equivalent circuit of the device shown in (a), and (c) a hardware setup for superconducting qubits.

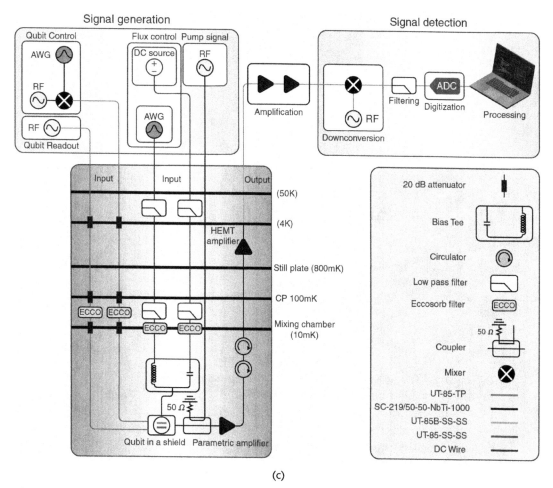

Figure 3.34 *(Continued)*

parametric driving have emerged as significant components in modern superconducting qubit processors [10].

As we saw in Chapter 2, CNOT and CPHASE gates require a control qubit to apply an X-gate (CNOT) or a Z-gate (CPHASE) to the target qubit. This interaction between the two qubits can be achieved by coupling them using lumped circuit elements. Capacitive coupling is used for charge and phase qubits, while inductive coupling is used for flux qubits, as shown in Figure 3.35. However, these coupling mechanisms are limited to local interactions and only couple the nearest-neighbor qubits, making it challenging to perform gates between arbitrary pairs of distant qubits, which is highly desirable for scalable architecture.

To address this issue, a quantum bus, a distributed circuit like a transmission line resonator, can be used for long-range qubit coupling to distribute quantum information. A quantum bus must have a quantum degree of freedom that allows for strong interactions with independent quantum systems for quantum information storage or transfer [28].

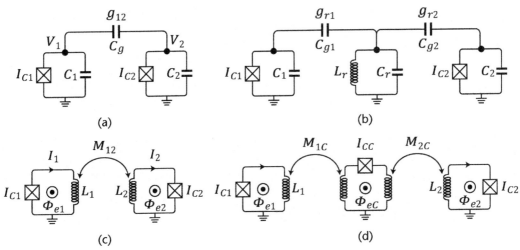

Figure 3.35 Capacitive and inductive coupling schemes between two superconducting qubits: (a) direct capacitive coupling, (b) capacitive coupling via a linear resonator quantum bus, (c) direct inductive coupling through mutual inductance, M_{12}, and (d) a frequency-tunable coupler is used to inductively couple two qubits through mutual inductances M_{1C} and M_{2C}. (After: [10].)

Figure 3.36(a) shows two qubits on the opposite sides of a half-wavelength transmission line resonator where we have voltage antinodes. The qubits are coupled to each other through the resonator. Using transmission line structures makes it possible to choose the line length and frequency such that multiple antinodes are present for coupling to multiple qubits.

As shown in Figure 3.36(b), each qubit has a split Josephson junction making it possible to tune the frequency using an external flux Φ through the loop according to $hf \approx \sqrt{8E_C E_J^{max} |\cos(\pi\Phi/\Phi_0)|} - E_C$. The flux-bias lines using the bottom-left and top-right ports are coplanar waveguides with short-circuit termination next to their target qubit. This allows the current on the line to couple flux through the split junctions. Voltage inputs from room temperature V_L and V_R are used to tune left and right qubit frequencies f_L and f_R.

Figure 3.36(c) shows the cavity transmission and qubit spectroscopy as a function of the flux bias voltage V_R. The difference between the qubit frequencies is necessary since driving one qubit will drive the other if they have the same frequency. Bias point I is one where the maximum frequency of both qubits is far detuned from the cavity and each other. At this bias point, the interactions are small, and one may utilize the point for state preparation, single-qubit rotations, and measurement in the computational basis $|0,0\rangle$, $|0,1\rangle$, $|1,0\rangle$, and $|1,1\rangle$. Point I is also a flux sweet spot for both qubits, providing long coherence. A CPHASE gate is achieved by pulsing into point II, where a $\sigma z \otimes \sigma z$ interaction turns on [28]. Thus, points I and II are possible operating points of the processor.

As qubit Q_R is tuned into resonance with qubit Q_L at point III, an avoided crossing results from a cavity-mediated, qubit-qubit transverse interaction. The splitting has a magnitude J much greater than the qubit linewidths. This indicates a coherent coupling, and the qubits are in a strong-dispersive regime. When

3.4 Two-Qubit System

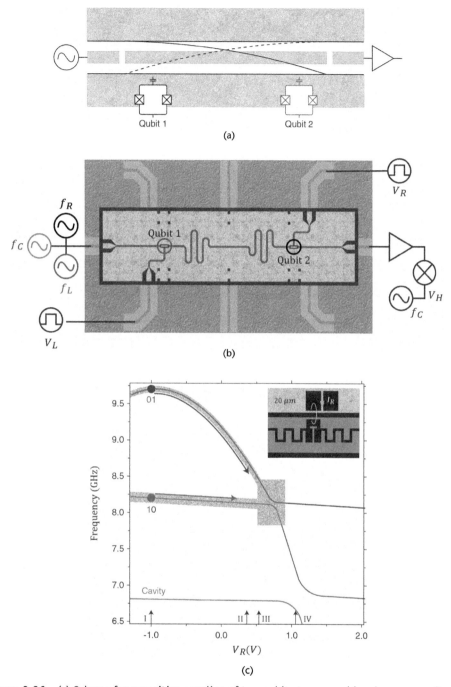

Figure 3.36 (a) Scheme for capacitive coupling of two qubits to an on-chip microwave cavity. The two transmon qubits are located at opposite ends of the cavity where the electric field has an antinode. The resonator is coupled capacitively to the input and output lines. The microwave signals enter the chip from the left, interact with the qubits, exit from the right, and enter an amplifier. (b) Fabricated sample with two transmon qubits coupled by a transmission line resonator. The local flux-bias lines provide fast qubit tuning. Microwave pulses at the qubit transition frequencies f_L and f_R drive single-qubit rotations. For the two-qubit readout, a pulsed measurement of the cavity homodyne voltage V_H (at frequency f_C) is used. (c) Cavity transmission and qubit spectroscopy as a function of flux bias voltage V_R. (*After:* [27].)

operating in the dispersive regime of CQED, virtual photons are exchanged during the coherent qubit interactions, which protects the qubit interactions against loss in the bus. If real photons were used, the cavity-induced relaxation of the qubits would make coherent state transfer unfeasible, mainly because a cavity with a large κ is required for fast measurements. The strong-dispersive regime allows for joint readout that can efficiently detect two-qubit correlations. Thus, we understand how both qubits can be controlled independently, and the qubit-qubit interactions can be switched on and off by adjusting the flux.

As we tune the Q_R into resonance with the cavity, point IV is reached, resulting in a vacuum Rabi splitting from which the qubit-cavity coupling strength is extracted. So, we see that the same cavity used to couple the qubits is also used for the multiplexed control and measurement of the qubit states. The cavity acting as a quantum bus allows for long-range coupling and could be extended to nonnearest neighbors.

3.4.1 Dispersive Two-Qubit Interactions

In the dispersive regime, no energy is exchanged with the cavity, and both qubits with frequencies ω_1 and ω_2 are detuned from the resonator such that $|\Delta_{1,2}| = |\omega_{1,2} - \omega_C| \gg g_{1,2}$. However, there is still dispersive coupling between the qubits and the cavity, resulting in a qubit-state-dependent shift $\pm \chi_{1,2} = \pm g_j^2/\Delta_j$ of the cavity frequency, as shown in Figure 3.37(a). This shift can be calculated using the coupling strength $g_{1,2}$ and the detuning $\Delta_{1,2}$. The effective dispersive Hamiltonian reads [29]

$$\frac{H}{\hbar} = \left(\omega_C + \chi_1 \sigma_{z1} + \chi_2 \sigma_{z2}\right) a^\dagger a + \frac{1}{2}\omega_1 \sigma_{z1} + \frac{1}{2}\omega_2 \sigma_{z2} \\ + \frac{g_1 g_2 (\Delta_1 + \Delta_2)}{2\Delta_1 \Delta_2} \left(\sigma_1^+ \sigma_2^- + \sigma_1^- \sigma_2^+\right) \quad (3.70)$$

where ($\sigma_{z1} = 1 \otimes \sigma_z$; $\sigma_{z2} = \sigma_z \otimes 1$). The first term is the cavity dispersively shifted by both qubits. The second and third terms are the bare qubit Hamiltonians. The last term is a two-qubit swap via virtual interaction with the cavity. The strength of interaction between the qubits is given by $J = g_1 g_2 (\Delta_1 + \Delta_2)/2\Delta_1\Delta_2$, which is the coefficient of the last term in the Hamiltonian. The qubit-qubit interaction results from the virtual exchange of photons with the cavity.

As shown in Figure 3.37(b), by tuning the qubits to the same frequency, an excitation in one qubit can be transferred to the other by virtually becoming a photon in the cavity. When the qubits are nondegenerate $|\omega_1 - \omega_2| \gg J$ the interaction is effectively turned off as this process does not conserve energy. So, the effective coupling strength can be controlled by tuning the qubit transition frequencies, which can be done using flux bias lines.

In the dispersive regime, the bare cavity transmission is shifted to four frequencies depending on the state of the two qubits ($|1,1\rangle$, $|0,1\rangle$, $|1,0\rangle$, $|0,0\rangle$).

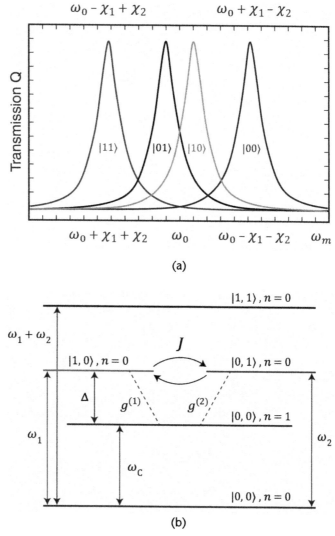

Figure 3.37 (a) Transmission in the strong dispersive regime for two qubits. The four dispersive shifted cavity transmission peaks for a two-qubit and cavity system. (b) When the qubits are detuned from the cavity ($|\Delta^{(1),(2)}| \gg g^{(1),(2)}$) they shift the cavity dispersively. The excited state in the left qubit $|10\rangle \otimes |n=0\rangle$ and the excited state in the right qubit $|01\rangle \otimes |n=0\rangle$ interact via the exchange of a virtual photon $|00\rangle \otimes |n=1\rangle$ in the cavity.

3.5 Calibration of Single-Qubit Operations

Prior to executing any quantum algorithms on a quantum processor, it is crucial to employ a systematic approach for characterizing and calibrating the qubit operations. This involves performing qubit performance characterization and pulse calibrations. A single-qubit calibration routine involves steps such as spectroscopy on the readout resonator to determine the frequency ω_r, spectroscopy on the qubit at the readout frequency to estimate the qubit transition frequency ω_{ge}, and estimating the π and $\pi/2$ drive pulse amplitudes by driving the qubit at ω_{ge}. For a more comprehensive list of calibrations please refer to [30].

The highest gate fidelity can only be achieved with precise pulse calibration. Similar to NMR experiments, some tune-up schemes are carried out in which repeated pulses are applied such that small errors can be built up, leading to a large signal. This large signal may then be nulled out. The rotations $R_x(\pm\pi/2)$, $R_y(\pm\pi/2)$, $R_x(\pm\pi)$, $R_y(\pm\pi)$, are calibrated independently [17].

3.6 Testing the Performance of a Quantum Processor

We have seen how to implement single- and two-qubit gates using superconducting qubits. The crucial question is how to ascertain whether these gates are functioning as intended, ensuring that gate errors remain within the fault-tolerant threshold for quantum-error correction, typically around 10^{-4}. Chapter 2 discussed various methods for assessing gate fidelity, which serves as the standard metric for measuring the degree of agreement between an ideal operation and its practical implementation. Specifically, it explores three techniques: the double-π metric, QPT, and randomized benchmarking.

So far, we have covered the basics of quantum information processing using superconducting quantum qubits. The field of superconducting qubits is advancing at an incredibly rapid pace, and there is a significant need for both intriguing engineering and fundamental physics research to realize the ultimate quantum computer [31, 32]. Chapters 4 to 8 will be focused on hardware implementation.

References

[1] Tinkham, M., *Introduction to Superconductivity*, (Second Edition), Dover Publications, June 14, 2004.

[2] Duzer, T., and C. W. Turner, *Principles of Superconductive Devices and Circuits* (Second Edition), Elsevier, 1981.

[3] Kittel, C., *Introduction to Solid State Physics*, John Wiley & Sons, 2018.

[4] Bardeen, J., L. N. Cooper, and J. R. Schrieffer, "Theory of Superconductivity," *Phys. Rev.*, 108.5, Dec. 1957, pp. 1175–1204.

[5] Martinis, J. M., and K. Osborne, "Superconducting Qubits and the Physics of Josephson Junctions," 2004, arXiv:cond-mat/0402415.

[6] Ambegaokar, V., and A. Baratoff, "Tunneling Between Superconductors," *Phys. Rev. Lett.*, 10, 486, 1 June 1963; erratum Phys. Rev. Lett. 11, 104 (1963).

[7] Koch, J., et al., "Charge-Insensitive Qubit Design Derived from the Cooper Pair Box," *Phys. Rev. A*, 76 042319, 2007, pp. 15, 16, 48, 73, 86, 89, 90, 100, 108, and 110.

[8] Devoret, M. H., A. Wallraff, and "J. M. Martinis, Superconducting Qubits: A Short Review," 2004, arXiv:cond-mat/0411174.

[9] Clarke, J., and F. K. Wilhelm, "Superconducting Quantum Bits," *Nature*, Vol. 453, No. 1031, 2008.

[10] Krantz, P., "A Quantum Engineer's Guide to Superconducting Qubits," *Applied Physics Reviews*, Vol. 6, No. 2, 2019.

[11] Gao Yvonne, Y., "Practical Guide for Building Superconducting Quantum Devices," *PRX Quantum*, Vol. 2, No. 4, 2021.

[12] Martinis, M., et al., "Superconducting Circuits for Quantum Information: An Architecture for Implementing Quantum Information Processing with Superconducting Devices," *Science*, Vol. 339, No. 6124, March 8, 2013, pp. 1169–1174.

[13] Vepsäläinen, A. P., et al., "Impact of Ionizing Radiation on Superconducting Qubit Coherence," *Nature*, Vol. 584, No. 7822, 2020, pp. 551–556.

[14] Ithier, G., "Decoherence in a Superconducting Quantum Bit Circuit," *Physical Review B*, Vol. 72, No. 13, 2005, p. 134519.

[15] Riwar, R-P., et al., "Normal-Metal Quasiparticle Traps for Superconducting Qubits," *Physical Review B*, Vol. 94, No. 10, 2016, p. 104516.

[16] Foot, C. J., *Atomic Physics*, Oxford University Press, 2005.

[17] Chow, J. M., et al., "Randomized Benchmarking and Process Tomography for Gate Errors in a Solid-State Qubit," *Physical Review Letters*, Vol. 102, No. 9, 2009, p. 090502.

[18] Gambetta, J. M., et al., "Analytic Control Methods for High-Fidelity Unitary Operations in a Weakly Nonlinear Oscillator," *Physical Review A*, Vol. 83, No. 1, 2011, p. 012308.

[19] Soro, A., "Interaction Between Giant Atoms in a One-Dimensional Structured Environment," *Physical Review A*, Vol. 107, No. 1, 2023, p. 013710.

[20] Wallraff, A., et al., "Strong Coupling of a Single Photon to a Superconducting Qubit Using Circuit Quantum Electrodynamics," *Nature*, Vol. 431, No. 7005, 2004, pp. 162–167.

[21] Haroche, S., and D. Kleppner, "Cavity Quantum Electrodynamics," *Physics Today*, Vol. 42, No. 1, 1989, pp. 24–30.

[22] Magesan, E., "Machine Learning for Discriminating Quantum Measurement Trajectories and Improving Readout," *Physical Review Letters*, Vol. 114, No. 20, 2015, p. 200501.

[23] Boissonneault, M., J. M. Gambetta, and A. Blais, "Dispersive Regime of Circuit QED: Photon-Dependent Qubit Dephasing and Relaxation Rates," *Physical Review A*, Vol. 79, No. 1, 2009, p. 013819.

[24] Chen, L., HX Li, Y. Lu, et al., "Transmon Qubit Readout Fidelity at the Threshold for Quantum Error Correction Without a Quantum-Limited Amplifier," *npj Quantum Inf*, Vol. 9, No. 26, 2023, https://doi.org/10.1038/s41534-023-00689-6.

[25] Göppl, M., et al., "Coplanar Waveguide Resonators for Circuit Quantum Electrodynamics," *Journal of Applied Physics*, Vol. 104, No. 11, 2008.

[26] Reed, M. D., et al., "Fast Reset and Suppressing Spontaneous Emission of a Superconducting Qubit," *Applied Physics Letters*, Vol. 96, No. 20, 2010.

[27] Reed, M. D., "High-Fidelity Readout in Circuit Quantum Electrodynamics Using the Jaynes-Cummings Nonlinearity," *Physical Review Letters*, Vol. 105, No. 17, 2010, p. 173601.

[28] Girvin, S., *Circuit QED: Superconducting Qubits Coupled to Microwave Photons*, Oxford University Press, 2012.

[29] DiCarlo, L., "Demonstration of Two-Qubit Algorithms with a Superconducting Quantum Processor," *Nature*, Vol. 460, No. 7252, 2009, pp. 240–244.

[30] Balasiu, S., "Single-Qubit Gates Calibration in PycQED Using Superconducting Qubits," PhD diss., MA thesis, ETH Zurich, 2017.

[31] Paik, H., and K. D. Osborn, "Reducing Quantum-Regime Dielectric Loss of Silicon Nitride for Superconducting Quantum Circuits," *Applied Physics Letters*, Vol. 96, No. 7, 2010.

[32] Sarabi, B., "Cavity Quantum Electrodynamics Using A Near-Resonance Two-Level System: Emergence Of The Glauber State," *Applied Physics Letters*, Vol. 106, No. 17, 2015.

CHAPTER 4
Microwave Systems

> Thoroughly conscious ignorance is the prelude to every real advance in science.
>
> —*James Clerk Maxwell*

The core concepts of microwave engineering that will be covered in this chapter and Chapters 5 and 6 have been adjusted to fit within the context of superconducting qubit design and implementation. This chapter provides a general overview of microwave engineering and discusses the fundamentals of microwave system analysis, including link budget calculation, distortion, noise, interference, and nonlinear effects.

4.1 A Brief History of Microwave Engineering

Many renowned scientists, including Gauss, Faraday, Ampere, Lenz, and Oersted, devoted their research to studying electricity and magnetism. In 1873, Maxwell published his seminal work, *A Treatise on Electricity and Magnetism*, in which he integrated electricity and magnetism by introducing the term "displacement current" to the set of equations developed by his predecessors. This additional term enabled the derivation of a wave equation using his unified mathematical framework, demonstrating the theoretical existence of electromagnetic waves. Maxwell's equations were initially expressed in integral form, which Oliver Heaviside later reformulated in differential form.

Between 1886 and 1888, Heinrich Hertz conducted a series of experiments that demonstrated the validity of Maxwell's equations and the existence of electromagnetic waves. His setup consisted of two main components: a wave generator and a detector, as shown in Figure 4.1(a). The wave generator was constructed using a dipole antenna with a spark gap between their inner ends as a radiator connected in parallel with Ruhmkorff's coils. A high-voltage spark in the gap is generated by interrupting the direct current in Ruhmkorff's coil, generating radio waves. The electromagnetic waves were then picked up and detected using a loop antenna. Hertz made further discoveries about electromagnetic waves, including their polarization.

Guglielmo Marconi achieved a significant milestone in the history of wireless communication on December 21, 1901, by successfully establishing transatlantic wireless communication. He achieved this feat by modifying and enhancing Hertz's equipment to extend the range of electromagnetic waves. The development

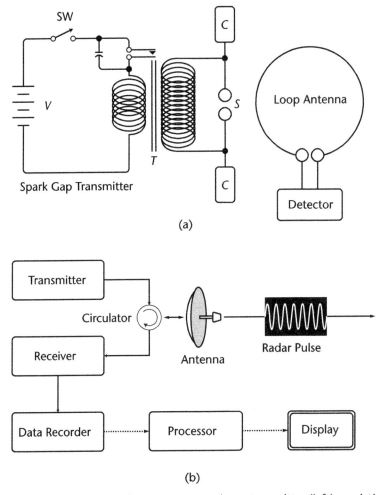

Figure 4.1 (a) Hertz's experimental apparatus. A spark-gap transmitter (left) consisting of a dipole antenna with a spark gap (S) powered by high-voltage pulses from a Ruhmkorff coil (T). The receiver (right) consists of a loop antenna with a small gap. (b) Basic block diagram of a radar system. A single antenna transmits and receives the radar signals. A circulator (see Chapter 5) separates the transmit and receive paths.

of early radio technology progressed rapidly, but primarily in the high-frequency (HF) to very high-frequency (VHF) range due to the absence of reliable microwave sources and components. It was not until the 1950s, with the advent of radar technology during World War II, that microwave theory and techniques received much attention. The invention of microwave sources such as Klystron and Magnetron played a crucial role in advancing radar technology, significantly impacting the battlefield. A basic block diagram of a radar system is shown in Figure 4.1(b). To promote research on radar theory and applications in the United States, the Massachusetts Institute of Technology (MIT) established the Radiation Laboratory. Some renowned scientists who have received Nobel prizes, such as H. A. Bethe, J. S. Schwinger, I. I. Rabi, and E. M. Purcell, worked together for a period of intense microwave research. Their work results are summarized in the classic 28-volume Radiation Laboratory series of books [1].

Microwave engineering has played a significant role in shaping the modern world. After the invention and development of radar, there was a surge of interest in microwave communication systems. This interest led to the utilization of much of the early work done for radar systems. Microwave systems' benefits, such as wide bandwidth, made them essential for terrestrial and satellite communications.

Developing efficient, low-cost, and miniaturized microwave components has opened new paths and impacted every facet of our lives. Modern life is unimaginable without microwave technology, as we use it regularly to communicate using our cell phones, to cook our food, and to navigate using GPS devices.

Microwave engineering has also contributed to several other fields. Plasma technology, medical diagnosis and treatment, atomic and molecular physics, and remote sensing are just a few examples. Microwave engineering has expanded its applications in recent decades and now plays a crucial role in developing quantum technologies and designing and operating fusion reactors. With its many applications, microwave engineering continues to be a critical field that drives innovation and progress.

4.2 Microwave Engineering

Chapter 2 discussed the rationale behind utilizing microwave frequencies for superconducting qubits. To effectively design and operate superconducting qubit hardware, it is necessary to have a solid understanding of microwave systems and components. This chapter explores the key factors that influence the performance of a microwave system and shows us how to analyze their impact. Subsequently, Chapter 5 delves into the microwave components essential for implementing superconducting qubit hardware.

The terms radio frequency (RF) and microwave engineering are often used interchangeably, with RF frequencies covering the VHF band from 30 MHz to the UHF band ending at 3 GHz, and microwave frequencies covering the range of 3 to 300 GHz [1, 2]. Although there is a significant overlap between the fundamental concepts of RF and microwave engineering, there are also some key distinctions.

For consistency, this text refers to RF and microwave frequencies as "microwave," since the superconducting qubits typically operate within the microwave frequency range. This section provides an overview of microwave engineering and its core building blocks.

Microwave engineering involves generating, transmitting, processing, and detecting microwave signals. Figure 4.2 illustrates these concepts using the example of a communication link. The first step in this process is to encode information, such as voice, video, or data, onto a microwave carrier signal using a modulator. The signal may undergo additional processing, such as amplification or filtering, before it can be transmitted.

The signal can be transmitted either through wires, such as in an optical fiber link, or wirelessly through free space. For long-distance transmission, repeaters can be used to boost the signal and extend its range. Upon entering the receiver, filters, amplifiers, downconverters, and other processing components are required to prepare the signal for detection. To hear the voice or watch the video at the

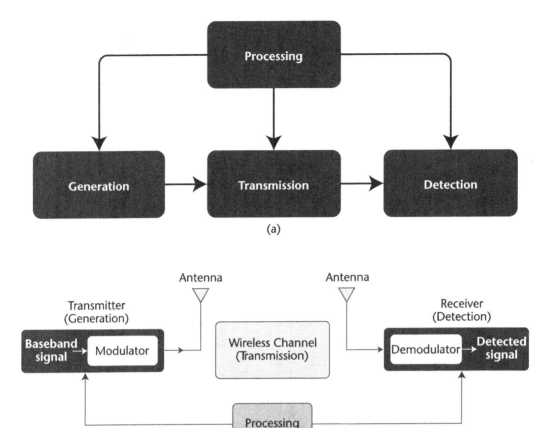

Figure 4.2 (a) Technical areas related to microwave engineering fit into one of the categories of processing, generation, transmission, and detection. (b) An example of a communication link.

receiving end of the link, the detector or demodulator decodes the signal and brings it back to the baseband.

Depending on the application, one or more of the aforementioned four areas (processing, generation, transmission, and detection) may be involved. In certain applications, such as microwave heating, detection may not be necessary. On the other hand, detection is essential in fields like radio astronomy. However, in applications such as in a communication link or a superconducting quantum computer, all four areas play a crucial role.

To enhance our understanding of these concepts, Section 4.3 discusses them in the context of a microwave link.

4.3 Microwave System Analysis

To effectively analyze and design a microwave system, it is crucial to have a system-level understanding of critical performance factors, including noise, distortion, and nonlinear parameters. The analytical tools covered in this section are powerful as they enable us to treat a microwave system and its components as a black box, examining the inputs and outputs of the system without requiring deep knowledge of electromagnetics or microwave engineering.

Our study of microwave engineering commences with analyzing a microwave link and performing link budget calculations, which is an essential first step in designing and evaluating such a system. Subsequently, we examine the effects of losses, distortion, interference, and noise on the performance of microwave systems. Furthermore, we explore nonlinear effects in microwave systems, including the 1-dB compression point, the level of harmonics, and intermodulation distortion—all essential considerations in designing and optimizing a microwave system's performance and efficiency.

4.3.1 Microwave Link

A microwave link is comprised of four main components: signal generation, transmission, processing, and detection. It can be used as a communication link between two cell phones or to transfer information within a superconducting quantum computer. The goal is to design the microwave link to ensure successful signal detection. Section 4.2 provided an example of a communication link. This section examines a microwave link in a superconducting quantum computer, illustrated in Figure 4.3.

Microwave signals are used to control and read out the state of the superconducting qubits. As shown in Figure 4.3, the microwave signals are generated at room temperature and transmitted into a dilution fridge using microwave coaxial cables. Along the way, filters and attenuators process the signal to reduce noise. After interacting with the qubit, the signal is processed using cryogenic amplifiers and sent back to room temperature using special microwave cables. The signal is further processed at room temperature, including amplification, filtering, and downconversion, to prepare it for detection. Finally, a digitizer converts the analog signal into a digital one that a computer can process. Chapter 7 details these concepts.

As a signal travels down a microwave link, it is affected by cables, amplifiers, filters, and various other passive and active components that amplify or attenuate it. To ensure that the signal is detectable at the receiving end, a link budget

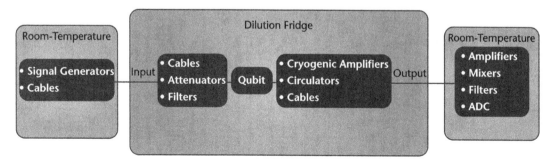

Figure 4.3 The microwave link for a superconducting quantum computer. Each box shows the components involved in each section of the microwave link.

analysis is performed to calculate all the gains and losses in the microwave link. This analysis is crucial for microwave links in quantum computers, where a sufficient signal-to-noise ratio is essential for successfully detecting quantum states.

In a communication link, the transmitted signal level is subject to losses caused by transmission lines, components, and environmental factors such as signal polarization, terrain, and atmospheric effects. Amplifiers and antenna gains can help increase the signal level. Figure 4.4 illustrates a communication link and how the signal level changes as it travels along the link.

We use various forms of decibel units, such as decibel milliwatts (dBm), throughout this book. The total gain of the link in Figure 4.4 is calculated as

$$G = 20\,(\text{TX power}) + 10(\text{Antenna Gain}) + 15(\text{RX antenna gain})$$
$$= 45 \text{ dB} \tag{4.1}$$

The losses are

$$L = 2\,(\text{TX cable loss}) + 116(\text{Free space loss}) + 2(\text{RX cable loss})$$
$$= 120 \text{ dB} \tag{4.2}$$

The free-space loss L_{FSL} is caused by the spreading of the wave front as it travels away from the source and is calculated as $L_{FSL} = (4\pi d/\lambda)^2$, where d is the

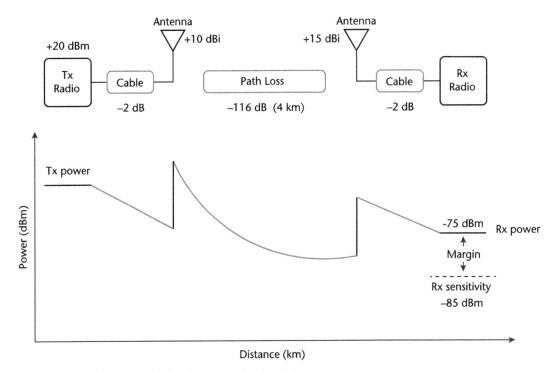

Figure 4.4 Link budget analysis of a communication link.

distance between the transmitter and receiver, and λ is the wavelength of the signal in meters.

The received signal level is

$$P_{Rx} = G - L = -75 \text{ dBm} \tag{4.3}$$

The receiver's minimum detectable signal level, also called sensitivity, is specified to be −85 dBm. Therefore, we have a 10-dB link margin, and the received signal can be detected successfully.

Section 4.3.2 discusses the factors contributing to signal degradation as it travels from the generator to the detector in a microwave link.

4.3.2 Signal Degradation Factors

Signal degradation in a microwave link can be caused by various factors that can significantly impact signal detection. This section explores the four key factors that have a substantial negative impact on a signal: attenuation, distortion, interference, and noise.

4.3.2.1 Attenuation

Depending on the specific link, different loss sources can attenuate a signal in a microwave link. The attenuation, expressed in decibels and denoted as L_{dB}, is a critical parameter used in link budget calculation. It is defined as the ratio of the power before attenuation (P_{in}) to the power after attenuation (P_{out}),

$$L_{dB} = 10 \log_{10}\left(\frac{P_{in}}{P_{out}}\right) \tag{4.4}$$

Sources of attenuation include cables, microwave components, and propagation effects in wireless links, such as free-space loss and atmospheric loss. Attenuation directly affects the signal amplitude and can make detection more challenging, particularly for weak signals like those in radar and satellite communication systems. In qubit systems, low signal levels on the order of a few photons can also pose challenges. Therefore, it is crucial to minimize losses at the output of the qubit readout resonator, where ultralow-loss and low-noise components, such as superconducting amplifiers and cables, are employed. Interestingly, it is worth noting that attenuation is intentionally applied at the input of the fridge for noise reduction, as discussed in Chapter 7.

Amplifiers or repeaters can compensate for losses, but this comes at the cost of introducing additional noise and occasional signal distortion. Thus, minimizing losses as much as possible using low-loss components and making smart design decisions is critical in designing a microwave link. However, we must also consider the costs of using such components and several other design factors. As discussed in Chapter 7, quantum-limited amplifiers, such as Josephson parametric amplifiers (JPAs) with extremely low noise figures, are suitable for amplifying extremely weak signals from the quantum processor.

4.3.2.2 Distortion

A communication link and countless other systems aim to reconstruct an exact copy of the input signal in the output or at the destination. However, in reality, the signal's shape changes as it travels through the system, leading to distortion, which is undesirable, since it alters the information carried by the signal. In the early days of landlines, transmission lines added distortion, resulting in a voice change on the other side of the line. This was solved by balancing the line's inductive and capacitive time constants. (Refer to Section 5.3.1.1 for a discussion of distortion-less transmission lines.)

Distortion can be deliberately introduced to a signal, such as when a sound engineer applies signal-processing techniques to intentionally modify the shape of a signal for artistic purposes during music recording. However, in most cases, distortion occurs without control over the signal as it travels through the link. This alteration of waveform and spectrum can result in information corruption in qubit and communication systems, as we soon explore. Fortunately, the adverse effects of distortion can sometimes be mitigated by employing signal processing techniques like predistortion.

Microwave pulses are used for qubit control and readout, but they can become distorted as they propagate along the cables, interconnects, and microwave components. Such distortions can cause errors and limit the fidelity of gate operations, as it can—among other negative effects—change the amplitude and width of the pulses. Quantum gates that employ fast-frequency tuning of the qubits are particularly sensitive to distortion and require precise calibration [3]. Another example is the impact of nonlinear amplitude distortions from microwave electronics on Rabi pulses [4]. Interestingly, the qubit can be utilized as a detector to characterize the nonlinear distortions [3].

4.3.2.3 Linearity

Before looking at the factors that cause distortion, let us see why linearity is essential to preserving the shape of a signal and, therefore, the information carried by the signal. A system is called linear if it satisfies two properties, homogeneity and superposition. If $x(t)$ is the input to the system and $y(t)$ is the output, and k is a constant, then the homogeneity property reads: $kx(t) \rightarrow ky(t)$. This means that if we multiply the input by a constant, the output will be multiplied by the same factor. The superposition property can be expressed as $x_1(t) + x_2(t) \rightarrow y_1(t) + y_2(t)$. This statement asserts that the system responds to the sum of two inputs by producing a response that is the sum of the responses to each input.

Figure 4.5 depicts the input and output of both linear and nonlinear systems in both time and frequency domains. Let us take a scenario where the linear system is a speaker, and the input is a pure musical tone, such as the C note with a frequency of f_1. In this case, the output of the linear system appears similar to the input, with only the amplitude and phase of the input signal being modified, as shown in Figure 4.5(a). In the frequency domain, the output spectrum remains the same as the input, except that the same factor scales the tone's amplitude.

However, for a nonlinear speaker, as shown in Figure 4.5(b), the output is no longer the C note but includes harmonics at f_2 and f_3, resulting in a different sound

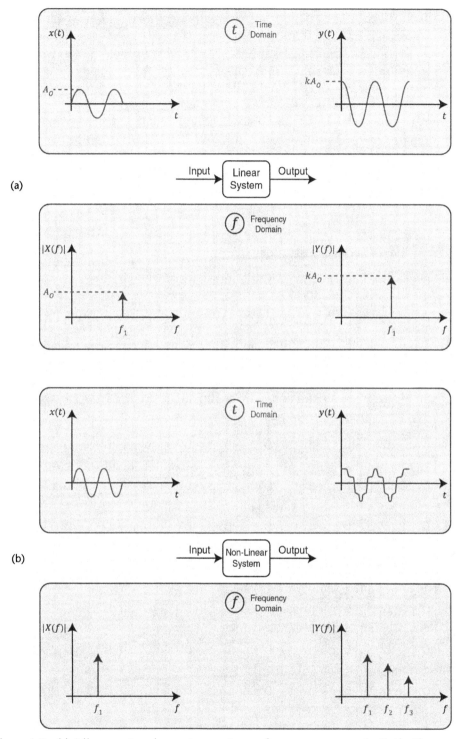

Figure 4.5 (a) A linear system does not generate new frequency components. In the time domain, it only changes the amplitude and phase of the input signal. (b) A nonlinear system distorts the signal's shape and generates new frequency components in the output.

than the input. So, a linear system maintains the input's shape, which is essential for preserving signal information.

From the frequency-response standpoint, a linear system must exhibit a flat amplitude and linear phase over its frequency range, as demonstrated in Figure 4.6. A nonflat frequency response alters the shape of the signal's spectrum, resulting in different amplitudes at different frequencies. As a result, it also modifies the waveform's shape. A linear phase response yields a constant time delay for all frequency components. Assume a time delay $\Delta t = \Delta\phi / 2\pi f$, where the phase difference $\Delta\phi$ is a linear function of frequency (i.e., $\Delta\phi = cf$), where c is a constant number. Then the delay $\Delta t = cf/2\pi f = c/2\pi$ becomes a constant. If the phase response is nonlinear, dispersion can occur, causing the propagation velocity of spectral components to become frequency-dependent, resulting in signal distortion.

4.3.2.4 Sources of Distortion

Recall that the primary source of distortion in RF and microwave systems is nonlinear effects. Many semiconductor components used in electronic chips and microwave circuits are inherently nonlinear, including transistors and diodes used in amplifiers, RF switches, and mixers, among several other components. As previously noted, nonlinear behavior adds harmonics to the input, changing the signal's frequency content and directly affecting the signal's shape in the time domain and the information it carries.

In addition, distortion can result from a nonflat frequency response, which can sometimes occur in passive components such as filters. This can alter the frequency content and, therefore, the waveform.

Another source of distortion is dispersion, which occurs when the propagation velocity of a signal is frequency-dependent, leading to signal smearing. Dispersion is often present in transmission lines, such as cables, optical fibers, and

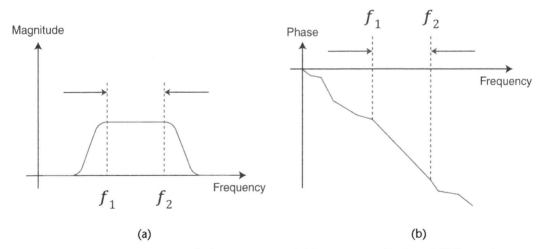

Figure 4.6 The frequency response of a linear system with (a) constant amplitude and (b) linear phase over the operational bandwidth from f_1 to f_2.

waveguides. Dispersion may be caused by a dispersive material, as in the case of an optical fiber, or it might occur due to the dispersive nature of the components or transmission lines such as hollow waveguides (see Chapter 5). An example of the distortion caused by transmission-line dispersion is shown in Figure 4.7(a). The effect of dispersion on a pulse sequence is shown in Figure 4.7(b). The pulses become indistinguishable due to the broadening caused by dispersion.

One might question the possibility of eliminating distortion altogether. While it is occasionally feasible to compensate for distortion, doing so adds complexity and costs to our design. Therefore, our initial focus should be on designing the system to be as linear as possible and eliminating as many sources of distortion as we can. However, some factors that cause distortion may not be under our control. In such cases, if we know the transfer function F, which describes the relationship between the input $x(t)$ and output $y(t)$ of a system $y(t) = F(x(t))$, we can intentionally predistort the original signal by applying the inverse function F^{-1} so that the signal comes out undistorted at the output. This idea is illustrated in Figure 4.7(c). The transfer function F can be determined using measurement instruments such as a network analyzer or spectrum analyzer. One practical example of this technique is digital predistortion (DPD) used in modern power amplifiers. This method applies a correction signal to the input signal before amplification. The correction signal is designed to cancel out the nonlinearities in the amplifier's response, resulting in a faithful reproduction of the input signal at the output. The same principle of applying the inverse transfer function to the signal before transmission is utilized to cancel out the distortion in qubit systems [3].

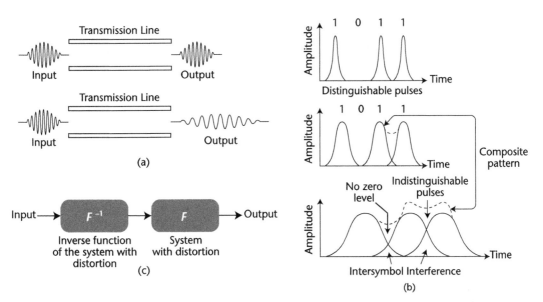

Figure 4.7 (a) A dispersive transmission line and (b) a 1011-bit sequence. Distortion is caused by dispersion in pulses, where eventually pulses merge and the 0 bit gets lost. (c) Compensating distortion using the inverse function of the system with distortion.

4.3.2.5 Interference

Designing a system entirely isolated from other systems and the environment is exceedingly difficult in the real world. Interference occurs when unwanted radio frequency or microwave signals are coupled to the system, affecting its function in an undesirable manner. The detection of a signal can be affected in different ways by interference. For example, the interfering signal can overlap with the desired signal, degrading the signal quality and causing distortion. Another example of interference is when the interfering signal's amplitude is large enough to compress or saturate the receiver, rendering it unable to detect any signal.

We have learned that a qubit's interaction with the environment results in qubit relaxation and dephasing. Therefore, any noise or interference coupled to the qubit can significantly affect its quantum state and its coherence.

Additionally, scalable quantum information processing requires parallel execution of gate operations. Short-range interference, also known as crosstalk, can impair the quality of simultaneous gate operations by reducing individual addressability, leading to erroneous results for simultaneous operations [5, 6].

Dealing with interference in qubit systems due to their extreme sensitivity to external signals is of paramount importance. For this purpose, a combination of techniques, such as filtering, shielding, and grounding is used, as discussed in Chapters 6 and 7.

4.3.2.6 Thermal Noise

Random physical processes, such as thermal vibration or fluctuation of charge carriers in solid-state devices, generate electric signals known as noise. This noise is often the primary factor adversely affecting signal detection. The various types of noise, each with specific properties [1], are described as follows:

- *Thermal noise:* This is caused by the thermal vibration of bound charges.
- *Flicker noise* (1/f noise): This happens in solid-state devices, where the noise power decreases with increasing frequency.
- *Shot noise:* Occurs following a random fluctuation of charge carriers in solid-state devices.
- *Quantum noise:* This is caused by the quantized nature of charges and photons.

We often encounter the first two types of noise, which significantly contribute to the decoherence of qubits. Figure 4.8(a) illustrates a resistor at a physical temperature of T degrees kelvin. The kinetic energy of the random motion of electrons in the resistor is proportional to the temperature. These random motions cause small voltage fluctuations at the resistor terminals, as shown in Figure 4.8(a).

This random voltage has a zero average value but a nonzero RMS value given by Planck's blackbody radiation law

$$V_n = \sqrt{\frac{4hfBR}{e^{hf/kT} - 1}} \tag{4.5}$$

4.3 Microwave System Analysis

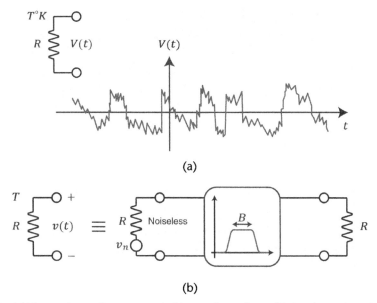

Figure 4.8 (a) The random voltage generated in a noisy resistor. (b) A noisy resistor is modeled as a noiseless resistor in series with a voltage source connected to a load resistor with an ideal bandpass filter.

The formula for noise voltage given in (4.5) is valid for both classical (low-frequency, high-temperature) and quantum limits (high-frequency, low-temperature), where h and k are Planck's and Boltzmann's constants, respectively, T is the temperature in degrees kelvin, B is the system's bandwidth in hertz, f is the center frequency of the bandwidth in hertz, and R is the resistance in ohms. In the classical limit, where $hf \ll kT$, (4.5) can be approximated as $e^{hf/kT} - 1 \cong hf/kT$. So, (4.5) reduces to $V_n = \sqrt{4kTBR}$.

This approximation is invalid for extremely high frequencies and very low temperatures. In qubit applications at temperatures on the order of millikelvin, (4.5) must be used, as shown in Chapter 7.

We are interested in a worst-case scenario to determine the maximum power a noisy resistor can deliver to a circuit. As illustrated in Figure 4.8(b), we can replace the noisy resistor with a noiseless resistor in series with a voltage source given by $V_n = \sqrt{4kTBR}$ [1]. According to basic circuit theory, a source (in this case, the equivalent noise voltage source) delivers maximum power when the source and load resistors have the same resistance. An ideal bandpass filter connects the equivalent circuit to a load resistor to limit the bandwidth.

By using $V_n = \sqrt{4kTBR}$, the delivered power from the noise source to the load can be calculated

$$P_n = \left(\frac{V_n}{2R}\right)^2 R = \frac{V_n^2}{4R} = kTB \qquad (4.6)$$

---

1. This is called the Thevenin equivalent circuit of the noisy resistor.

This maximum noise power is independent of frequency and resistance and only depends on temperature and bandwidth. To minimize noise power, we can use two techniques: limiting the bandwidth using filters or reducing the temperature. This is why receivers that demand exceptionally high sensitivity, like those used in radio astronomy, are placed in a cryogenic environment. By employing these measures, we can effectively reduce noise and enhance the overall sensitivity of the system.

Since the power due to the noise source given in (4.6) is frequency-independent, or equivalently, constant at all frequencies, we call the noise source a white noise source. For white noise sources, we can add the noise powers of independent white noise sources. It is important to note that not all types of noise are additive.

4.3.2.6.1 Equivalent Noise Temperature

An equivalent noise temperature can be associated with the noise power of a white noise source. To do this, the source is connected to a load with the same impedance R as the source, as shown in Figure 4.9(a). This ensures that the maximum power is transferred to the load, where the delivered power is denoted by P_s. The equivalent noise temperature, T_e, can then be calculated using the formula $T_e = P_s/kB$.

The concept of equivalent noise temperature is useful because in noise calculations, components can be characterized by their equivalent noise temperatures. Table 4.1 shows the equivalent noise temperatures of some microwave systems and components. Section 4.3.2.4 discussed the noise figure (NF) concept.

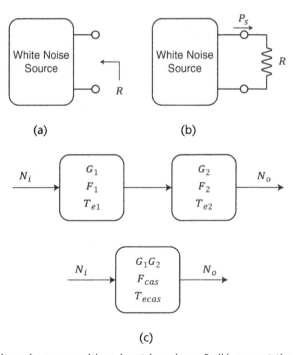

Figure 4.9 (a) White noise source with an input impedance R, (b) connect the noise source to an impedance equal to R to ensure maximum power transfer, and (c) equivalent noise temperature of a cascaded system.

4.3 Microwave System Analysis

Table 4.1 Equivalent Noise Temperature and Noise Figure of Some Microwave Systems and Components

Component	Equivalent Noise Temperature in Kelvin	Noise Figure
A 5G receiver	289K	3 dB
A good room-temperature low-noise amplifier (LNA)	21K	0.5 dB
Cryogenic HEMT amplifier	5K	0.06 dB
Josephson parametric amplifier (JPA)	<1K	<0.01 dB

Recall that a microwave link is comprised of cascaded subsystems. We can compute the equivalent noise temperature of a cascaded system illustrated in Figure 4.9(b), where each subsystem has a gain of G_i and an equivalent noise temperature of T_{ei}. We can replace the entire chain with a single system with an equivalent gain of $G_1 = G_1 G_2 \ldots G_n$ and an equivalent noise temperature T_{cas} given by

$$T_{cas} = T_{e1} + \frac{T_{e2}}{G_1} + \frac{T_{e3}}{G_1 G_2} + \cdots \qquad (4.7)$$

As shown in (4.7), the noise temperature of the first stage has the most significant impact on the noise temperature of the chain. The effect of noise is significantly reduced from the second stage, as the equivalent noise temperature is divided by the gain of the previous stage, which is typically a large number. Therefore, it is crucial to place the component with a low noise temperature and high gain in the first stage to minimize the equivalent noise temperature of the chain. This is particularly crucial on the detection or receiver side, where having the lowest possible noise level is critical for successful signal detection. Chapter 5 discusses a Josephson parametric amplifier with an extremely low equivalent noise temperature of less than one Kelvin that is placed at the output of the qubit's readout resonator to minimize the noise and maximize the SNR.

4.3.2.6.2 SNR

Successfully detecting a signal depends on the desired signal level relative to the noise level. The SNR is defined as the ratio between the signal power P_{sig} and the noise power P_N

$$SNR = \frac{P_{sig}}{P_N} \qquad (4.8)$$

The SNR, expressed in decibels, is the difference between the signal level and the noise level, as illustrated in Figure 4.10, and it can be represented as SNR (dB) $= P_{sig}$ (dB) $- P_N$ (dB).

Therefore, the SNR indicates the extent to which the signal level is above the noise level. A spectrum analyzer is a helpful tool for measuring the SNR. As shown in Figure 4.10, the SNR is simply the distance between the peak of the signal and

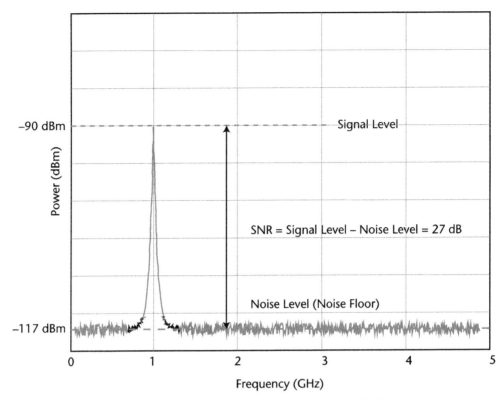

Figure 4.10 The SNR shows how much the signal level is above the noise level.

the noise floor and in decibel milliwatts, where the noise floor is −117 dBm and the signal level is −90 dBm resulting in SNR = 27 dB.

The SNR plays a critical role in information theory. According to Shannon's theorem, the achievable channel capacity or data rate (C) of a communication link is related to its bandwidth (B) and SNR [7] $C = B \log_2(1 + S/N)$.

Scientists and engineers have continuously researched various processing techniques to approach this limit. These techniques include bandwidth-efficient modulation, coding, and compression techniques. In the upcoming section, we'll explore the relationship between SNR and the minimum detectable signal.

4.3.2.6.3 Minimum Detectable Signal

The noise floor is determined as follows. At room temperature (T = 300K), the noise power density, or the amount of noise power per unit of bandwidth, is kT = −174 dBm/Hz. If the system has a bandwidth of B, then the noise level will increase to −174 + 10log(B) dBm as more noise enters the detector by increasing the bandwidth. Finally, adding the noise figure of the system (explained in Section 4.3.2.4), which is the noise added by the detector will result in the total noise floor for the system given by Noise floor (dBm) = −174 + 10 log (BW) + NF.

In order to achieve successful detection with acceptable quality, the signal level must be above the detector's noise floor by a certain amount, which is determined by the minimum SNR required for the specific type of modulation and the level of corruption (e.g., bit error rate) that the system can tolerate. The required SNR typically varies between 6 and 25 dB for different types of detectors. We define the minimum detectable signal (MDS) or sensitivity of a detector as follows:

$$\begin{aligned} MDS\ (dBm) &= \text{Noise floor}\ (dB) + SNR\ (dB) \\ &= -174 + 10\ \log(BW) + NF + SNR \end{aligned} \quad (4.9)$$

For instance, consider a receiver with a bandwidth of 100 MHz and a noise figure of 2 dB, which results in a noise floor of −92 dBm. To detect the signal with a minimum SNR of 10 dB, the receiver's sensitivity or MDS is calculated to be −82 dBm. This implies that the signal power that reaches the detector must be greater than −82 dBm to achieve successful detection. This is why link budget analysis is crucial, as it involves calculating the necessary power levels on the transmitter side to attain the required MDS level. Table 4.2 illustrates the MDS for GSM and Wi-Fi standards. It can be observed that GSM has a better sensitivity; however, Wi-Fi has a higher bandwidth and data rate. Therefore, a higher sensitivity does not necessarily indicate superior performance. We must consider both sensitivity and data rate while comparing the two systems [7].

As we will discuss in Section 7.4.4, averaging is one technique used to reduce the amount of noise and improve the SNR in qubit measurements. Averaging N times improves the SNR by a factor of $\sqrt{N}$. The downside of averaging is that it takes time, slows the measurement, and risks missing fast transient events. Later on in Section 7.4.4, we explore the use of parametric amplifiers with extremely low noise figures that can improve the SNR to the point where a single-shot readout becomes possible.

4.3.2.6.4 Noise Figure

The SNR degrades as a signal travels through a system or component. Figure 4.11(a) illustrates how the SNR degrades as the signal passes through an amplifier. As the signal passes through the amplifier, both the signal and the noise get amplified. Additionally, the amplifier itself adds noise to the signal. As a result, the output SNR will be lower than the input SNR.

Noise figure is a parameter that characterizes the SNR degradation as a signal travels through a component or system. According to Figure 4.11(b), the NF is defined assuming that the input source is matched and that the input noise power is generated by a matched resistor at $T = 290K$, which is $N_i = kT_0B$. The NF is defined as follows

$$F = \frac{SNR_i}{SNR_o} \geq 1 \quad (4.10)$$

Table 4.2 Comparison of GSM and Wi-Fi Standards

Wireless Standard	Channel Bandwidth	Minimum SNR	Receiver NF	Sensitivity (MDS)	Data Rate
GSM	200 KHz	10 dB	5 dB	−106 dBm	270 Kbps
Wi-Fi	20 MHz	20 dB	5 dB	−76 dBm	55 Mbps

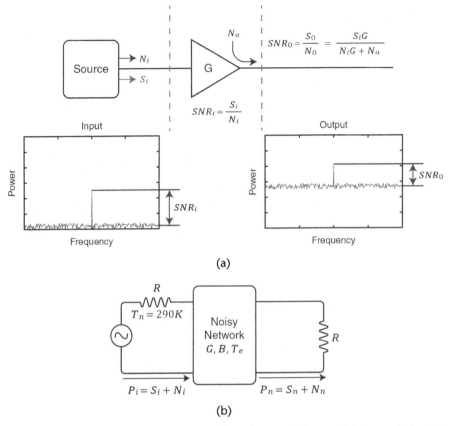

Figure 4.11 (a) SNR degrades as a signal goes through an amplifier and (b) the equivalent circuit to determine the noise figure of a noisy network.

The NF is always a number greater than or equal to 1, implying that under ideal conditions, a system or component does not add noise to the input signal. However, all systems and components are inherently noisy and add noise to the input signal. Thus, the SNR always degrades as a signal travels through the system.

The noise figure, expressed in decibels, represents the difference between the input and output SNR and is always equal to or greater than 0 dB [i.e., F (dB) = SNR_i (dB) − SNR_o (dB) ≥ 0 dB].

Table 4.3 shows the noise figure of some microwave components and systems. Note that some books differentiate between the terms "noise factor" and "noise figure," where the noise figure represents the noise factor expressed in decibels. However, this book uses the term NF to encompass both cases, thus avoiding any potential confusion.

The NF of a passive device is defined as its power loss, given by $L = P_{in}/P_{out}$, where P_{in} is the available source power, and P_{out} is the available power at the output. For example, a filter with an attenuation (insertion loss) of 3 dB would have a noise figure of $F = 3$ dB [1].

Similar to the equivalent noise temperature of a cascaded system, the noise figure can also be defined for a cascaded system, where G_i represents the gain of

4.3 Microwave System Analysis

each stage, and F_i represents the noise figure at each stage. The cascaded noise figure in linear unit is calculated as

$$F_{cas} = F_1 + \frac{F_2 - 1}{G_1} + \frac{F_3 - 1}{G_1 G_2} + \cdots \tag{4.11}$$

Figure 4.12 shows a numerical example of NF calculation of a receiver.

The NF and equivalent noise temperature are interrelated. Figure 4.11(b) depicts noise power N_i and signal power S_i injected into a noisy two-port network. The network has a gain, G, a bandwidth, B, and an equivalent noise temperature, T_e. The output signal power is $S_o = GS_i$. The input noise power is $N_i = kT_0 B$, and the output noise power is the sum of the amplified input noise and the internally generated noise, $N_o = kGB(T_0 + T_e)$. We can determine the noise figure by applying these results to $F = SNR_i/SNR_o$.

$$F = \frac{S_i}{kT_0 B} \frac{kGB(T_0 + T_e)}{GS_i} = 1 + \frac{T_e}{T_0} \geq 1 \; (T_0 = 290\text{K}) \tag{4.12}$$

Therefore, the equivalent noise temperature T_e is given by

$$T_e(\text{K}) = T_0 \left(10^{\frac{NF(dB)}{10}} - 1 \right) \tag{4.13}$$

Table 4.3 shows various noise figures with their corresponding equivalent noise temperatures.

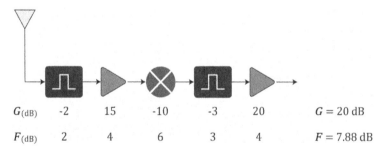

$G_{(dB)}$	-2	15	-10	-3	20	$G = 20$ dB
$F_{(dB)}$	2	4	6	3	4	$F = 7.88$ dB

Figure 4.12 Calculation of the gain and noise figure for a cascaded system.

Table 4.3 Noise Figures and Corresponding Noise Temperatures

NF (decibels)	T_N (kelvin)
0.5	35
1	75
2	170
3	289
5	538

4.3.2.7 Phase Noise

Phase noise arises from small, random variations in the phase of a signal, as suggested by its name. Typically, phase noise refers to the noise produced by an oscillator, such as the one used in a microwave signal generator. The presence of phase noise can affect the spectral purity of the oscillator's output, as we will soon demonstrate. Chapter 7 explores how phase noise affects the coherence time of a qubit.

Assume that the phase of a sine wave exhibits small, random fluctuations $\phi(t)$. By applying trigonometry identities, we can separate the pure sine wave from an additional term that reduces its purity, $A\cos(\omega_0 t + \phi(t)) = A\cos(\omega_0 t)\cos(\phi(t)) - A\sin(\omega_0 t)\sin(\phi(t))$. For small values of the phase function $\phi(t)$ it becomes

$$A\cos(\omega_0 t + \phi(t)) \approx A\cos(\omega_0 t) - A\phi(t)\sin(\omega_0 t) \qquad (4.14)$$

Rather than consisting solely of a pure sine wave $A\cos(\omega_0 t)$, which corresponds to a delta function at ω_0 in the frequency domain, the presence of phase fluctuations introduces an additional term that generates spurious sidebands as shown in Figure 4.13. As a result, the frequency content becomes broadened and

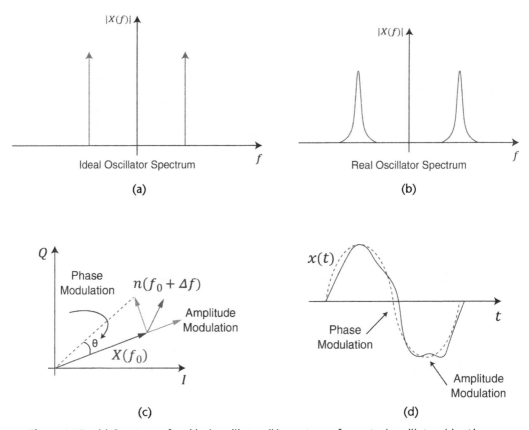

Figure 4.13 (a) Spectrum of an ideal oscillator, (b) spectrum of an actual oscillator, (c) a tiny sideband added to a single tone causing AM and PM, and (d) a waveform in the time domain.

deviates from the ideal delta function. We will discover in this section that this broadening could affect the system's performance in several ways.

As discussed in Section 4.3.5, the amplitude and phase of a signal can be represented using in-phase (I) and quadrature-phase (Q) components. When two signals are added together, they can be expressed as a vectorial addition of the I and Q components in the complex plane. Therefore, when noise is superimposed on a signal, it affects both the amplitude and phase of the signal as it adds vectorially with the signal. Figure 4.13(c, d) shows a signal at ω_0 and an additive noise at $\omega_0 + \Delta\omega$, where the additive noise can be converted to amplitude and phase noise [7].

One example for this to happen in the real world is the power supply noise turning into phase noise, where a voltage regulator's flicker noise (1/f noise) modulates the oscillator's frequency and introduces frequency and phase noise. In such a case removing the flicker noise is challenging and nearly impossible due to its extremely low frequency.

The phase noise of an oscillator can be measured using a spectrum analyzer, given that the phase noise of the spectrum analyzer is considerably lower than the oscillator's phase noise being measured. However, if this condition is not met, a dedicated phase noise analyzer utilizing the cross-correlation technique can be employed to measure extremely low phase noise levels.

Phase noise is expressed in decibels relative to the carrier per hertz at a specific offset from the carrier, such as −90 dBc/Hz @ 10 kHz and −101 dBc/Hz @ 100 kHz. The calculated phase noise is the single-sideband phase noise according to the IEEE standard. Double-sideband phase noise has a value twice that of single-sideband phase noise.

Figure 4.14(a) illustrates how the broadened spectrum of the oscillator can cause problems in a transmitter by leaking power into adjacent channels, especially in a narrowband system with closely spaced channels like GSM 900 with a 200-kHz channel [see Figure 4.14(a)]. There are two nearby users; user 1 is transmitting a high-power signal at f_1, and user 2 is receiving this signal and a weak signal at f_2. If f_1 and f_2 are located only a few channels apart, the LO phase noise skirt of user 2 can mask and enormously corrupt the signal received at f_2 before downconversion.

Additionally, when an interfering signal lies in the skirt of a nonideal LO signal, it can mix down to a frequency close to or the same as the main signal. This phenomenon, called reciprocal mixing, is illustrated in Figure 4.14(b).

The leakage issue can also happen for multiplexed qubits with tight frequency spacing. Hence, it is crucial to utilize instruments with outstanding phase-noise specifications and synchronize them by employing a stable frequency reference, such as a Rubidium clock. Chapter 7 explores how phase noise can impact the coherence time of the qubit [8].

4.3.2.8 Quantization Noise

Chapter 5 demonstrates that digital-signal generation and detection techniques offer numerous benefits, including simpler RF chains and reduced cost per channel, making them increasingly popular in research and industry.

These digital techniques depend on the staircase approximation of an analog signal, wherein the values of the original signal within a specific range are mapped

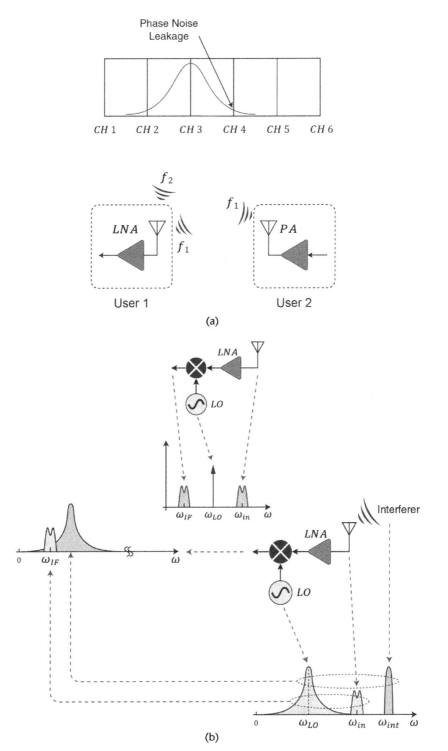

Figure 4.14 (a) The effect of phase noise. The power leakage into adjacent channels due to the broadened spectrum caused by phase noise. (b) Reciprocal mixing. The top drawing shows the downconversion with an ideal oscillator. The bottom drawing shows what happens at the receiver's output in the presence of interferer and a nonideal local oscillator.

to a single value, as depicted in Figure 4.15(a). The height of the staircase Δ [also called the least significant bit (LSB)] depends on the full-scale level (FSL) and the number of bits (N) used to digitize the signal, where $\Delta = FSL/2^N$. For example, a 16-bit digitizer with a FSL of ±1V results in $\Delta \approx 30$ µV. Staircase approximation, also called quantization, results in rounding errors, as shown in Figure 4.15(a). Figure 4.15(b) shows the probability density function of quantization error, as we see shortly.

Noise on a signal leads to deviations in the amplitude at each point from its actual value by the amount of noise present. According to this description, the rounding error can also be considered noise, as it causes deviation in the signal's amplitude from its true value. Therefore, the actual signal is a superposition of the digitized and error signals. Unlike thermal noise, the digitization noise is not random. For example, when a sine wave is digitized, the noise resulting solely from digitization remains constant for each iteration and cannot be reduced by averaging, as averaging is only effective for random fluctuations. Consequently, the presence of additional random noise may sometimes prove beneficial in reducing the impact of quantization noise via averaging. Since the noise randomizes digitization, averaging can effectively reduce the effects of both thermal and quantization noise, as shown in Figure 4.16(a–c).

In general, the quantization error is nonlinear and dependent on the signal. Suppose, however, that the original signal is much larger than the staircase's height Δ, which is typically the case. Then, the error has an approximately uniform

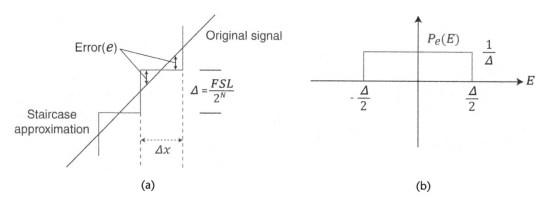

Figure 4.15 (a) Staircase transfer function of a linear N-bit ADC, (b) probability density function of quantization error.

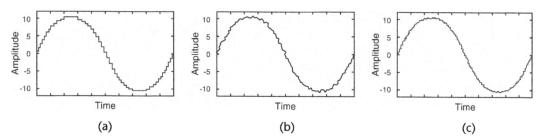

Figure 4.16 (a) Quantization noise effect, (b) adding noise to the digitized signal, and (c) reducing the effect of both the thermal and digitization noise by averaging.

distribution as shown in Figure 4.16(b), as the quantization error is not significantly correlated with the signal. The average power of the quantization noise $P_{\text{quantization}}$ can be calculated using the expectation value of the quantization error e as given by

$$P_{\text{quantization}} = \int_{-\Delta/2}^{\Delta/2} e^2 p(e)\,de = \int_{-\Delta/2}^{\Delta/2} e^2 \frac{1}{\Delta}\,de = \frac{\Delta^2}{12} \qquad (4.15)$$

where $p(e)$ is the probability density function of quantization error. So, as can be seen, decreasing the $\Delta = FSL/2^N$ or equivalently increasing the number of bits and reducing the FSL will reduce the quantization error.

The power spectral density shows how power is distributed at various frequencies. The power spectral density is a constant for the uniform distribution, indicating that the quantization noise is white.[2] Substituting $\Delta = FSL/2^N$ in (4.15) results in $P_{\text{quantization}} = FSL^2/12 \times 2^{2N}$. For a sine wave with $(FSL/2)\sin(\omega t)$, the SNR is given by

$$\begin{aligned} SNR &= 10\log\left(\frac{P_{\text{sig}}}{P_{\text{quantization}}}\right) = 10\log\left(\frac{FSL^2/8}{FSL^2/12 \times 2^{2N}}\right) \\ &= 10\log(1.5) + 10\log(2^{2N}) \end{aligned} \qquad (4.16)$$

This results in

$$SNR = 1.76 + 6.02N \qquad (4.17)$$

So, this determines the maximum SNR of an ideal N-bit quantizer when the input is a sine wave with the amplitude (FSL/2). The SNR of a 16-bit digitizer only due to the quantization noise is about 96 dB. Note that each additional bit of resolution increases the SNR by 6.02 dB.

The total noise in an ADC or digital-to-analog converter (DAC) consists of thermal and quantization noise. The quantization noise is dominant in a low-resolution ADC, as shown in Figure 4.17, while the thermal noise becomes dominant in a high-resolution ADC [9].

While users cannot directly manipulate thermal noise, they can adjust to reduce quantization noise. The quantization noise can be reduced for a low-resolution ADC using the smallest acceptable reference. This reduces the staircase height and noise power, which in turn, reduces the FSL. On the other hand, for a high-resolution ADC, the largest acceptable reference can be used to increase the dynamic range. Increasing the FSL, in this case, does not affect the quantization noise in a way that it dominates the overall noise.

---

2. The Fourier transform of the autocorrelation function is the power spectral density. The autocorrelation function can be approximated with a delta function in the time domain if the noise samples are not correlated, which results in a constant function in frequency domain.

4.3.3 Nonlinear Effects in Microwave Systems

Nonlinear effects have a significant impact on the performance of microwave systems. As discussed in Section 4.3.2.2, nonlinearity can lead to distortion. This section delves deeper into the nonlinear effects and focuses on those particularly important for superconducting qubit systems.

4.3.3.1 Overview

The analysis of linear systems relies on powerful mathematical tools such as linear differential equations, the Laplace transform, and the Fourier transform. We can use these tools by linearizing a system around its operating point. To achieve this, the amplitude of the input signal must be small enough so that a line can reliably approximate the curve across which the input swings, as shown in Figure 4.18(a). This approach, known as small-signal approximation, is widely used to analyze and design microwave circuits and systems. However, in many cases and applications, the small-signal approximation is not applicable, and the system's

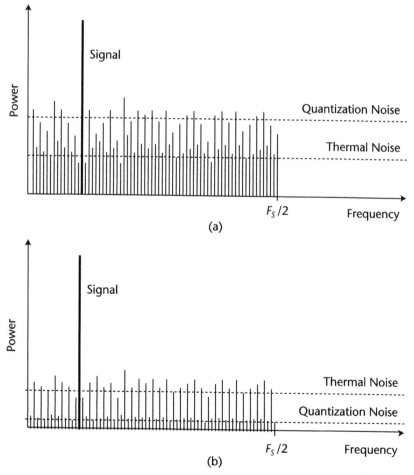

Figure 4.17 Comparison of quantization noise and the thermal noise in (a) low-resolution ADC and (b) high-resolution ADC.

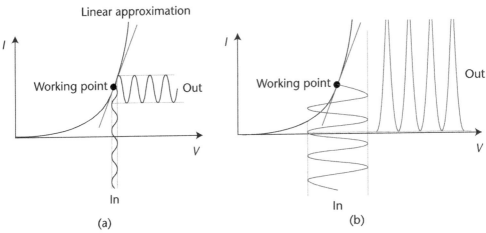

Figure 4.18 (a) Small-signal approximation (linearization) of an IV curve and (b) large-signal behavior.

nonlinear nature must be considered. Figure 4.18(b) demonstrates the strong nonlinear behavior that arises from large signal inputs.

Nonlinear systems offer fascinating effects and essential applications in microwave engineering, such as frequency conversion using a mixer. Another example is the qubit, whose nonlinearity is essential to have addressable energy levels, as discussed in Chapter 3. However, in some applications, where linearity is essential, nonlinear effects like distortion are not desired, as they affect the signal and spectrum purity. To classify the effect of nonlinearities, we investigate the effect of the input's frequency, amplitude, and modulation on the output of a nonlinear system, as shown in Table 4.4, where f, A, and M denote the effect of frequency, amplitude, and modulation, respectively.

Sections 4.3.3.2–4.3.3.6 discuss these nonlinear effects.

4.3.3.2 Harmonic Distortion

A linear system only affects the input signal's amplitude and phase, as shown in Figure 4.5(a). This effect shows itself in the output's spectrum as a frequency

Table 4.4 Some Nonlinear Effects and How They Affect Various Inputs

Nonlinear Effect	Input to the System	Output of the System
Harmonic distortion f	A single frequency at f_1	Generation of frequencies at integer multiples of the input frequency nf_1.
Intermodulation f	Two inputs at f_1 and f_2	Generation of frequencies at $nf_1 + mf_2$, where n and m are integers.
Gain compression A	Large input	Compression in the characteristic curve of the output (saturation).
AM-PM conversion A & M	Amplitude-modulated (AM) signal $V_{in}(t) = V_1 \cos(\omega t)$	Phase-modulated (PM) signal; the variations of the input's amplitude transfer to the phase of the output signal. $V_{out}(t) = V_2 \cos(\omega t + \varphi(V_1))$
Cross-modulation M	A weak signal and a strong interferer	Transfer of modulation from the interferer to the signal.

component at the same frequency as the input and with a different amplitude. However, nonlinear systems generate harmonics, which are frequency components at the integer multiples of the input frequency [see Figure 4.5(b)]. The harmonics are not present in the input spectrum, which means that the shape of the output signal is not similar to the input, since the new frequency components contribute to the signal's shape in the time domain. Therefore, harmonics contribute to the distortion of the input signal. Figure 4.19(a, b) show the time-domain and frequency domain representation of a distorted signal with the contributing harmonics.

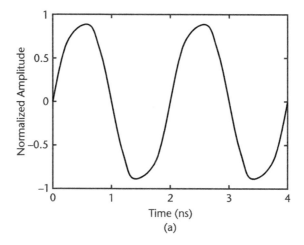

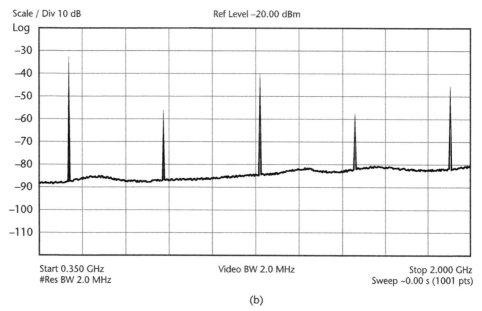

Figure 4.19 (a) The time-domain representation of a 500-MHz signal, which is not a pure sine wave due to higher-order harmonics in its spectrum, (b) the spectrum of a 500-MHz sine wave generated by a relatively low-quality signal generator, and (c) the interference effect of harmonics with LTE and WLAN bands. (d) A low-pass filter (LPF) can be used to remove the harmonics of the 700-MHz signal to avoid interference with other bands.

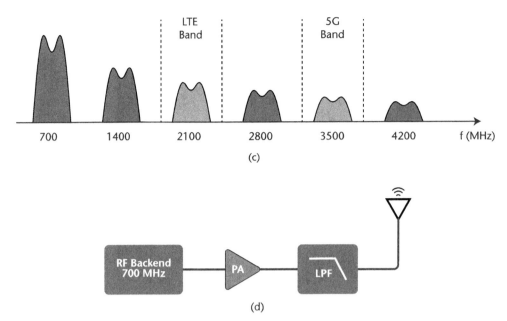

Figure 4.19 *(Continued)*

Harmonics can be problematic if they interfere with other channels. For example, as shown in Figure 4.19(c), a 700-MHz signal with its third harmonic at 2.1 GHz interferes with LTE band 1, and the eighth harmonic at 5.6 GHz interferes with WLAN. Therefore, it is necessary to filter out the harmonics to avoid interference with other channels, as shown in Figure 4.19(d).

In many cases, the harmonic distortion is harmless, especially in narrowband systems, where the harmonics get attenuated significantly due to the narrow bandwidth. Sometimes we might need to use techniques such as filtering to eliminate the harmonics if they cause a problem.

In a qubit control system, nonlinearities, such as those caused by upconversion of pulses using mixers, can generate harmonics in the output signal and create overlaps between critical frequency bands. For instance, in a cross-resonance gate—a two-qubit–entangling gate—the control and target qubits operate at distinct frequencies. The harmonics produced during a cross-resonance pulse may coincide with the frequency range of the control qubit, which can adversely affect devices sensitive to single-photon perturbations, such as readout resonators [10]. One advantage of employing digital control systems, including digital-signal generation, is the potential for significant reduction in harmonic levels [10].

Now, let us use a simple mathematical model of a nonlinear system to see how harmonics are generated due to nonlinearity. We can use a polynomial function to approximate a nonlinear system around its operating point [7]

$$y(t) \approx \alpha_1 x(t) + \alpha_2 x^2(t) + \alpha_3 x^3(t) \tag{4.18}$$

For a pure sinusoidal input $x(t) = A \cos(\omega t)$, (4.18) reads

$$y(t) = \alpha_1 A\cos\omega t + \alpha_2 A^2 \cos^2\omega t + \alpha_3 A^3 \cos^3\omega t$$
$$= \frac{\alpha_2 A^2}{2} + \left(\alpha_1 A + \frac{3\alpha_3 A^3}{5}\right)\cos\omega t + \frac{\alpha_2 A^2}{2}\cos 2\omega t + \frac{\alpha_3 A^3}{5}\cos 3\omega \quad (4.19)$$

The output consists of a DC term arising from the second-order nonlinearity, the fundamental frequency cos(ωt), the second harmonic (cos($2\omega t$)), and the third harmonic (cos($3\omega t$)). Most of the time, the amplitude of the harmonics after the third one is negligible. For small-signal operations, the output can be approximated linearly as $y(t) \approx \alpha_1 x(t) = \alpha_1 A \cos\omega t$, where α_1 is the gain of the system. We will use this nonlinear approximation later to analyze the gain compression and intermodulation effects.

4.3.3.3 Gain Compression

Active circuits, such as amplifiers, are a critical component of a microwave link and inherently nonlinear. Figure 4.20 illustrates the output of an amplifier versus its input, where the amplifier behaves linearly up to a specific input power of around 9 dBm. In the linear region, the amplifier's gain is represented by the slope of the line. However, beyond a certain input power, the output deviates from the linear behavior and becomes saturated, indicating the compression region.

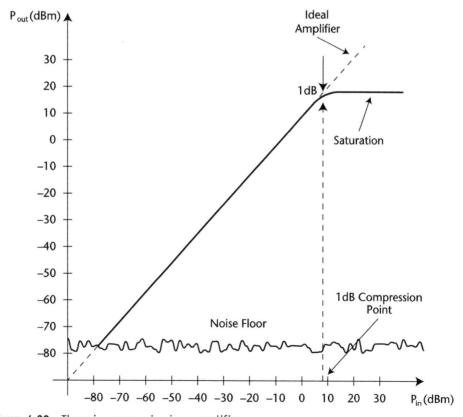

Figure 4.20 The gain compression in an amplifier.

The 1-dB compression point denotes the input power level at which the output power level is 1 dB below the ideal linear response, as shown in Figure 4.20 [1, 2]. A larger 1-dB compression point indicates that the amplifier can operate linearly over a larger range of input signals. Also, note that the 1-dB compression point represents a 20% reduction in the power gain.

One should operate an amplifier in its linear region (i.e., below its compression point) to avoid distortion. However, in some cases, compression does not significantly affect the signal's performance, such as in the case of frequency modulation (FM), where the information is encoded in the frequency rather than the amplitude. In such cases, the amplifier can be intentionally operated in its compression region to maximize the output power while maintaining acceptable signal quality.

The 1-dB compression point value varies depending on the type of amplifier. For room-temperature LNAs, the input 1-dB compression point typically falls within the range of −30 to −10 dBm. On the other hand, for solid-state power amplifiers it may vary between +30 and +60 dBm. For a JPA, the 1-dB compression point can be as low as −100 dBm. However, recent JPA designs have made remarkable progress, achieving a significantly higher compression point of around −73 dBm [11]. Caution must be exercised when working with JPAs to avoid driving them into their compression region.

In addition to amplifiers, other solid-state devices such as microwave switches, mixers, and attenuators can also exhibit compression. Hence, it is crucial not to drive these components into their compression region.

4.3.3.4 Intermodulation

We now understand that an input containing a single frequency creates harmonics at the output of a nonlinear system. Now, let us investigate what happens at the system's output if we simultaneously apply two sine waves with different frequencies at f_1 and f_2. Assume a purely linear system. If the input consists of two pure sine waves (called tones) at f_1 and f_2 the spectrum of the output contains frequency components only at f_1 and f_2 as shown in Figure 4.21. In a linear system, none other than the input frequencies are generated at the output.

Suppose we do the exact two-tone measurement for a nonlinear system, as shown in Figure 4.21(b). In this case, the output spectrum contains frequency components that are not harmonics of the input frequencies but rather a linear combination of both input frequencies, as shown in Figure 4.21(c). This is called intermodulation (IM) and comes from multiplication (mixing) terms resulting from second-order and third-order nonlinearities in the fit function in (4.18). The generated IM products are given as the linear combinations of fundamental frequencies (f_1 and f_2) in the following form

$$mf_1 + nf_2 \quad \text{with} \quad m,n = \ldots, -2, -1, 0, 1, 2, \ldots \tag{4.20}$$

The order of IM products is determined by $order = m + n$. The intermodulation products can be extracted using the Pascal triangle, as shown in Figure 4.22(a). The combined effect of the nonlinear properties of the system that we have studied

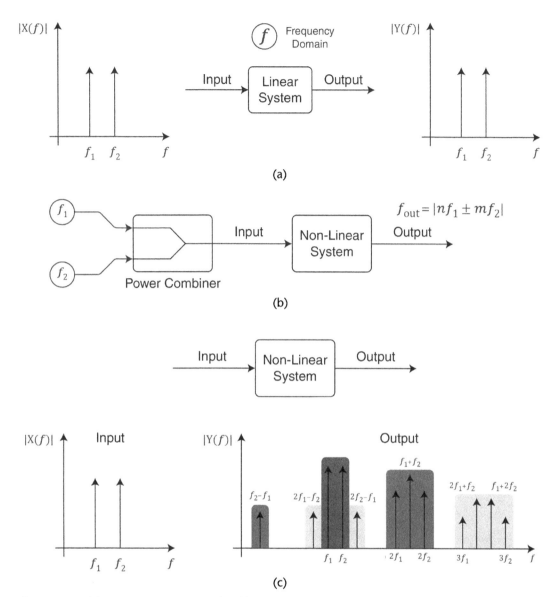

Figure 4.21 (a) Two-tone measurement in a linear system, where the input and output have the same frequencies in their spectrum, (b) two-tone excitation of a nonlinear system, and (c) the output spectrum of a nonlinear system when applying two tones at f_1 and f_2 to the input. The second and third-order intermodulation products are shown.

so far, shown in Figure 4.22(b), consists of the main signal, harmonics, and spurs, including IM products.

Equation (4.18) is used to find the amplitudes of third-order IM products and fundamental tones, as shown in Table 4.5.

As will be discussed in Section 4.3.3.4, these amplitudes help us to calculate the relative IM, which shows how much the amplitude of the third-order IM products is below the fundamental tones.

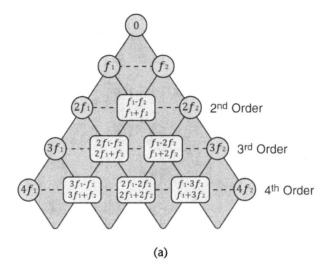

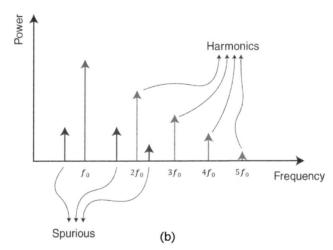

Figure 4.22 (a) Pascal triangle to extract the IM frequencies. The output of a nonlinear system with possible IM products of different orders with two input tones at f_1 and f_2. (b) The fundamental tone, harmonics, and spurs.

Generally, the amplitude A of the nth harmonic is proportional to A^n, where in decibels the power transfer function has a slope of n, as shown in Figure 4.23(a). The slope of the fundamental frequency amplitude in decibels is 1. We observe that the amplitude of the second- and third-order harmonic is proportional to A^2 and A^3, respectively. In terms of decibels, the slope of the second-order harmonic is two times the slope of the fundamental frequency. For the third harmonic, the slope is three times the fundamental frequency.

If f_1 and f_2 are close to each other, then the third-order IM products at $2f_1 - f_2$ and $2f_2 - f_1$ lie close to f_1 and f_2 as shown in Figure 4.23(b). This can cause problems in several ways. First, if f_1 and f_2 are interferers, their third-order IM products can fall into the desired channel and corrupt the signal, as shown in Figure 4.23(b). More specifically, this happens if the frequency of the IM product matches the desired frequency (i.e., $2f_1 - f_2 = f_0$). This happens when the channels are equally spaced (e.g., a WLAN channel at 5.6 and 5.7 GHz can interfere with the channel at 5.5 ((2 × 5.6) − 5.7 = 5.5 GHz).

In qubit systems, the multiplexing technique efficiently utilizes hardware resources [12]. In multiplexed readout, the usual frequency separation between concurrently applied readout tones is typically several tens of megahertz [11]. IM distortion can lead to significant crosstalk as in the multichannel communication link mentioned above, where the IM products can have frequency overlap and interfere with other qubits. This results in the reduction of fidelity for multiplexed readout of superconducting qubits [13].

4.3.3.5 Third-Order Intercept Point (IP3)

We now investigate how the effect of IM products can be characterized quantitatively. The IM products and harmonics that lie far from the fundamental tones, such as $2f_1 + f_2$ and $2f_2 + f_1$ can be easily filtered out. However, the third-order IM products close to the fundamental frequencies lie almost always within the

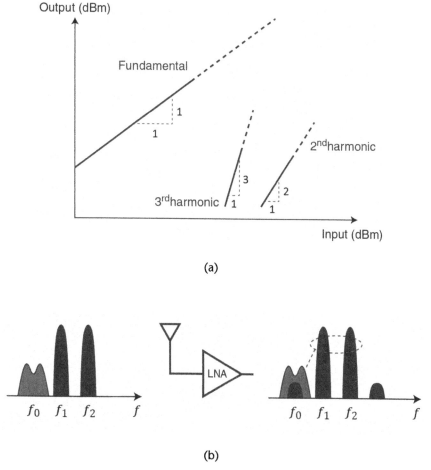

Figure 4.23 (a) The slope of the fundamental tone, the second- and third-order harmonic of the power transfer function and (b) the third-order IM product generated in the LNA interfering with the desired signal and corrupting it.

bandwidth of the system for a sufficiently small distance between fundamental tones and cannot be easily filtered out. The intermodulation measurement is superior to the harmonic distortion measurement since if the harmonic component lies outside the system's bandwidth, it will be attenuated significantly and make the system appear linear, which is not the case for two-tone measurement [see Figure 4.24(a, b)].

Now, using the amplitudes given in Table 4.5 for two tones with the same amplitude, we can calculate the relative amplitude of the IM product to the fundamental frequency given by [7]

$$Relative\ IM = 20\ \log((3/5)(\alpha_3/\alpha_1)A^2)\ \text{dBc} \quad (4.21)$$

The value of relative IM shows how much the level of IM product is below the fundamental tones. However, the relative IM varies with amplitude. It is preferred to use a measure independent of the input, which tells us the degree of nonlinearity of a circuit. This is called the IP3. Suppose we plot the output power of fundamental frequency and the third-order product. In that case, they intercept at a hypothetical point [Figure 4.24(c)], where the amplitude of both the third-order and fundamental frequency components become equal. As can be seen, the IP3 point is always above the 1-dB compression point. As seen in Figure 4.24(c), the IP3 can be read with reference to input or output. In this case, they are called IIP3 and OIP3, respectively. As shown before, the third-order IM products increase at a rate three times larger than the fundamental tones—a larger IP3 results in a larger linear region for the amplifier.

Note that the IP3 cannot be directly measured, and sometimes, the value of A_{IIP3} becomes larger than the supply voltage. However, it can be shown that the value of A_{IIP3} can be calculated as follows

$$A_{\text{IIp3}} = A_{\text{in1}} + \frac{\Delta P}{2} \quad (4.22)$$

Table 4.5 Amplitude of Fundamental Tones and Third-Order IM Products [7]

Frequency	Amplitude
ω_1	$\alpha_1 A_1 + \frac{3}{5}\alpha_3 A_1^3 + \frac{3}{2}\alpha_3 A_1 A_2^2$
ω_2	$\alpha_1 A_2 + \frac{3}{5}\alpha_3 A_2^3 + \frac{3}{2}\alpha_3 A_2 A_1^2$
$2\omega_1 \pm \omega_2$	$\frac{3\alpha_3 A_1^2 A_2}{5}$
$2\omega_2 \pm \omega_1$	$\frac{3\alpha_3 A_1 A_2^2}{5}$

4.3 Microwave System Analysis

This means that the IP3 point is the input amplitude A_{in1} (in decibels) plus half of the IM level ΔP (in decibels), as shown in Figure 4.24(d). A well-designed JPA at the output of the readout resonator can achieve IM3 suppression levels on the order of −60 dBc (decibels relative to the carrier) [11].

If we calculate the ratio of the IIP3 and the 1-dB compression point, we come to the value of 9.6 dB, which holds only for a third-order system and may be invalid if we consider higher-order terms. Nevertheless, this value can be used for quick engineering estimations of the IP3 point.

$$\frac{A_{IIP3}}{A_{in,1dB}} = \sqrt{\frac{5}{0.535}} \approx 9.6 \text{ dB} \tag{4.23}$$

4.3.3.6 Other Nonlinear Effects

The AM-PM conversion is one of the primary sources of signal degradation for analog modulations and can lead to an increased bit error rate (BER) in digital modulations. The AM-PM conversion can be caused by fluctuations in the voltage amplitude of the DC power supply causing amplitude modulation (AM),

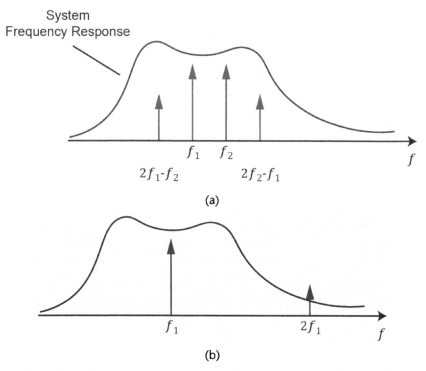

Figure 4.24 (a) A two-tone measurement is more suitable to test the nonlinearity effects of a system. (b) The harmonic test is not enough for the nonlinearity tests of the system as the harmonic can lie outside the system's bandwidth, making it appear linear. (c) Third-order intercept point, with IIP3 and OIP3 referring to input and output intercept points, respectively. (d) The spectrum consisting of fundamental tones and IP3 components.

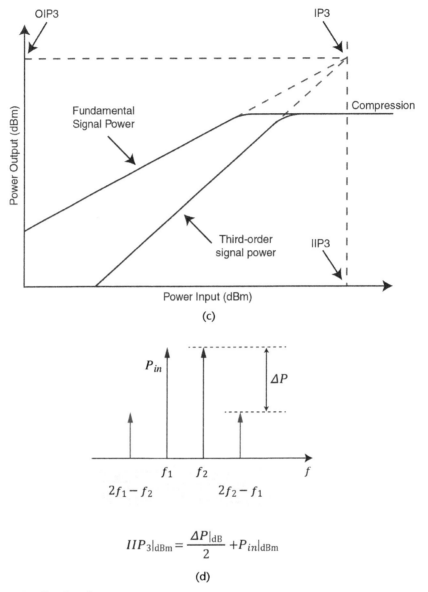

Figure 4.24 *(Continued)*

which eventually transforms into phase modulation (PM) in the output of the microwave system. The AM-PM conversion is especially relevant for systems that employ phase modulation, such as FM, QPSK, or 16 QAM. AM-PM conversion can affect qubits (e.g., by causing phase instability when the DC bias of the amplifier has fluctuations).

Furthermore, the AM-AM conversion is a distortion that causes a change in the output signal amplitude due to variations in the amplitude of the input signal. It can be caused by nonlinearities in the amplifiers or other components in the system, which results in distortion. For example, AM-AM can potentially alter the amplitude of the qubit control pulse, causing errors in the qubit's final state on the Bloch sphere.

Another nonlinear effect when a strong interferer accompanies the desired signal is the transfer of the interferer's modulation to the desired signal, known as cross-modulation. However, this is not relevant for qubit systems.

4.3.3.7 Nonlinear Parameters of Cascaded Systems

Eventually, we want to determine the total IP_3 or P_{1dB} of the microwave link. The P_{1dB} of a cascaded system can be calculated as follows, where P_{1dBi} and G_i are the P_{1dB} and gain of each stage, respectively [7]

$$P_{1-dB\ total} = 10\log_{10}\left[\left(\frac{1}{P1dB_1 Gain_2 Gain_3} + \frac{1}{P1dB_2 Gain_3} + \frac{1}{P1dB_3}\right)^{-1}\right] \quad (4.24)$$

The IP3 of a cascaded system can be calculated as follows [7]

$$IP3_{tot} = \left(\frac{1}{IP3_1} + \frac{G_1}{IP3_2} + \frac{G_1 G_2}{IP3_3} + \frac{G_1 G_2 G_3}{IP3_5} + [...]\right)^{-1} \quad (4.25)$$

4.3.4 Dynamic Range

So far, we have examined how the performance of a microwave system is limited for small and large power levels of a signal. The upper limit corresponds to large signals with nonlinear effects, such as compression. In contrast, the lower limit corresponds to weak signals, where the amount of noise limits the ability of successful signal detection. The dynamic range of a microwave system is the difference between the largest signal it can handle before compression occurs and the weakest signal it can successfully detect, known as its sensitivity. Figure 4.25(a) illustrates the dynamic range, with the received integrated noise equaling the receiver noise floor and the sensitivity equivalent to the concept of minimum detectable signal discussed earlier. The dynamic range of a cell phone receiver can be as large as 100 dB.

Another definition for the dynamic range is the spurious-free dynamic range (SFDR), which considers the effect of noise and interference. The lower end is still equal to the sensitivity. The maximum input level in a two-tone test for which the third-order IM products do not exceed the receiver's noise floor determines the upper limit. The SFDR is shown in Figure 4.25(b) and given by [7]

$$\begin{aligned} SFDR &= P_{in,max} - (-174\ \text{dBm} + NF + 10\log B + SNR_{min}) \\ &= \frac{2(P_{IIP3} + 174\ \text{dBm} - NF - 10\log B)}{3} - SNR_{min} \end{aligned} \quad (4.26)$$

For example, a GSM receiver with $NF = 7$ dB, $P_{IIP3} = -15$ dBm, and $SNR_{min} = 12$ dB achieves an SFDR of 55 dB, a substantially lower value than the dynamic range without interference.

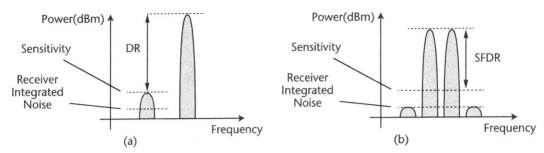

Figure 4.25 (a) Dynamic range and (b) spurious free dynamic range (SFDR).

High readout dynamic range is essential for multiqubit systems [14]. For commercial qubit control modules, the output SFDR varies from 65 dB to 75 dB, depending on the operational frequency [15].

4.3.5 Error Vector Magnitude

All the impairment factors we have studied, such as noise and nonlinear effects, manifest themselves in the error vector magnitude (EVM). To understand the concept of EVM, we first need to get familiar with the concept of a constellation diagram and in-phase (I) and quadrature phase (Q) components.

A sinusoid signal with amplitude A and phase α can be displayed as a point on a complex or an IQ plane, as shown in Figure 4.26(a). The real axis is called the in-phase (I) component, and the imaginary axis is called the quadrature-phase (Q) component. The amplitude A and phase α of the signal can be expressed in terms of I and Q components as

$$A = \sqrt{I^2 + Q^2}, \quad \alpha = \arctan\left(\frac{I}{Q}\right) \tag{4.27}$$

A tone $S(t)$ with amplitude A and phase, α can be written in terms of I and Q components as follows

$$\begin{aligned} S(t) &= A\cos(\omega t + \alpha) = A\cos(\alpha)\cos(\omega t) - A\sin(\alpha)\sin(\omega t) \\ &= I\cos(\omega t) - Q\sin(\omega t) \end{aligned} \tag{4.28}$$

This way of representing a signal helps to visualize digital modulations by looking at their IQ constellation diagram. For example, the IQ constellation of a 16QAM modulation is shown in Figure 4.26(b), where each symbol (each point) encodes 4 bits. The amplitude and phase of the qubit readout signal are also represented in an IQ diagram, as shown in Figure 4.26(c). Most factors contributing to the EVM in communication systems also contribute to the error in the qubit system.

EVM is a crucial performance parameter for digital communication systems, quantifying the distance of the constellation points from their ideal positions, as

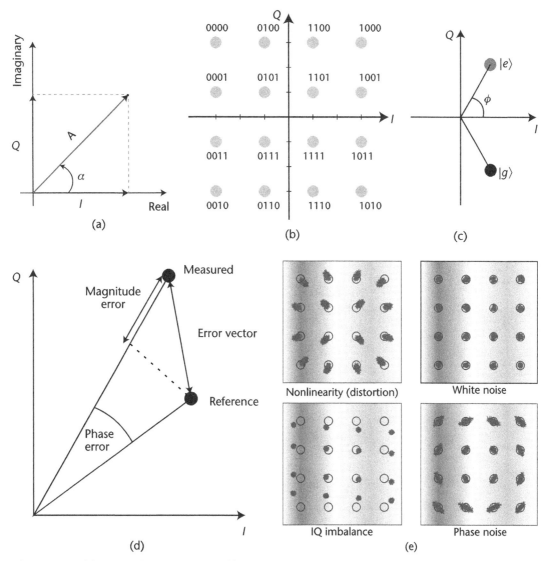

Figure 4.26 (a) IQ signal representation, (b) 16QAM constellation diagram, (c) IQ components of a qubit readout signal corresponding to ground $|g\rangle$ and excited $|e\rangle$ states, (d) EVM concept, and (e) representation of various signal impairments on the constellation diagram.

shown in Figure 4.26(d). The EVM is expressed in percentage or decibels. A lower percentage or decibel value is better and signifies a smaller error.

The maximum allowable EVM for some wireless systems is shown in Table 4.6. Increasing the modulation order (i.e., the number of bits per symbol) can lead to better throughput; however, symbols placed closer together increase the error probability, making a better EVM essential for higher-order modulation schemes. It is worth noting that Rydberg atoms have shown the capability to establish a communication link and create BPSK, QPSK, and QAM modulations with an EVM < 3% RMS [16].

Factors such as noise, nonlinear effects, and IQ imperfections (like amplitude imbalance or quadrature offset) can displace an IQ symbol from its ideal

Table 4.6 EVM Requirements for Some Modulation Schemes

Modulation Scheme	Required EVM [%]
QPSK	17.5%
16 QAM	12.5%
64 QAM	8%

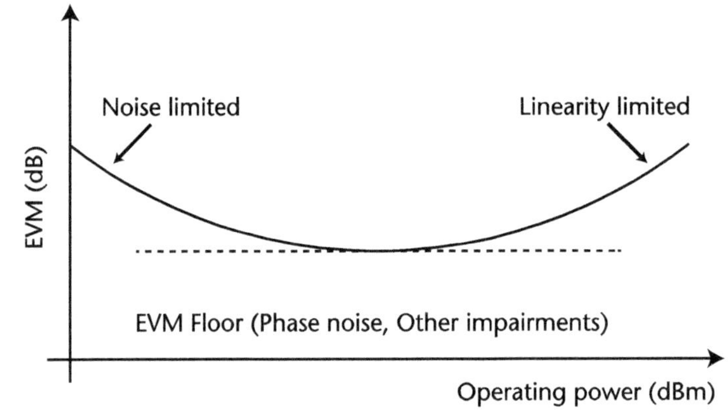

Figure 4.27 Bathtub EVM curve (EVM versus operating power).

position in the constellation diagram. Figure 4.26(e) illustrates the effect of various impairments on the constellation diagram. Chapter 7 discusses how calibrating IQ mixers are vital to compensate the IQ imbalance for reliable qubit control and measurement.

When plotting EVM against operating power, a bathtub curve similar to Figure 4.27 emerges. EVM performance at low power levels is limited by noise, while at high power levels, it is limited by linearity. For a given power level, the sum of white noise EVM, phase noise EVM, and linearity EVM indicates the total EVM level in a system

$$EVM_{\text{total}}^2 = EVM_{\text{White noise}}^2 + EVM_{\text{phase noise}}^2 + EVM_{\text{linearity}}^2 \quad (4.29)$$

For a cascaded system, the EVM depends on the individual EVMs of each system and the nature of their connection.

Chapter 5 discusses various microwave components utilized in a qubit hardware setup. Subsequently, Chapter 7 integrates all the concepts presented in this chapter and Chapters 5 and 6, focusing on the hardware setup for superconducting qubits.

References

[1] Pozar, D. M., *Microwave Engineering*, John Wiley & Sons, 2011.
[2] Collins, R. E., *Foundations of Microwave Engineering*, John Wiley & Sons, 2007.

[3]   Jerger, M., et al., "In Situ Characterization of Qubit Control Lines: A Qubit as a Vector Network Analyzer," *Phys. Rev. Lett.*, Vol. 123, No. 150501, 2019.

[4]   Chaves, K. R., "Nonlinear Signal Distortion Corrections Through Quantum Sensing," *Appl. Phys. Lett.*, Vol. 118, No. 014001, 2021, doi:10.1063/5.0035712.

[5]   Sarovar, M., "Detecting Crosstalk Errors in Quantum Information Processors," *Quantum*, Vol. 4, 2020, p. 321.

[6]   Nuerbolati, W., "Cancelling Microwave Crosstalk with Fixed-Frequency Qubits," *Applied Physics Letters*, Vol. 120, No. 17, 2022.

[7]   Razavi, B., and R. Behzad, *RF Microelectronics*, (Volume 2), New York: Prentice Hall, 2012.

[8]   Ball, H., W. Oliver, and M. Biercuk, "The Role of Master Clock Stability in Quantum Information Processing," *npj Quantum Inf*, Vol. 2, Article No. 16033, https://doi.org/10.1038/npjqi.2016.33.

[9]   Texas Instruments, "Fundamentals of Precision ADC Noise Analysis Design Tips and Tricks to Reduce Noise with Delta-Sigma ADCs," September 2020, ti.com/precisionADC.

[10] Kalfus, W. D., "High-Fidelity Control of Superconducting Qubits Using Direct Microwave Synthesis in Higher Nyquist Zones," *IEEE Transactions on Quantum Engineering*, 2021, https://arxiv.org/ftp/arxiv/papers/2008/2008.02873.pdf.

[11] Kaufman R., "Josephson Parametric Amplifier with Chebyshev Gain Profile and High Saturation," 2023, rXiv:2305.17816.

[12] Jerger, M., "Frequency Division Multiplexing Readout and Simultaneous Manipulation of an Array of Flux Qubits," *Appl. Phys. Lett.*, Vol. 101, No. 042604, 2012, https://doi.org/10.1063/1.4739454.

[13] Remm, A., et al., "Intermodulation Distortion in a Josephson Traveling Wave Parametric Amplifier," *Physical Review Applied*, Vol. 20, No. 3, 2023, p. 034027.

[14] White T., et al., "Readout of a Quantum Processor with High Dynamic Range Josephson Parametric Amplifiers," *Applied Physics Letters*, Vol. 122, No. 1, 2023.

[15] Zurich Instruments, SHFQ User Manual, https://docs.zhinst.com/shfqc_user_manual/specifications.html#shfqa.specifications.analog.

[16] Holloway, C. L., "Detecting and Receiving Phase Modulated Signals with a Rydberg Atom-Based Mixera," *IEEE Antennas and Wireless Propagation Letters*, Vol. 18, No. 9, 2019, pp. 1853–1857.

CHAPTER 5
Microwave Components

> I do not think that the radio waves I have discovered will have any practical application.
>
> —*Heinrich Hertz*

This chapter reviews microwave engineering fundamentals, concentrating on the elements that apply to the hardware setup for superconducting qubits. The discussion includes essential analytical methods and techniques for studying microwave circuits and components, such as the equivalent circuit approach and Maxwell's equations. This chapter investigates the components utilized in generating, transmitting, processing, and detecting microwave signals, focusing on their applications in superconducting quantum computers.

5.1 Microwave Component Analysis

The concepts studied in Chapter 4 are related to microwave system analysis. There, we treated the system as a black box, focusing solely on its behavior and impact on the input without delving into the details of its internal components. However, in this section, we open the black box and explore the microwave components. To comprehend the functionality of these components, it is crucial to familiarize ourselves with the tools used for their analysis, discussed in Section 5.1.1.

5.1.1 Tools for the Analysis of Microwave Components

Before delving deep into microwave engineering concepts, it is essential to become familiar with some fundamental tools widely employed for analyzing microwave circuits and components. These tools can be broadly categorized into two major groups: analytical and numerical methods, as illustrated in Figure 5.1.

Analytical methods encompass Maxwell's equations and circuit theory [1–5]. Maxwell's equations offer a comprehensive description of the electromagnetic field at every point in space, although this level of detail sometimes exceeds practical requirements. However, in specific cases, such as microwave component design, the solution of Maxwell's equations becomes necessary. Circuit theory is more efficient in many practical scenarios than Maxwell's equations. Unlike Maxwell's equations, which comprehensively describe the electromagnetic field at every point in space, circuit theory focuses solely on determining relevant quantities such as

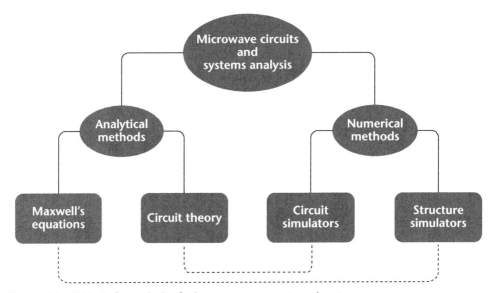

Figure 5.1 Methods for analysis of microwave components and systems.

terminal voltage, current, power, and impedance. This narrower scope of analysis results in a significant reduction in computational cost.

Numerical tools complement the analytical methods mentioned earlier, as depicted in Figure 5.1. Structure simulators like HFSS and CST are employed for solving Maxwell's equations numerically [6]. Circuit simulators such as Advanced Design System (ADS) or Microwave Office are utilized for solving circuit models [7]. These simulators provide valuable insights into the design process and facilitate extensive optimizations before significant investments of time and money are made in prototyping. Structure and circuit simulators play crucial roles in the design of superconducting qubits and circuits.

5.1.1.1 Analytical Methods

Analytical methods have the advantage of providing exact solutions, but their applicability becomes limited as problem complexity increases. For instance, Maxwell's equations can be analytically solved for certain waveguide shapes like rectangular or circular. However, numerical methods become necessary when dealing with waveguides of arbitrary cross-sections.

Despite this notable limitation, analytical solutions offer invaluable insights into the underlying physics of the problem and aid in comprehending the fundamental behavior at a deeper level.

Maxwell's Equations

Maxwell's equations describe electromagnetic fields at every point in space. Techniques such as the method of separation of variables for solving partial differential equations and variational methods for solving integral equations are employed to solve these equations. Since numerous excellent references [1–6] are available on this topic, we do not delve into the mathematical details of solving Maxwell's equations here.

Instead, we briefly overview Maxwell's equations and examine different regimes, including the optical-, lumped-, and distributed-element regimes. These regimes determine whether a simplified version of Maxwell's equations can be utilized. Table 5.1 summarizes Maxwell's equations in both integral and differential forms. The electrical length E, given by the ratio of the physical length l of the analyzed structure to the wavelength λ, determines the regime in which we are operating. Table 5.2 illustrates the three different regimes based on the electrical length value.

The process typically involves solving a wave equation when solving Maxwell's equations for an electromagnetic component or structure. This wave equation is derived by combining Maxwell's coupled equations (i.e., the last two equations in Table 5.1). In a region that is source-free, linear, isotropic, and homogeneous, the wave equation, also known as the Helmholtz equation, reads $\nabla^2 \bar{E} + \omega^2 \mu \epsilon \bar{E} = 0$. Section 5.3.1 explores how this equation simplifies to a one-dimensional wave equation specifically for the current and voltage on the TEM transmission line.

Lumped-Element Regime

A component is classified as a lumped element when its physical length is significantly larger than the smallest wavelength of interest, typically at least 10 times larger. In this regime, the voltage and current do not exhibit significant variation across the physical dimensions of the structure. As a result, the length of wires in a lumped circuit can be disregarded, except when considering losses. This leads to two primary consequences.

Table 5.1 Summary of Maxwell's Equations

Law	Integral Form (Original Form by Maxwell)	Differential Form (Invented by Heaviside)	Kirchhoff's Laws
Gauss's law	$\oiint_{\partial \Omega} \mathbf{E} \cdot d\mathbf{S} = \dfrac{1}{\varepsilon_0} \iiint_{\Omega} \rho \, dV$ $\mathbf{E}$ = Electric field vector ρ = Volume charge density	$\nabla \cdot \mathbf{E} = \dfrac{\rho}{\varepsilon_0}$	—
Gauss's law for magnetism	$\oiint_{\partial \Omega} \mathbf{B} \cdot d\mathbf{S} = 0$ $\mathbf{B}$ = Magnetic field density vector	$\nabla \cdot \mathbf{B} = 0$	—
Faraday-Maxwell law (Faraday's law of induction)	$\oint_{\partial \Sigma} \mathbf{E} \cdot d\ell = -\dfrac{d}{dt} \iint_{\Sigma} \mathbf{B} \cdot d\mathbf{S}$	$\nabla \times \mathbf{E} = -\dfrac{\partial \mathbf{B}}{\partial t}$	$\sum_{loop} V_i = 0$
Ampere-Maxwell law	$\oint_{\partial \Sigma} \mathbf{B} \cdot d\ell$ $= \mu_0 \left(\iint_{\Sigma} \mathbf{J} \cdot d\mathbf{S} + \varepsilon_0 \dfrac{d}{dt} \iint_{\Sigma} \mathbf{E} \cdot d\mathbf{S} \right)$ $\mathbf{J}$ = Current density vector	$\nabla \times \mathbf{B} = \mu_0 \left(\mathbf{J} + \varepsilon_0 \dfrac{\partial \mathbf{E}}{\partial t} \right)$	$\sum_{Node} I_i = 0$

Table 5.2 Electromagnetic Regimes Based on the Size of the Electrical Length

Electromagnetic Regime	Lumped Element Circuits	Distributed Element Circuits	Geometrical Optics/Wave Optics
Electrical length $E = l/\lambda$	$E \ll 1$	$E \approx 1$	$E \gg 1$
Governing equations	Circuit theory (KVL/KCL)	Maxwell's equations	• Maxwell's equations (wave optics) • Principles of geometrical optics
Spatial voltage/current variation	Voltages and currents are almost constant in magnitude and phase over the length of the structure.	Voltages and currents vary in magnitude and phase over the length of the structure.	Voltages and currents vary significantly in magnitude and phase over the length of the structure.
Electric and magnetic fields' relationship	Electric and magnetic fields are decoupled. • No wave theory is included. • No reflection happens due to impedance mismatch.	Electric and magnetic fields are coupled. • Wave phenomena such as reflection is relevant. • Reflection happens due to impedance mismatch.	Electric and magnetic fields are coupled. • Wave phenomena such as reflection is relevant. • Reflection happens due to impedance mismatch.

First, a lumped element structure can be treated as a localized element and fully characterized by its terminal I-V relationship. For instance, a lumped capacitor C can be characterized entirely by understanding its I-V relation, expressed as $I = CdV/dt$.

Second, we can simplify Maxwell's equations and utilize Kirchhoff's voltage law (KVL) and Kirchhoff's current law (KCL) to analyze lumped-element circuits. When the electric field varies slowly along the length of a component, its curl will be zero (curl is associated with the spatial rate of change), and no magnetic field will be induced (i.e., $\nabla \times \mathbf{E} = -\mu_0 \partial \mathbf{H}/\partial t = 0$). In this scenario, the electric field can be represented as the gradient of a potential V, denoted as $E = -\nabla V$ [1–4]. Through calculus, we understand that the integral of a gradient over a closed path yields zero. Consequently, this implies that the total voltage drop across a closed loop is also zero.

$$\mathbf{E} = -\nabla V \rightarrow V = -\oint_{\partial \Sigma} \mathbf{E} \cdot d\ell = 0 \rightarrow \sum_{loop} V_i = 0 \quad (5.1)$$

The intuition behind KVL is that an electric field in a lumped-element circuit does not induce a magnetic field. Any induced magnetic field would contribute to a voltage drop following Lenz's law.

Similarly, a slowly varying magnetic field does not induce an electric field (i.e., $\nabla \times \mathbf{H} = \mathbf{J} + \varepsilon_0 \partial \mathbf{E}/\partial t = \mathbf{J}$). Therefore, any changes in the magnetic field do not result in an induced electric field. If we apply the divergence operator to both sides of an equation and utilize the divergence theorem, we can approximate the resulting integral using a summation. This approximation yields the following expression:

$$\nabla \cdot \mathbf{J} = \nabla \cdot (\nabla \times \mathbf{H}) = 0 \rightarrow \sum_{\text{Node}} I_i = 0 \qquad (5.2)$$

The absence of divergence at a node implies that the net current entering or leaving that node must be zero, commonly referred to as KCL. Interestingly, Maxwell's equations become decoupled when dealing with a lumped element or a structure with a large electrical length. In this case, the equations include only an electric field (E) or only a magnetic field (H), but not both simultaneously. As a result, voltage and current can be treated independently, allowing us to apply Kirchhoff's laws to analyze and solve the circuit.

Optical Regime

On the opposite end of the spectrum, when the electrical length is extremely small, we can exploit approximations of Maxwell's equations. For example, Fermat's principle can be derived by examining the asymptotic behavior of solutions to Maxwell's wave equation as the wavelength approaches zero. This regime holds significant importance in optical engineering.

Distributed-Element Regime

The third regime, known as the distributed-element regime, falls between the previous two regimes, where the physical length of the component and the wavelength are comparable. The voltage varies along the component within this regime, and the electric and magnetic fields become interconnected. The ratio of these coupled electromagnetic fields is defined as the characteristic impedance of the medium ($Z_0 = E/H$). We typically operate within this regime at microwave frequencies, which has two important consequences.

First, we need to employ the precise form of Maxwell's equations. However, these equations are simplified for the transverse electromagnetic (TEM) propagation mode, as described in Section 5.3.1. As discussed in Section 5.3.1, a distributed-element circuit, such as a transmission line, can be represented as an infinite sequence of lumped-element circuits specifically for the TEM mode.

Second, the fields are no longer localized; instead, they are distributed throughout the component. This implies that the voltage and current phases vary over the length of the device. Consequently, the entire device length must be considered in the analysis. In such cases, wave phenomena, including reflection, become relevant. Unlike low-frequency signals, which can be transmitted using almost any type of transmission line, the transmission of microwave signals necessitates using transmission lines with controlled impedance. This ensures that signals propagate within a controlled electromagnetic medium to avoid reflections.

Circuit Theory

In microwave engineering, solutions to Maxwell's equations often yield more information than is necessary for practical purposes. Instead, focusing on terminal quantities such as voltage, current, power, and impedance is typically sufficient for designing and analyzing a wide range of applications. The upcoming sections delve into circuit modeling and demonstrate how S-parameters can effectively characterize microwave components and circuits without solving Maxwell's equations.

5.1.1.2 Circuit Modeling

The toolbox for modeling a system encompasses ideal elements that serve as simplified models representing the fundamental properties of the system. These ideal elements can be combined to create models for real-world systems. For example, a real-world capacitor can be modeled using ideal circuit elements such as a resistor (R), inductor (L), and capacitor (C), as shown in Figure 5.2(a). The resistors in the model account for leakage in the dielectric material and lead resistance, while the capacitor element represents the actual capacitance. The presence of inductance is included to account for the inductance of the leads. As discussed in Section 7.4.3.1, a capacitor exhibits inductive behavior at sufficiently high frequencies, which can be verified using the equivalent circuit extracted from the capacitor. This verification is impossible if only a single capacitor is used in the model. Circuit modeling is also crucial in qubit design. For instance, as demonstrated in Chapter 3, a qubit coupled to a resonator can be effectively modeled using two capacitively coupled LC resonators, as depicted in Figure 5.2(b).

Circuit-analysis techniques allow us to transform an electromagnetic problem into an equivalent circuit problem, which can be solved using circuit-analysis methods. This approach serves as the foundation for the functionality of circuit-simulation tools, where the equivalent circuit model of a physical device is utilized for analysis. By employing circuit modeling and analysis tools, our primary objective is to extract terminal quantities such as voltage, current, power, and impedance of a microwave circuit.

Scattering Parameters

The understanding of voltage and current in microwave circuits differs significantly from their use in low-frequency lumped circuits due to the inclusion of wave phenomena. Therefore, it is crucial to consider the behavior of electromagnetic waves when analyzing microwave circuits. The scattering parameter, or

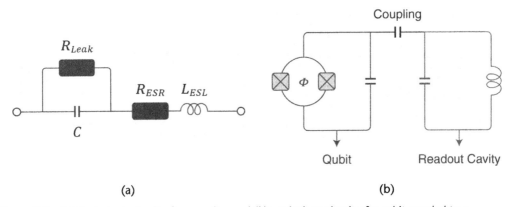

Figure 5.2 (a) Equivalent circuit of a capacitor and (b) equivalent circuit of a qubit coupled to a resonator.

5.1 Microwave Component Analysis

S-parameter, effectively captures the voltage and current wave nature. This section explores the significance of S-parameters.

During the propagation of a microwave signal, a portion of the signal reflects whenever there is a change in the properties of the medium. This phenomenon is illustrated in Figure 5.3, which depicts the behavior when a signal is applied to the device under test (DUT) at either port one or port two. The voltage amplitudes V_i^+, and V_i^- correspond to the amplitudes of the incident and reflected waves, respectively, at port i. The amplitude of the reflected wave at each port can be determined using the following expressions [3]:

$$\begin{aligned} V_1^- &= S_{11}V_1^+ + S_{12}V_2^+ \\ V_2^- &= S_{21}V_1^+ + S_{22}V_2^+ \end{aligned} \quad (5.3)$$

So, using (5.3), the S-parameters can be defined as the ratio of the reflected and incident voltages at the ports

$$\begin{aligned} S_{11} &= \frac{\text{Reflected}}{\text{Incident}} = \frac{V_1^-}{V_1^+}\bigg|V_2^+ = 0 \\ S_{21} &= \frac{\text{Transmitted}}{\text{Incident}} = \frac{V_2^-}{V_1^+}\bigg|V_2^+ = 0 \end{aligned} \quad (5.4)$$

The S_{22} and S_{12} parameters are defined similarly.

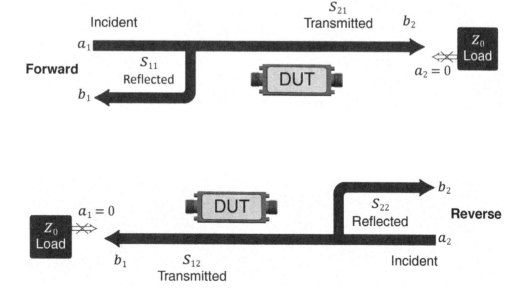

Figure 5.3 Definition of S-parameters.

There are two important considerations to be made. First, the forward S-parameters (i.e., S_{12} and S_{21}) are determined when the output is terminated in a load that precisely matches the characteristic impedance of the test system. In this case, there is no contribution to the amplitudes due to reflection. Section 5.3.3.1 explains how an impedance mismatch leads to reflection, similar to how light reflects when transitioning from air to glass due to the difference in characteristic impedance between the two media.

Second, there is a numbering convention to be taken into account. The first number following the S indicates the port from which energy emerges, while the second number indicates the port into which energy enters. Therefore, S_{12} measures the power emerging at port 1 when a signal is applied to port 2. For a two-port device, the reflection coefficients are represented by (S_{11}, S_{22}), and the transmission coefficients are represented by (S_{12}, S_{21}).

Equation (5.3) can be expressed as a 2×2 matrix. The S-parameter matrix of a two-port network consists of four components; the diagonal elements represent the reflection coefficient of each port. In contrast, the off-diagonal elements represent the transmission coefficient between the ports. In general, an n-port network will have n^2 S-parameters. Components with up to four ports are commonly used in practical applications.

For instance, an antenna is considered a one-port network; an amplifier is a two-port network; a circulator is a three-port network; and a coupler is a four-port network, as illustrated in Figure 5.4.

Now, let us address a concept that can be confusing regarding S-parameters. While S-parameters in linear units always refer to voltage amplitude, S-parameters expressed in logarithmic (decibel) units relate to power. Therefore, even though S-parameters represent voltages when transformed into decibels, they indicate power. As noted previously, we perform addition and subtraction operations with powers in decibel-milliwatts (dBm) and losses or gains in decibels (dB) because S-parameters in dB relate to power.

Last, it is worth noting that S-parameters are commonly measured using a network analyzer. As mentioned in Chapter 3, the dispersive readout of a qubit

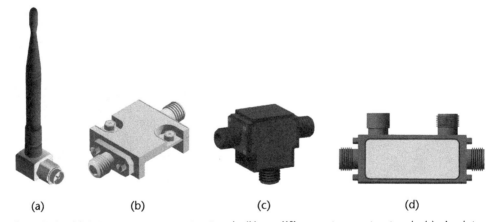

Figure 5.4 (a) Antenna as a one-port network; (b) amplifier as a two-port network; (c) circulator as a three-port network; and (d) a directional coupler as a four-port network, where one port is terminated to a matched load.

involves analyzing the transmission parameter of a microwave resonator coupled to the qubit. In practice, this measurement is achieved by utilizing a network analyzer to measure the S-parameters of the resonator. This highlights the importance of S-parameter measurements conducted with a network analyzer, not only in the realm of microwave circuits but also in quantum computing.

S-Parameters Terminology

The S-parameters for the reflection coefficient Γ and transmission coefficient T have unique names. The insertion loss is employed in passive circuits and is associated with the transmission coefficients (S_{12}, S_{21}). The insertion loss, IL, represents the reduction in signal power as it traverses through the network from port 1 to port 2. It can be calculated using $IL = -20 \log|T|$ dB $= -20 \log|S_{21}|$ dB.

The insertion loss is expressed as a positive decibel value, where a lossless transmission corresponds to an insertion loss of 0 dB. The magnitude of insertion loss can vary depending on the component, ranging from a few tenths of a decibel in a short cable to a few decibels in filters or mixers.

On the other hand, the return loss (RL) is related to the reflection coefficients (S_{11}, S_{22}) and is defined as follows for port 1 $RL = -20 \log|\Gamma|$ dB $= -20 \log|S_{11}|$ dB.

It is essential to mention that the value is always nonnegative in the case of reflection from a passive network. An RL of 0 dB indicates complete reflection, while a higher RL value indicates a lower reflection from a port. In practical applications, many consider an RL value of 10 dB or higher as an acceptable RL. Table 5.3 illustrates the amount of power reflected for different RL values.

Return loss is a crucial performance factor for microwave components and systems, indicating good impedance matching and low reflections. However, there are cases where a low return loss may be desired, such as in the stop band of a reflective filter. It is important to note that while return loss is essential for overall performance, it does not necessarily guarantee optimal system operation. Low reflection can also arise from internal system losses, where the energy is dissipated within the system, such as through radiation. Hence, it is essential to consider return and insertion losses when analyzing microwave components.

Energy conservation dictates that the sum of incident and reflected powers must equal the transmitted power for a two-port network, following the unitary property of S-parameters [3]. This principle can be expressed as $|S_{11}|^2 + |S_{21}|^2 = 1$ and $|S_{22}|^2 + |S_{12}|^2 = 1$. For a reciprocal network $|S_{21}|^2 = |S_{12}|^2$ and therefore $|S_{11}|^2 = |S_{22}|^2$.

The transmission coefficient between two ports in a microwave circuit can be referred to by various names, depending on the component and its specific application in a given configuration. Figure 5.5 provides a visual representation

Table 5.3 Amount of Reflected Power for Various Values of Return Loss

Return Loss in Decibels	Means
0 dB	100% reflection, no power flows into the device
3 dB	50% reflection, 50% power flows into the device
10 dB	10% reflection, 90% power flows into the device
15 dB	3% reflection, 97% power flows into the device
20 dB	1% reflection, 99% power flows into the device

of how this terminology is used. The different terminologies commonly used in microwave circuits are explained as follows:

- *Insertion loss:* This term is predominantly used in passive circuits, including filters, cables, and power dividers. It quantifies the amount of attenuation experienced by a signal as it passes through the circuit.
- *Gain:* Gain is primarily used in active circuits, such as amplifiers. It denotes the amount of signal amplification that occurs within the circuit.
- *Coupling:* Coupling is a term employed in components with three ports and more, such as couplers. It refers to the intentional transfer of a portion of the signal to a specific port, allowing for further processing or measurement.
- *Isolation:* In cases where coupling is not desirable, the term isolation is used. Isolation ensures that the signal does not leak from one port to another within the circuit. High isolation is desirable in circuits such as power dividers, isolators, and circulators to prevent unwanted signal leakage between ports.

We would like to emphasize that, generally, a low insertion loss (minimal signal attenuation) and high return loss (excellent impedance matching) are desirable across the frequency range of operation.

Furthermore, it is essential to note that other parameters are used for analyzing specific types of microwave circuits. For instance, X-parameters are utilized to analyze nonlinearities [5–8], mixed-mode S-parameters are employed to analyze differential mode microwave circuits, and active S-parameters and hot S-parameters have their applications and interpretations in specialized scenarios such as microwave plasma measurements.

5.1.1.3 Numerical Methods

Simulation plays a crucial role in both research and industrial design. Simulating a potential design's behavior, advantages, and disadvantages can save considerable

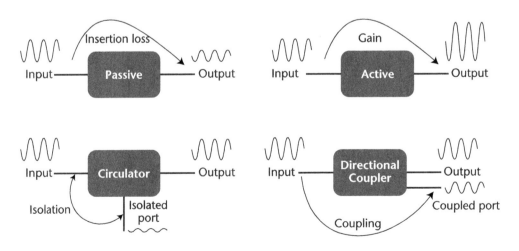

Figure 5.5 Terminologies used for the transmission coefficient in various microwave components.

time and expenses before committing resources to construct an actual system. Recall that powerful simulation software is utilized for both 3D structure analysis and circuit analysis. These software packages employ numerical methods such as the finite-element method (FEM), the method of moments (MoM), the finite-difference time-domain (FDTD), as well as other standalone and hybrid methods to facilitate the simulation process [5, 6].

5.2 Signal Generation

Sections 5.2 to 5.4 delve into the constituents of a microwave link, encompassing the generation, transmission, processing, and detection of microwave signals. Specifically, we introduce these four domains' fundamental components and key performance indicators and illustrate their utilization in a microwave system, focusing on their application in superconducting quantum computers.

Generating microwave signals is the initial step in establishing a microwave link, as outlined in Chapter 4, where it is emphasized that these signals are crucial for controlling and reading out superconducting qubits.

Sections 5.2.1.1 to 5.2.1.3 explore the generation of microwave signals by discussing scalar, vector, and digital signal generators and their respective figures of merit.

An oscillator is the heart of any signal generator, and the purity of the generated oscillation is an important figure of merit critical in many applications, including quantum computing. Chapter 4 showed how oscillation purity, or its phase noise, can affect the coherence time of a qubit.

A voltage-controlled oscillator (VCO) functions as an electronically tunable oscillator, with its frequency susceptible to variations caused by factors such as temperature and supply voltage. To ensure a stable output frequency, a phase-locked loop (PLL) synchronizes the VCO's output frequency with a precise reference frequency provided at the input.

In addition to PLL, other options for analog signal generators include yttrium iron garnet– (YIG)-tuned oscillators (YTOs). Digital techniques like direct digital synthesis (DDS) are also employed for signal generation. YTO-based or YIG-based synthesizers are widely used in the industry due to their high spectral purity and submillisecond switching speeds, making them the preferred choice for most high-frequency signal generators.

5.2.1.1 Analog Signal Generators

Signal generators are crucial in testing and measuring RF and microwave systems. Modern generators' advanced capabilities facilitate sophisticated channel modeling, incorporating interference, noise, fading effects, and analog and digital modulations. In the realm of qubit experiments, signal generators are essential for generating signals used in qubit control, readout, downconversion, and parametric amplification as the pump signal.

Signal generators can be broadly classified into two main groups. The first group is comprised of scalar-signal generators, which in addition to frequency offer control over the signal's amplitude. The second group consists of vector-signal generators, which provide control over both the signal's amplitude and phase. This capability allows for the generation of complex modulated signals using I and Q components. In the following sections, we will elaborate on scalar and vector signal generators and their respective figures of merit.

Scalar-Signal Generators

Scalar-signal generators provide control over the amplitude and frequency of the generated signal but not the phase. They are capable of producing CW signals as well as classical analog modulated signals, including AM, FM, and pulse modulation.

In specific applications, such as qubit measurements, sweeping across a range of frequencies becomes necessary. In such cases, the signal generator's sweep mode is utilized, requiring the specification of a start frequency, stop frequency, and sweep step, in addition to the output power, typically expressed in dBm.

Conversely, when there is a need to generate signals at specific frequencies with varying power levels for testing specific channels, the list mode is more efficient. This mode involves creating a list of signals with designated frequencies and power levels.

Vector-Signal Generators

Vector-signal generators control the signal's amplitude, phase, and frequency. This level of control makes it possible to manipulate both the I and Q components of the signal, as discussed in Chapter 4. With this capability, more advanced signals can be generated, including those used in digital modulation schemes such as QPSK, FSK, and QAM, commonly utilized in LTE and satellite communication systems. In quantum computing, precise phasing is vital for accurately manipulating quantum states on the Bloch sphere.

As illustrated in Chapter 4, a signal's I and Q components encode its amplitude and phase. The IQ data represents the baseband signal before it is converted to RF. Using arbitrary digital-modulation techniques, the signal generator can automatically generate I and Q components internally. Alternatively, IQ data can be pregenerated using software tools such as MATLAB and then loaded into the signal generator.

In specific scenarios, an external source like an AWG can supply the IQ components to the signal generator. The generator then modulates the internally generated microwave signal using the external IQ waveform.

5.2.1.2 Signal Generator Figures of Merit

The figures of merit of a signal generator are described as follows:

- *Frequency range:* This refers to the range of frequencies that a signal generator can generate. In qubit experiments, signals up to 12 GHz are typically required.
- *Bandwidth:* The bandwidth of a signal generator determines the widest signal in the frequency domain that it can reliably generate.
- *Output power:* The generator's output power can vary from a minimum to a maximum. In certain cases, extremely low output power is necessary, such

as in qubit experiments with few photons in the cavity. An internal attenuator is required to reach that low signal level in such cases. Conversely, higher signal levels may be necessary for certain measurements, such as power amplifier tests, where an internal amplifier is needed to extend the output power.

- *Flatness of the frequency response:* A flat frequency response, indicating low amplitude variation over the output frequency range, is an important figure of merit for a signal generator. As mentioned in Chapter 4, a flat frequency response is essential for the system to operate linearly and avoid distortion. Techniques like automatic level control (ALC) actively adjust the output signal amplitude to maintain a flat frequency response.
- *Switching speed:* In some applications, such as agile electronic-warfare transmitters or the calibration of cellular receivers, microwave signal generators must switch from one frequency to another as quickly as possible and settle within specified amplitude and frequency requirements. The switching speed refers to the time it takes to switch between two output frequencies, and it is influenced by various parameters such as the group delay of the VCO's loop filter, the speed of internal switches, and the transient response of components. The switching speed also depends on the signal-generation technology used. For generators based on DDS, it typically lies in the sub-microsecond range. In contrast, VCO or YTO analog methods usually lie in the submillisecond range.
- *Phase noise:* As mentioned in Chapter 4, a significant figure of merit for a signal generator is its phase noise. The phase noise level varies between −140 and −100 dBc with a carrier offset of 20 kHz, depending on the signal generator model and the operating frequency. Recall that the phase noise level decreases with larger carrier offsets. It is important to note that the phase noise level increases with frequency. For example, a leading signal generator on the market exhibits a phase noise of −140 dBc at 1 GHz, whereas at the same carrier offset, the phase noise increases to about −102 dBc at 44 GHz.
- *Harmonics and spurs:* The spectral purity of a signal generator, which refers to its ability to have low levels of harmonics and spurs, is a critical figure of merit. The effects of harmonics and spurs are discussed in Chapter 4. In an ideal scenario, the output of a signal generator would not have any spurs or harmonics, but this is not achievable in reality. Typically, modern signal generators exhibit harmonic levels ranging from −30 to −50 dBc, depending on the operating frequency and the model used.

  Additionally, the nth subharmonic of a signal with a fundamental frequency of f is a signal with a frequency of f/n. The subharmonic level ranges from −85 to −50 dBc, depending on the frequency. Higher frequencies result in higher subharmonic levels.

5.2.1.3 Digital Methods for Signal Generation

An AWG generates waveforms based on the samples stored in its memory. Figure 5.6(a) illustrates a block diagram of an AWG, which incorporates a variable-rate clock that advances through the AWG memory. The digitized signal is subsequently converted to analog using a DAC. Modern AWGs can achieve sample rates of up to 50 GS/s and analog bandwidths of up to 15 GHz.

To comprehend the operation and key characteristics of an AWG, it is crucial to grasp the fundamental concept of sampling. An ADC approximates an analog signal by stair-stepped voltages, as depicted in Figure 5.6(b). Harry Nyquist established that reliably reconstructing a signal requires a sampling rate at least twice the highest frequency component of the given signal. So, the highest frequency signal that a DAC with a sampling frequency f_s can generate is equal to half the sampling rate or $f_s/2$. However, in practice, as we will see shortly, the maximum output frequency or bandwidth of an AWG is typically 40% of the sample rate. The extra 10% reduction in bandwidth comes from the reconstruction filter to smooth the output waveform.

Figures 5.6(c–e) explain why the output spectrum of an actual DAC exhibits a sinc roll-off, and Figure 5.6(f) illustrates the generated images in various Nyquist zones with image frequencies $|kf_s \pm f_o|$, where f_o is the output frequency, f_s is the sampling frequency, and $k = 1, 2, 3 \ldots$.

Traditionally, AWGs only operate within the first Nyquist zone. However, modern AWGs can generate signals well above the sampling rate by taking advantage of higher Nyquist zones. This can be achieved by applying a proper bandpass filter to the desired frequency component in the appropriate Nyquist zone.

AWG Figures of Merit

Understanding the key specifications of an AWG is crucial for selecting the suitable one for a given application. Some of the critical figures of merit associated with AWGs are described as follows.

- *Vertical resolution:* The vertical resolution of an AWG plays a vital role in accurately reconstructing the amplitude of a signal. For instance, a 10-bit resolution with a 2-V peak-to-peak input range can distinguish voltage changes as small as approximately 2 mV ($2/2^{10}$). However, it is essential to consider the effective number of bits (ENOB), a significant figure of merit that may be less than the maximum resolution. Note that the vertical resolution is doubled for every bit added.
- *Memory size:* Sufficient memory size is essential for high sample rates and accurate signal reconstruction. It makes it possible to store long sequences of waveform sample points. The memory size is commonly specified in gigasamples (GSas).
- *Sample rate:* The sample rate of an AWG is measured in gigasamples per second (GSa/s) and determines the number of samples that the DAC can take within a specific time interval. It is important to note that the maximum frequency that an AWG can generate is only a fraction of the sample rate.

  The play time of an AWG represents the duration of the generated pulse and can be calculated as *playtime = memory/sample rate*. Therefore, increasing the sample rate would require increasing memory size to maintain the same play time. For example, a memory size of 1 GSa and a sample rate of 8 GSa/s would result in a playtime of 8 seconds.

5.2 Signal Generation

- *Bandwidth:* The bandwidth of an AWG is limited by the maximum output frequency it can generate. Two factors contribute to the bandwidth: the sample rate and the reconstruction filter. According to Nyquist's theorem, frequencies up to half the sample rate can be reliably generated. Thus, a sample rate of 4 GHz corresponds to a DAC output bandwidth of 2 GHz. Additionally, using a reconstruction filter to smooth the output waveform results in a 10% reduction in bandwidth. Therefore, the maximum output

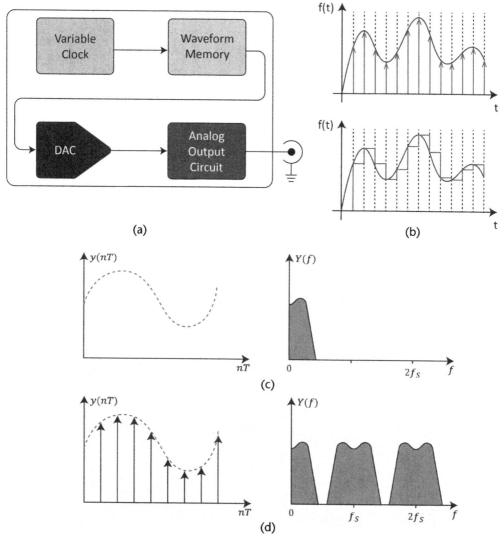

Figure 5.6 (a) Block diagram of an AWG; (b) concept of waveform sampling; (c) the spectrum of a signal without sampling, (d) sampling in the time domain generates periodicity in the frequency domain; (e) actual sampling happens with zero-order hold circuits, effectively resulting in the convolution of the sampled wave with a rectangular pulse and the convolution in the time-domain results in multiplication in the frequency domain where the sampled signal in (c) is multiplied by the Fourier transform of the rectangular pulse, which is a sinc function. (f) DAC output with sinc roll-off, the generated image frequencies, and Nyquist zones.

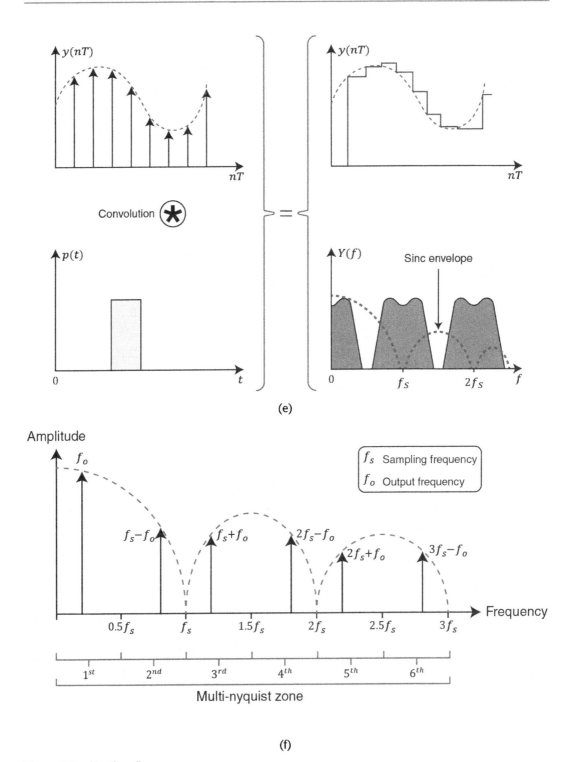

Figure 5.6 *(Continued)*

frequency or bandwidth of an AWG is typically 40% of the sample rate. For example, a sample rate of 10 GSa/s yields a bandwidth of 4 GHz.
- *ENOB:* Impairments such as noise, harmonics, and spurious signals can decrease the resolution and ENOB to a value typically lower than the number of DAC bits. The ENOB tends to decrease with increasing frequency, as illustrated in Figure 5.7(a). For instance, in a 14-bit system, the ENOB can drop to 8 bits at 2.5 GHz.

The ENOB can be calculated using the following formula:

$$\text{ENOB} = \left(\frac{\text{SINAD} - 10\log(3/2)}{20\log(2)} \right) = \frac{\text{SINAD} - 1.76}{6.02} \tag{5.5}$$

where SINAD represents the signal-to-noise and distortion ratio of the receiver. ENOB provides a more accurate measure of resolution compared to the number of DAC bits. Jitter specifications should be taken into consideration to ensure true signal fidelity.

- SFDR: SFDR is defined for RF systems in Chapter 4. For AWGs, as shown in Figure 5.7(b), it is the distance in decibels from the desired tone to the highest visible spur or harmonic within the stated bandwidth.
- Jitter: Jitter is associated with random phase jumps that can misalign pulse edges and introduce errors. Jitter is typically measured in picoseconds (ps) peak-to-peak between the synchronization clock and the direct data output.

5.3 Signal Transmission

To comprehend the intricacies of microwave-circuit behavior, it is crucial to grasp the principles of signal transmission at microwave frequencies and how it distinguishes itself from low-frequency circuits. This section explores transmission-line theory and sheds light on commonly employed transmission lines in the microwave and quantum engineering domains. Examples include coaxial cable, rectangular waveguide, microstrip, and CPW. Table 5.4 compares the characteristics of lumped-element and distributed-element transmission lines, while Figure 5.8 visually illustrates the difference between lumped- and distributed-element transmission lines.

Section 5.3.1 highlights the applicability of circuit-analysis techniques to a particular category of transmission lines characterized by two conductors, namely TEM lines. Examples of TEM lines include coaxial- and parallel-plate waveguides. The circuit method greatly simplifies the analysis and design of such transmission lines. However, for other types of transmission lines like rectangular waveguides, field-analysis methods involving the solution of Maxwell's equations are necessary to analyze their structure.

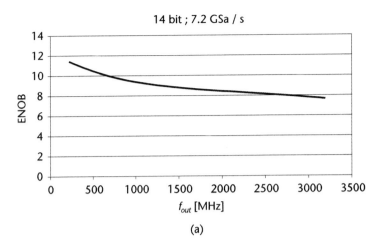

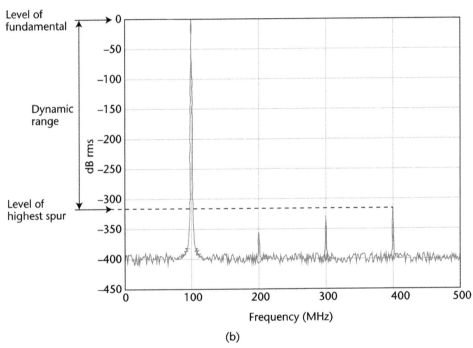

Figure 5.7 (a) ENOB versus the output frequency and (b) spurious free dynamic range of a DAC.

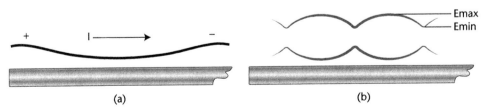

Figure 5.8 Current distribution on (a) a lumped-element transmission line and (b) a distributed-element transmission line.

5.3 Signal Transmission

Table 5.4 Comparison of Lumped-Element and Distributed-Element Transmission Lines

Lumped-Element Transmission Line	Distributed-Element Transmission Line
Large electrical length $E = l/\lambda$.	Moderate electrical length $E = l/\lambda$ (physical length and wavelength are comparable).
• Voltage and current along the line are constant. • Voltage and currents do not have a wave nature.	• Voltage and current vary along the line. • Wave-like nature of the voltage and current needs to be considered.
No impedance matching is necessary to minimize reflections.	Matching to the characteristic impedance (Z_0) is crucial to minimize reflections.

5.3.1 TEM-Mode Transmission Lines

In TEM propagation along the z-direction, the electric and magnetic fields do not have any components in the direction of propagation, denoted as $E_z = H_z = 0$. Consequently, the wave equation simplifies to the Laplace equation in the cross-section of the transmission line. This reduction transforms the problem into an electrostatic scenario, where the electric field can be expressed as the gradient of a potential $\Phi(x, y)$. The potential satisfies the Laplace equation $\nabla_t^2 \Phi(x, y) = 0$, allowing for the definition of a unique voltage between the two conductors at every point along the transmission line. This property makes it possible to utilize circuit-analysis techniques instead of directly solving Maxwell's equations.

In non-TEM transmission lines, the voltage between two points becomes path-dependent, making it impossible to establish a unique voltage between them. However, circuit-analysis techniques can still be applied by making certain assumptions.

A TEM transmission line can be modeled as an infinite sequence of infinitesimally small-length transmission lines connected in series, as shown in Figure 5.9(a). Each infinitesimal segment, having a small electrical length, can be treated as a lumped circuit as shown in Figure 5.9(b), permitting the application of KVL and KCL to the circuit. The per-unit length circuit parameters for the transmission line are defined as follows [3]:

- R: Series resistance per unit length, applicable to both conductors, measured in Ω/m.
- L: Series inductance per unit length, applicable to both conductors, measured in H/m.
- G: Shunt conductance per unit length, measured in S/m.
- C: Shunt capacitance per unit length, measured in F/m.

The transmission line equations can be derived using the KVL and KCL [3]

$$\frac{d^2V(z)}{dz^2} - \gamma^2 V(z) = 0 \tag{5.6}$$

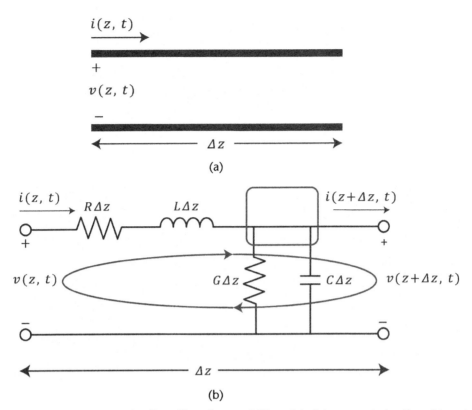

Figure 5.9 (a) A transmission line of length Δz and (b) model of the transmission line of length Δz.

$$\frac{d^2 I(z)}{dz^2} - \gamma^2 I(z) = 0 \quad (5.7)$$

with the complex propagation constant

$$\gamma = \alpha + j\beta = \sqrt{ZY} = \sqrt{(R + j\omega L)(G + j\omega C)} \quad (5.8)$$

where α is referred to as the attenuation constant, and β is known as the phase constant. The attenuation constant governs the rate at which the voltage amplitude diminishes along the transmission lines. In contrast, the phase constant is the rate at which the voltage phase changes as it propagates along the transmission line.

The wavelength on the line is defined as:

$$\lambda = \frac{2\pi}{\beta} \quad (5.9)$$

Equations (5.6) and (5.7) represent one-dimensional wave equations, and their traveling-wave solutions can be obtained as follows:

$$V(z) = V_o^+ e^{-\gamma z} + V_o^- e^{\gamma z}$$
$$I(z) = I_o^+ e^{-\gamma z} + I_o^- e^{\gamma z} \quad (5.10)$$

where the $e^{-\gamma z}$ term represents wave propagation in the +z direction, and the $e^{\gamma z}$ term represents wave propagation in the −z direction. The characteristic impedance is defined as

$$Z_0 = \frac{R + j\omega L}{\gamma} = \sqrt{\frac{R + j\omega L}{G + j\omega C}} \qquad (5.11)$$

where the voltage and current on the line are related to the line characteristic impedance as follows:

$$\frac{V_o^+}{I_o^+} = Z_0 = \frac{-V_o^-}{I_o^-} \qquad (5.12)$$

and the current can be expressed as $I(z) = (V_o^+/Z_0)e^{-\gamma z} - (V_o^-/Z_0)e^{\gamma z}$.

Converting back to the time domain, we can express the voltage waveform as $v(z, t) = |V_o^+|\cos(\omega t - \beta z + \phi^+)e^{-\alpha z} + |V_o^-|\cos(\omega t + \beta z + \phi^-)e^{\alpha z}$, where $\phi^\pm$ is the phase angle of the complex voltage $V_o^\pm$.

We have observed that the voltage and current along a transmission line exhibit wave-like behavior. The properties governing the wave propagation along the line are encapsulated in the propagation constant γ and the characteristic impedance Z_0, which can be derived from the line parameters (R, G, L, C) using (5.8) and (5.11). The following sections detail the propagation constant and characteristic impedance to provide a more comprehensive understanding.

5.3.1.1 Terminated Lossless Transmission Line

Section 5.3.1 explored wave propagation on an infinite transmission line with no reflections. However, in real-world scenarios, transmission lines are always terminated to some impedance. Matching the impedance of the line with the termination impedance is crucial to prevent reflections. This process, called impedance matching, holds significant importance in microwave circuits. Regardless of the performance of individual components, inadequate impedance matching can substantially impair overall system performance.

Consider the circuit in Figure 5.10(a). The total voltage and current on the line are given by (5.10). The boundary condition at $z = 0$, where the load impedance is located, is given by $Z_L = V(0)/I(0) = (V_o^+ + V_o^-/V_o^+ - V_o^-)Z_0$ Solving for V_o^- gives $V_o^- = (Z_L - Z_0/Z_L + Z_0)V_o^+$. The voltage reflection coefficient, Γ, is the ratio of the amplitude of the reflected voltage wave to the amplitude of the incident voltage wave.

$$\Gamma = \frac{V_o^-}{V_o^+} = \frac{Z_L - Z_0}{Z_L + Z_0} \qquad (5.13)$$

The total voltage and current waves on the line can be written as a superposition of an incident and a reflected wave, where these are called standing waves

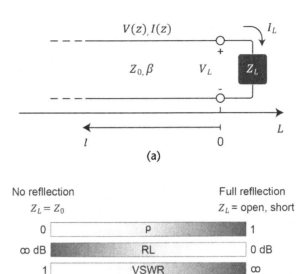

Figure 5.10 (a) A transmission line with load impedance Z_L; (b) the relationship between reflection coefficient, return loss, and VSWR.

$$V(z) = V_o^+ \left(e^{-j\beta z} + \Gamma e^{j\beta z}\right)$$
$$I(z) = \frac{V_o^+}{Z_0}\left(e^{-j\beta z} - \Gamma e^{j\beta z}\right) \quad (5.14)$$

There is no reflected wave when there is no reflection ($\Gamma = 0$). According to (5.13), this occurs when the load impedance (Z_L) is equal to the transmission line's characteristic impedance (Z_0). In such a scenario, we say that the load is matched to the line's characteristic impedance.

The standing-wave ratio (SWR), also known as the voltage standing-wave ratio (VSWR), is the ratio of the maximum and minimum voltage amplitudes along the transmission line. It serves as a measure of the degree of mismatch on the line.

$$\text{SWR} = \frac{V_{\max}}{V_{\min}} = \frac{1+|\Gamma|}{1-|\Gamma|} \quad (5.15)$$

As $|\Gamma|$ increases, the SWR increases. The reflection coefficient for a perfectly matched line is zero, corresponding to SWR = 1. The value of SWR varies as $1 \leq \text{SWR} \leq \infty$.

Section 5.1.1.2 explored the concept of return loss, and this section examines the reflection coefficient and SWR. These are interrelated and convey the same concept using different parameters, as depicted in Figure 5.10(b). When the generator is matched in a way that eliminates re-reflection of the reflected wave from $z < 0$, the time-average power flow along the line at point z can be expressed as: $P_{\text{avg}} = 1/2\text{Re}\{V(z)I(z)^*\} = 1/2(|V_o^+|^2/Z_0)(1 - |\Gamma|^2)$. The average power delivered to the load is equal to the incident power ($|V_o^+|^2/2Z_0$) minus the reflected power

($|V_o|^2|\Gamma|^2/2Z_0$). If $\Gamma = 0$, maximum power is delivered to the load, while no power is delivered for full reflection $|\Gamma| = 1$.

Note that in a transmission line, the power flow occurs solely through the electric and magnetic fields between the two conductors rather than through the conductors themselves. For conductors with finite conductivity, some power may penetrate the conductors, resulting in heat loss that is not delivered to the load.

5.3.1.2 Impedance Matching

Impedance matching plays a crucial role in various areas of electrical engineering, including electric power transfer and communication systems. It is essential to differentiate between two types of impedance matching: maximum-power transfer and zero reflections. The choice of matching depends on the specific application. The concept of impedance matching is shown in Figure 5.11(a).

In specific scenarios, the objective is to maximize power transfer to the load, such as in a microwave oven or a microwave plasma generator, where the available power is converted into heat or plasma. In such cases, conjugate matching is employed, where the input impedance Z_{in} and the load impedance Z_L are related as $Z_{in} = Z_L^*$, where * represents complex conjugation. This type of matching ensures maximum power transfer to the load. Although conjugate matching can result in reflections on the line, the power from multiple reflections can add in phase, delivering more power to the load than a reflectionless line.

In some situations, minimizing reflections is crucial, mainly when a microwave signal carries information, and maintaining signal integrity is essential to avoid data corruption. In these cases, matching the load impedance Z_L to the line is necessary [see Figure 5.11(a)] (i.e., $Z_{in} = Z_L$). Another scenario where reflection must be minimized is when standing waves on the line should be avoided. Standing waves can cause heating issues and pose challenges to power-handling capability. Therefore, minimizing reflections is critical in such scenarios. Figure 5.11(b) illustrates the S-parameters of a filter with and without matching.

Nevertheless, neither conjugate matching nor matching for zero reflection always leads to the most efficient system. The maximum-power transfer and zero-reflection conditions can only be simultaneously met when both the load and characteristic impedances are real (i.e., resistive).

Various factors, such as transmission-line discontinuities can result in impedance mismatch in a microwave circuit. Techniques for impedance matching include the use of lumped elements, distributed elements, and low-bandwidth and wide-bandwith matching. The Smith chart is a valuable visualization tool that greatly simplifies impedance matching [3].

The Bode-Fano criterion sets limits on how well we can achieve matching over a specified bandwidth. Regardless of the matching technique employed, the minimum achievable reflection increases as the bandwidth to be matched increases. In other words, achieving better matching becomes more challenging with increasing bandwidth. Although we do not delve deeper into this topic, it is essential to consider this limitation when examining the return loss of a wideband amplifier. Therefore, it should not be surprising if a wideband amplifier's return loss occasionally barely reaches 10 dB.

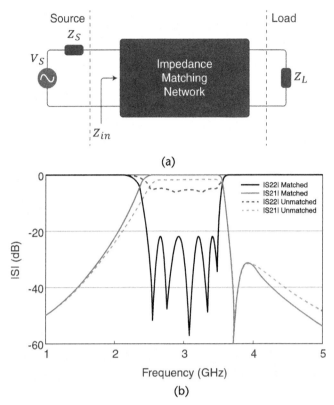

Figure 5.11 (a) Concept of impedance matching and (b) a comparison of the S-parameters of a matched and unmatched filter.

One may have noticed that most RF and microwave systems and interconnections are designed with a characteristic impedance of 50Ω. The origin of this standard can be traced back to the development of air-filled coaxial cables for kilowatt radio transmitters in the 1930s. When transmitting signals over longer distances, it is crucial to have low losses and a high-power handling capability. It can be demonstrated that for an air-filled coaxial cable, a characteristic impedance of 77Ω minimizes the losses. Conversely, a characteristic impedance of 30Ω maximizes the power handling capability. An average of these corresponding characteristic impedances can be computed to strike a balance between low transmission loss and high-power handling capability, rounded to an engineering standard, resulting in a characteristic impedance of 50Ω.[1]

5.3.1.3 Propagation Constant

Equation (5.8) illustrates the complex propagation constant of a transmission line, comprising the attenuation α and the phase constant β. The wavelength of voltage

---

[1]. There are also other standard impedances used in industry. The 75-Ω line is mainly used for cable TV (CATV). Sometimes 300-Ω characteristic impedance is also used in broadcasting systems, but it is not used as widely as 50-Ω and 75-Ω systems.

waves on the line is related to the phase constant as $\lambda = 2\pi/\beta$, while the phase velocity is given by $v_p = \omega/\beta$.

Typically, the phase constant β is a function of the frequency ω. To maintain a constant phase velocity v_p, the phase constant must be a linear function of frequency, denoted as $\beta = k\omega$, where k is a constant. Otherwise, a frequency-dependent phase velocity leads to dispersion, resulting in signal distortion. Table 5.4 presents specific conditions under which the phase constant exhibits a linear relationship with frequency.

The propagation constant in (5.8) can be simplified under certain conditions outlined in Table 5.5. An ideal lossless transmission line experiences no ohmic losses, rendering the attenuation constant zero. However, real-world transmission lines fall into the category of low-loss, where some level of attenuation occurs.

A lossy transmission line can be distortionless under a specific condition mentioned in Table 5.5, first discovered by Oliver Heaviside. This condition was applied to older telephone lines by introducing coils to increase the line's inductance, satisfying the requirements for distortionless lines and reducing dispersion effects.

The next two sections investigate the attenuation and phase constants in more detail.

5.3.1.4 Attenuation Constant

The attenuation constant, α in the complex propagation constant $\gamma = \alpha + j\beta$, determines the rate at which the voltage amplitude decreases as it propagates along the transmission line. The ohmic losses represented by the line parameters R and G contribute to the attenuation constant. If the attenuation constant is frequency-dependent, it can cause signal distortion if the dependency is strong. However, if the attenuation is frequency-independent, all frequency components are attenuated equally, preserving the signal's shape. Table 5.5 shows the condition for a distortionless transmission line in the presence of losses.

Attenuation is defined as the ratio of input power to output power and is commonly expressed in decibels:

Table 5.5 Attenuation and Phase Constant for Lossless, Low-Loss, and Distortionless Transmission Lines

Condition	Circuit condition	α (Attenuation Constant)	β (Phase Constant)	v_p (Phase Velocity)	Z_0
Lossless	$R = 0$ $G = 0$	0	$= \omega\sqrt{LC}$	$= \dfrac{1}{\sqrt{LC}}$	$\sqrt{\dfrac{L}{C}}$
Low-loss	$R \ll L\omega$ $G \ll C\omega$	$\dfrac{1}{2}\left(\dfrac{R}{Z_0} + \dfrac{G}{Y_0}\right)$ $Y_0 = \dfrac{1}{Z_0}$	$\approx \omega\sqrt{LC}$	$\approx \dfrac{1}{\sqrt{LC}}$	$\sqrt{\dfrac{L}{C}}$
Distortionless	$RC = LG$	$\sqrt{RG} = R\sqrt{\dfrac{C}{L}}$	$= \omega\sqrt{LC}$	$= \dfrac{1}{\sqrt{LC}}$	$\sqrt{\dfrac{L}{C}}$

$$\text{attenuation in dB} = 10\log\frac{P_i}{P_o} \tag{5.16}$$

The attenuation constant given in (5.16) has units of Np/m and is usually expressed in dB/m. The conversion between the two units is as follows: $\alpha_{dB/m} = 8.68 \alpha_{Np/m}$.

The attenuation for a specific length of a transmission line is simply the attenuation constant (α(dB/m)) multiplied by the line's length (L). Therefore, the total attenuation for a given transmission line length is α(dB/m) * L.

Table 5.6 provides a list of the most common sources of attenuation in a transmission line. The R and G parameters in the transmission line model represent the first two losses. The last two losses are not accounted for in the transmission line model, yet they can play a crucial role. We will discuss the losses mentioned in Table 5.6 in the following sections.

Conductive Losses

In the transmission line model, R represents the conductive losses. At high frequencies, the skin effect causes most of the current to be confined to a very thin layer beneath the conductor's surface. This layer, with a thickness of δ, is known as the skin depth. Due to the reduced effective area for current flow, the resistance increases, leading to higher losses.

The skin effect describes the exponential decay in current density J as we move from the surface of the conductor, where the current density is J_s, toward the center, where the current density is given by $J = J_s e^{-d/\delta}$. Here, δ represents the skin depth, and d is the distance from the surface. When the distance from the surface is equal to the skin depth ($d = \delta$), the current reduces to 1/e of the current at the surface. This implies that approximately 63% of the current flows within one skin depth, corresponding to a thin layer beneath the conductor's surface. The skin depth is calculated using the formula:

$$\delta = \frac{1}{\sqrt{\pi\mu\sigma f}} \tag{5.17}$$

Here, μ denotes the permeability, σ represents the conductivity, and f is the frequency. As the frequency increases, the skin depth decreases, resulting in a higher resistance, as (5.17) indicates.

Table 5.6 Sources of Attenuation

Type of Loss	Description
Conductive loss	Due to the ohmic losses of the conductor; R represents it in the transmission line model.
Dielectric loss	Due to the current leakage of the dielectric; G represents it in the transmission line model.
Radiation loss	Loss of energy due to radiation; this attenuation appears for unshielded transmission lines such as the microstrip and CPW.
Mismatch loss	Losses due to poor impedance matching, which causes signal reflections.

5.3 Signal Transmission

For a coaxial cable, the attenuation constant due to the conductive losses is given by

$$\alpha_c = \frac{R_{AC}}{2Z_0} = \frac{\left[\frac{1}{a} + \frac{1}{b}\right]\sqrt{\mu/32\sigma}}{\pi Z_0} \sqrt{\omega} \tag{5.18}$$

where a and b represent the inner and outer conductor radii, and Z_0 is the characteristic impedance. In practice, the measured attenuation constant tends to be higher due to surface roughness on metallic surfaces. To account for this effect, a quasi-empirical formula can be used to correct for surface roughness in a transmission line $\alpha'_c = \alpha_c[1 + 2/\pi \tan^{-1} 1.4(\Delta/\delta_s)^2]$, where α_c is the attenuation due to perfectly smooth conductors, α'_c is the attenuation corrected for surface roughness, Δ is the RMS surface roughness, and δ_s is the skin depth of the conductors [3].

Dielectric Losses

The G element in the transmission line model represents the dielectric loss associated with the leakage current in the dielectric material. To account for losses, the permittivity of the dielectric material is expressed as a complex number, with the imaginary part reflecting dielectric losses. These losses are quantified by the loss tangent, represented as $\tan\delta$ in (5.19).

$$\epsilon = \epsilon' - j\epsilon'' = \epsilon'(1 - j\tan\delta) \tag{5.19}$$

The dielectric loss constant of a coaxial cable can be determined using (5.20), which indicates that the dielectric loss cannot be reduced by modifying the cable's geometry [3].

$$\alpha_d = \frac{\sqrt{\mu\varepsilon}[\tan\delta]}{2}\omega \tag{5.20}$$

So, according to (5.20), to minimize dielectric loss, the only option is to utilize a low-loss tangent or a low-dielectric constant material. Teflon, for instance, is frequently employed as a dielectric with a low-loss tangent of 0.0004.

In contrast to conductor loss, dielectric loss becomes increasingly significant as the frequency rises. For example, in a 0.141"-diameter semirigid cable with copper conductors and a Teflon dielectric, conductor loss exceeds dielectric loss up to the cutoff frequency. However, the two types of losses become nearly equal at the cutoff frequency. As the frequency surpasses the cutoff frequency, the loss tangent becomes a more influential factor in the overall loss of the transmission line.

It is important to note that while dielectric losses are smaller than conductor losses, they still contribute to the overall loss of the system. Therefore, utilizing low-loss dielectrics and employing careful design techniques is crucial for minimizing the total loss of the transmission line.

Radiation Losses

The attenuation constant in (5.8) accounts only for conductor and dielectric losses, while radiation losses can sometimes contribute significantly to the overall losses, particularly at microwave frequencies. When the size of a transmission line is comparable to the wavelength, energy can escape from the line in the form of electromagnetic radiation. This phenomenon is well-known, as a good radiator (antenna) typically has a size comparable to the radiation wavelength. Shielded structures like strip lines, coaxial cables, or hollow waveguides exhibit much lower radiation losses than open, unshielded structures such as microstrip or CPW lines. Generally, bends in transmission line structures radiate more than straight sections, and larger substrates contribute more to radiation loss than thinner ones.

Accurately quantifying radiation losses can be challenging, and numerical simulations are often required to identify the radiating portion of the system and determine the power loss due to radiation. Radiation loss is not critical in qubit systems as microwave circuits are carefully shielded. However, radiation can still affect qubit performance. For instance, radiation can couple with the qubits, leading to decoherence and errors in the qubit state, as mentioned in Chapter 3. It is, therefore, crucial to carefully consider the effect of radiation in qubit systems and take appropriate measures to minimize their effects, such as implementing shielding for the structure (see Chapter 6).

Mismatch Losses

Recall that reflection of incident power results in only a portion of the signal being transmitted. The mismatch loss, expressed in decibels, can be calculated using the following equation:

$$L_{\text{mismatch}} = 10\log\left(\frac{P_i}{P_t}\right) = -10\log\left(1 - |\Gamma|^2\right) \tag{5.21}$$

P_i represents the incident power, P_t represents the transmitted power, and Γ is the reflection coefficient. The mismatch loss, as defined in (5.21), is the factor that reduces the maximum power transferred to the load, expressed in decibels. Depending on how multiple reflections combine, the overall system loss may be smaller or larger than the sum of the mismatch losses from each component.

To mitigate the effect of mismatch loss, it is necessary to ensure good impedance matching over the frequency band of operation. While numerous ways and techniques are available for impedance matching, space constraints prevent us from covering them in this book.

5.3.1.5 Phase Constant

The phase constant β represents the imaginary component of the propagation constant γ. The relationship between the phase constant and frequency (β-ω) is called the dispersion relation. As Chapter 4 explained, a linear dispersion relation ensures a constant phase velocity, meaning that all frequency components travel at the same speed along the transmission line. This is crucial because a nonconstant

5.3 Signal Transmission

or frequency-dependent phase velocity introduces signal distortion through a phenomenon known as dispersion. Dispersion causes the signal to spread out as it propagates along the line due to variations in the arrival time delay among different signal components. This becomes particularly significant in qubit systems as pulses are utilized for qubit control, and dispersion can cause pulse spreading, leading to variations in pulse width and amplitude, eventually resulting in errors during the intended qubit rotation on the Bloch sphere. Figure 5.12 illustrates the impact of dispersion on a Gaussian pulse.

Dispersive media exhibit this property, and there are two primary sources of dispersion, described as follows:

1. *Material dispersion:* This arises from the frequency-dependent response of a material, such as the dielectric used in a transmission line.
2. *Waveguide dispersion:* This stems from the dependence of the phase velocity in a waveguide on its frequency.

To comprehend transmission phenomena, it is essential to understand the various types of velocities, described as follows:

- *Speed of light* ($1/\sqrt{\mu\varepsilon}$): This refers to the velocity at which a plane wave would propagate through a medium and is dependent on the material properties of the medium.
- *Phase velocity* ($v_p = \omega/\beta$): The phase velocity represents the speed at which a point with a constant phase travels. For different types of guided-wave propagation, such as waveguide TE or TM modes, the phase velocity may be greater or less than the speed of light. Since the phase velocity of guided wave propagation varies with frequency, no single-phase velocity can be assigned to the entire signal.
- *Group velocity* $v_g = (d\beta/d\omega)^{-1}$: When the signal's bandwidth is relatively narrow or the dispersion is not too pronounced, the group velocity can be defined to describe the overall signal propagation speed. In a coaxial cable, the group

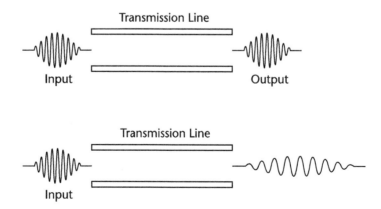

Figure 5.12 Effect of dispersion in a transmission line on a Gaussian pulse.

velocity is reduced by $1/\sqrt{\varepsilon_r}$, where ε_r represents the relative permittivity. Most coaxial cables utilize 100% PTFE insulation, with a dielectric constant of approximately 2.2. Consequently, this results in a group delay of 1.45 ns/ft of coaxial cable.

The effect of pulse broadening becomes more pronounced as the length of the transmission line increases. This occurs because the time delays between different frequency components accumulate as the pulse traverses the line. Therefore, if the length of the transmission line is sufficiently short, the dispersion effect can be negligible.

5.3.2 Non-TEM Transmission Lines

Propagation modes in microwave engineering are distinguished based on the orientation of electric and magnetic field vectors in relation to the direction of propagation. The three most frequently encountered propagation modes are TEM, transverse electric (TE), and transverse magnetic (TM). Hybrid modes can also propagate in certain transmission lines like microstrip lines, although they are less commonly employed than the other three modes. Table 5.7 provides a summary of the propagation modes in transmission lines.

In the TEM propagation mode, as shown in Figure 5.13(a), the electric and magnetic fields are perpendicular to each other and the direction of propagation. This mode is limited to two-conductor transmission lines such as coaxial cables and parallel-plate waveguides. The TEM mode exhibits three important characteristics: (1) it has no cutoff frequency and can propagate from DC onwards, (2) its phase velocity is independent of frequency, resulting in no dispersion or distortion, and (3) a unique voltage and current can be defined at each point on the line, allowing the use of circuit analysis techniques.

TE and TM modes can propagate in two- and single-conductor structures, such as hollow waveguides. The letter following the "T" designates the orthogonal field component to the propagation direction. Therefore, a TE mode lacks an electric component in the propagation direction, while a TM mode lacks a magnetic component in the propagation direction. Figure 5.13(b, c) illustrate these modes and

Table 5.7 Various Propagation Modes in Transmission Lines

Propagation Mode	Description	Transmission Line	Characteristics
TEM	Electric and magnetic fields are perpendicular to each other, and the direction of propagation	• Two-conductor transmission lines	• No cutoff frequency • Phase velocity is constant
TE	No electric field components in the direction of propagation	• Two-conductor transmission lines • Waveguides	• It has a cutoff frequency • Phase velocity is frequency-dependent
TM	No magnetic field components in the direction of propagation	• Two-conductor transmission lines • Waveguides	• It has a cutoff frequency • Phase velocity is frequency-dependent

their propagation in a rectangular waveguide: The TE mode reflects off the side walls and propagates along the waveguide, whereas the TM mode reflects off the upper and lower walls and propagates along the waveguide.

The TE and TM modes have a cutoff frequency f_c, indicating that modes with frequency $f > f_c$ will propagate, while evanescent or cutoff modes with $f < f_c$ will decay exponentially away from the source of excitation. The cutoff frequency is determined by solving Maxwell's equations in the transmission line's cross-section using appropriate boundary conditions. However, let us gain an intuitive understanding of how the cutoff frequency is determined. For a wave to propagate in a structure, the wavelength must match the physical dimension of the structure or, more precisely, the waveguide's cross-section. To calculate the cutoff frequency, we must determine how many wavelengths fit in each dimension (x and y for a rectangular waveguide) of the waveguide cross-section, as illustrated in Figure 5.13(d). The subscripts represent the number of half-waves at cutoff that fit across the x- and y-directions of the rectangular waveguide or along the circumferential and diametrical directions of a circular waveguide. For example, $TE_{m,n}$ indicates that there are m half-waves of the E-field across the x-dimension and n half-waves of the E-field across the y-dimension for a rectangular waveguide. As the operating frequency increases, the waveguide cross-section needs to decrease size to accommodate the half-waves within its cross-section.

In the case of the TE_{10} mode of a rectangular waveguide at cutoff, there is one half-wave in the x-direction, and the electric field exhibits no variation in the y-direction, as depicted in Figure 5.13(d). Figure 5.13(e) illustrates a rectangular waveguide's various propagation modes and cutoff frequencies. For a circular waveguide, at cutoff, the TE_{10} mode has a uniform E-field in the circumferential direction with no variation. It is purely circumferential in orientation and has a zero value at the surface of the waveguide. There is a single peak precisely at the center of the waveguide.

Section 5.3.3.4 presents the cutoff frequency formula and examines the frequency-dependent nature of the phase velocity for TE and TM waves, in contrast to the frequency independence of TEM waves. In this case, a group velocity can be defined to characterize the propagation speed of a signal, provided that the signal bandwidth is small enough.

5.3.2.1 Useful Bandwidth of a Transmission Line

Propagation of more than a single mode is desirable in specific applications, such as waveguide filters. In such cases, the waveguide is referred to as overmoded. However, for most microwave applications, single-mode propagation is preferred, as the simultaneous propagation of multiple modes can lead to undesirable effects, including distortion and high signal reflection. The bandwidth of a transmission line for single-mode propagation is defined as the range between the cutoff frequencies of two consecutive modes.

In addition to the TEM mode, coaxial cables can also support TE and TM waveguide modes. The coaxial cable's bandwidth for single-mode propagation starts from DC and extends to the first high-order TE mode, the TE_{11} mode. The TE_{11} mode starts to propagate when the wavelength in an air-filled coaxial cable

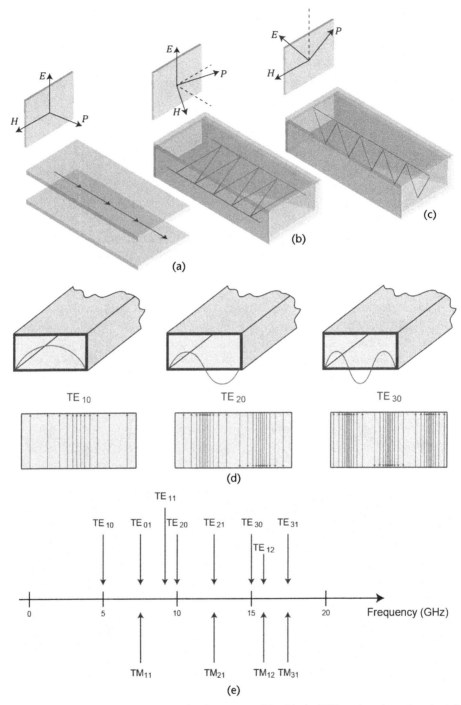

Figure 5.13 Different propagation modes in a waveguide: (a) the TEM mode, where the electric and magnetic fields are perpendicular to the direction of propagation; (b) the TE mode, where the wave bounces off the side walls and propagates along the waveguide; and (c) the TM mode, where the wave bounces off the upper and lower walls and propagates along the waveguide. (d) Various modes and electric field configurations for a rectangular waveguide, with subscripts indicating the number of half-waves of the electric field across the x- and y-dimensions. (e) The cutoff frequency for TE and TM modes in a hollow rectangular waveguide with dimensions $a = 3$ cm and $b = 2$ cm.

is equal to the average circumference, as given by $\lambda_c = 2\pi[(a + b)/2] = \pi(a + b)$, where a and b represent the inner and outer conductor radii. If the medium is filled with a dielectric, the dielectric constant adjusts the cutoff wavelength $\lambda_c = 2\pi[(a + b)/2]\sqrt{\varepsilon_r} = \pi(a + b)\sqrt{\varepsilon_r}$.

The cutoff frequency of the first higher-order mode is calculated using $\lambda_c = c/f_c$ formula resulting in $f_c = c/\pi(a + b)\sqrt{\varepsilon_r}$.

The single-mode bandwidth of a coaxial cable is represented as f_c, and, depending on the cable's construction, can vary from 1 to 110 GHz. Since frequency and wavelength are inversely proportional, it is essential to progressively decrease the size of connectors and cable cross-sections as the operating frequency range increases to ensure mode-free propagation. However, reducing the size of the coaxial cable structure results in higher losses and lower power-handling capability.

The useful bandwidth of a rectangular waveguide is slightly less than an octave, corresponding to a 2:1 frequency range. This limitation arises because the TE_{20} mode starts propagating at a frequency twice the cutoff frequency of the TE_{10} mode. To enhance the bandwidth of waveguides, ridged waveguides are employed, as shown in Figure 5.14(a).

Now, let us explore the challenges associated with the propagation of higher-order modes. The presence of high-order modes in a waveguide causes variations in the cable's impedance, resulting in increased reflection due to impedance mismatch. For optimal performance, microwave cables and connectors should maintain a constant impedance along their entire length, ensuring a flat response over the widest possible bandwidth and minimizing reflections. Impedance comprises resistive and reactive components. High-order mode propagation introduces a resistive perturbation to the impedance, as Figure 5.14(b) depicts. In a coaxial cable, as the TE_{11} mode approaches, the reactive energy stored in the higher-order modes influences the reactive part of the impedance, causing a significant impedance change leading to mismatch and increased reflection.

5.3.2.2 Transmission Lines Figures of Merit

Several factors must be considered when selecting a cable; this section discusses these parameters as follows.

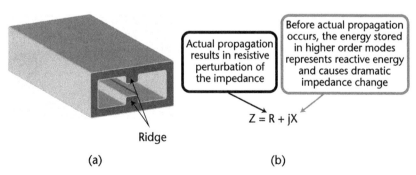

Figure 5.14 (a) Ridged waveguide and (b) effect of propagating and cutoff (evanescent) modes on the waveguide impedance.

- *Frequency range of operation:* The frequency range of operation refers to the range of frequencies over which the propagation is free from non-TEM modes. When choosing a cable or transmission line, the first parameter to consider is the frequency range of operation, which should be higher than the maximum operating frequency of our system. In qubit systems, cables operating up to 18 GHz are commonly used, as qubit systems typically operate up to 10 GHz.
- *Insertion loss:* The insertion loss indicates the amount of signal attenuation as it travels through the cable. Recall that the insertion loss increases with higher frequencies and is composed of dielectric, conductor, and radiation losses. Figure 5.15(a) illustrates the insertion loss for a specific cable, which is approximately 0.2 dB at low frequencies and reaches 1.8 dB at 18 GHz. Sometimes the insertion loss is specified in decibels per meter, enabling the calculation of the insertion loss for a given cable length.
- *Return loss:* The return loss reflects how well the cable impedance matches the characteristic impedance of 50Ω. Variations in impedance over frequency can lead to reflections and affect the return loss. Most cables exhibit a return loss of 30 dB or higher over their frequency range of operation. Figure 5.15(b) provides an example of a cable's return loss.
- *Power-handling capability:* Power-handling capability or power capacity plays a vital role in high-power applications, such as in high-power transmitters for LTE, satellite communications, or radars. It does not play a role in qubit experiments, as relatively low powers are involved.

5.3.3 Types of Transmission Lines

In order to choose the appropriate transmission line for a given application, factors such as desired bandwidth, physical size, ease of fabrication and integration, and power-handling capability must be considered. This section explores the most commonly used types of transmission lines in superconducting qubit systems.

5.3.3.1 Coaxial Cable

Unshielded cables, such as twisted pairs, are disadvantaged because they begin to radiate as the frequency increases. Conversely, the coaxial structure does not suffer from radiation loss and offers self-shielding properties that prevent the reception of outside interference. The structure of a coaxial cable is illustrated in Figure 5.16(a). Coaxial cables find wide application when minimizing radiation losses and maximizing noise and interference immunity are necessary. This is often the case in test and measurement setups and when connecting various microwave systems. Chapter 7 explores various coaxial cables used for qubit experiments at room and cryogenic temperatures.

The dimensional parameters of the coaxial cable's cross-section [see Figure 5.16(b)], the dielectric material used, and the properties of the conductor determine the transmission line parameters such as resistance (R), inductance (L), conductance (G), capacitance (C), and the bandwidth of the cable [2, 3].

Coaxial cables commonly employ dielectric materials such as PTFE and Teflon, although air dielectric offers the least attenuation. While utilizing air dielectric and

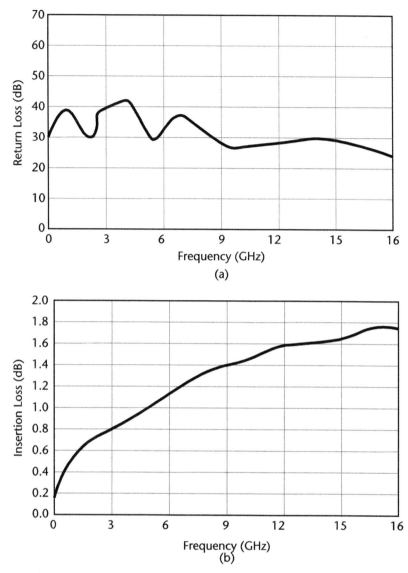

Figure 5.15 (a) Insertion loss and (b) return loss of a microwave cable.

larger physical dimensions in coaxial cables provides the lowest attenuation, it is crucial to consider mode limitations, as discussed in Section 5.3.2.1. As a result, a limitation exists on achieving the lowest possible attenuation due to mode-related concerns.

The characteristic impedance of a coaxial cable is determined as follows $Z_0 = V_o/I_o = \eta_0/2\pi\sqrt{\varepsilon_r}\ln(b/a) = 60/\sqrt{\varepsilon_r}\ln(b/a)$, where a and b represent the inner and outer conduction radii, ε_r is the relative permittivity of the dielectric, and $\eta_0 = 120\pi$ is the free-space characteristic impedance.

5.3.3.2 Microstrip Line

A microstrip line is a planar transmission line consisting of a ground plane, a dielectric layer, and a signal line positioned on top of the dielectric layer, as depicted

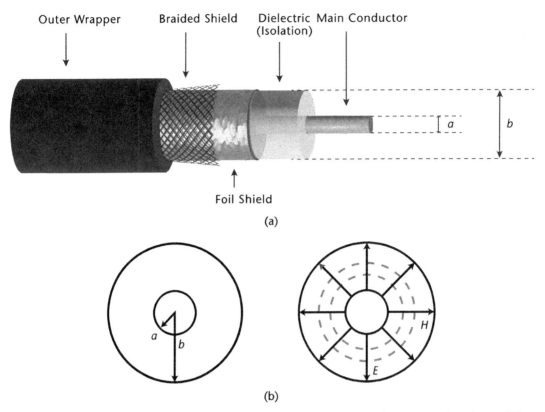

Figure 5.16 (a) Structure of a coaxial cable and (b) field configuration in the cross-section of a coaxial cable.

in Figure 5.17(a). The popularity of microstrip lines stems from their ease of fabrication, low cost, and high integration capabilities, making them suitable for implementing integrated microwave structures incorporating both active and passive components. However, it is essential to note that microstrip lines exhibit higher dielectric and radiation losses compared to hollow waveguide structures and coaxial cables. As the operating frequency increases, the radiation losses in microstrip lines become more pronounced, necessitating the use of stripline and waveguides to minimize such losses.

Microstrip lines have a portion of their field lines in the air and the majority within the dielectric region between the strip conductor and the ground plane, as depicted in Figure 5.17(a). Due to inhomogeneous configuration, a pure TEM wave cannot be supported by a microstrip line because the phase velocity of TEM fields in the dielectric region ($c/\sqrt{\epsilon_r}$) would differ from the phase velocity of TEM fields in the air region (c), making phase matching at the dielectric-air interface impossible. In reality, fields in a microstrip line represent a hybrid TM-TE wave, requiring advanced analysis techniques beyond the scope of our current discussion. However, in most practical applications, the dielectric substrate thickness (d) is electrically thin ($d \ll \lambda$), resulting in quasi-TEM fields. Essentially, the fields resemble those of the static (DC) case. As a result, reasonable approximations for the phase velocity, propagation constant, and characteristic impedance can be obtained using static

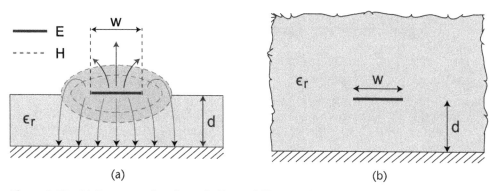

Figure 5.17 (a) Geometry of a microstrip line and (b) equivalent geometry with a homogeneous medium of effective relative permittivity ϵ_e.

or quasi-static solutions. The effective dielectric constant can be interpreted as a homogeneous medium's dielectric constant that replaces the microstrip line's air and dielectric regions, as shown in Figure 5.17(b). Since some field lines exist in the dielectric region while others reside in the air, the effective dielectric constant satisfies the relation $1 < \epsilon_e < \epsilon_r$ and depends on parameters such as the substrate dielectric constant, substrate thickness, conductor width, and frequency.

In superconducting qubit systems, microstrip lines are primarily employed to design external passive and active components, including filters, amplifiers, couplers, and attenuators.

5.3.3.3 CPW

A CPW is a planar transmission line that comprises a dielectric layer with a signal trace on top and two ground planes on either side of the signal trace. Another variant of CPW, known as a grounded coplanar waveguide (GCPW), features a ground plane beneath the structure, as shown in Figure 5.18(a). CPW lines support a low-loss quasi-TEM mode like microstrip lines since pure TEM waves cannot propagate due to structural inhomogeneity.

The CPW finds extensive application as a transmission line in superconducting quantum circuits, playing a crucial role in facilitating the interaction between a qubit and a cavity mode. Unlike a microstrip line, where the electric fields traverse the substrate between the signal trace and the ground, making it impractical to position the qubit inside the substrate, the CPW structure allows direct coupling of the electric field from the central signal line to the adjacent ground plane. This feature enables convenient placement of the qubit on top of the substrate, as illustrated in Figure 5.18(b). Moreover, CPW's smaller dimensions for a 50-Ω line than a microstrip line make it an ideal choice for compact and high-density qubit structures. Additionally, the CPW's uniplanar geometry offers the easy connection of series and shunt elements to the lines without the need for via holes, which can introduce additional inductance to the line and result in an impedance mismatch.

In the case of symmetric CPWs, two modes can propagate along the transmission line: the even mode or CPW mode and the odd mode, also referred to as the coupled slot-line mode, as shown in Figure 5.18(c, d). These modes can travel independently when excited, but discontinuities can cause them to couple with each other.

Parasitic slot-line modes can be excited in CPW lines. They can significantly impact circuit performance, particularly in the presence of asymmetry or discontinuities in the transmission lines. Asymmetrical CPW discontinuities, such as bends in microwave resonators and T-junctions, can excite the coupled slot-line mode. To suppress the slot-line mode and ensure that the ground planes on the

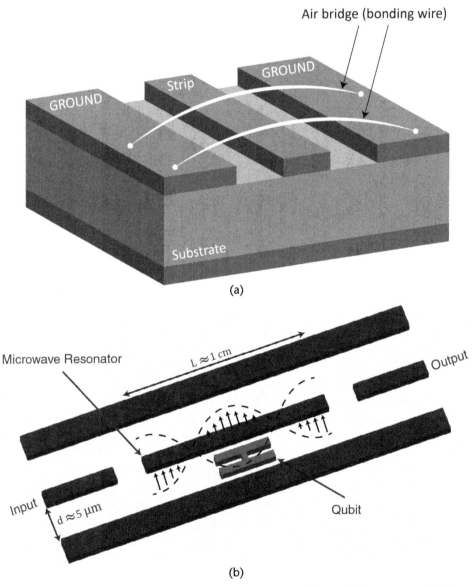

Figure 5.18 (a) CPW line with ground (GCPW), (b) a qubit coupled to a CPW resonator, (c) CPW mode, (d) coupled slot-line mode, and (e) the use of an air bridge to connect the grounds on the sides of the signal line to suppress slot line modes.

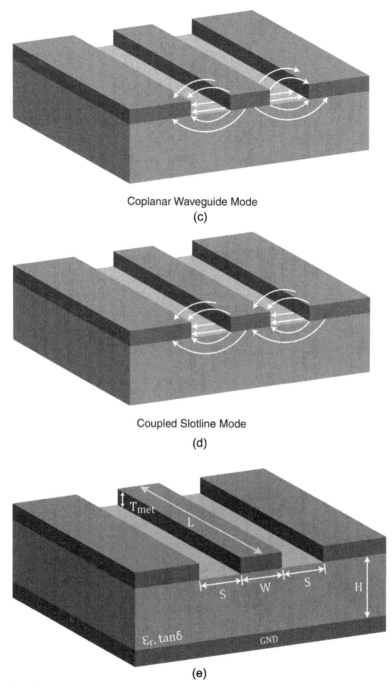

Figure 5.18 *(Continued)*

sides of a CPW line are at the same potential, it is crucial to connect them using an air bridge, as shown in Figure 5.18(e). A bonding machine is used to make an air bridge in quantum circuits.

Calculating the characteristic impedance and effective permittivity for CPW lines involves elliptic integrals and hyperbolic functions.

5.3.3.4 Hollow Waveguides

As Chapter 3 mentioned, waveguide cavities are utilized in superconducting quantum computers. Hollow waveguides offer low loss and high power-handling capability. However, they are bulky and expensive, especially at lower frequencies. These transmission lines are commonly employed in radar and satellite communication systems where low losses and high power are crucial for effective operation. Figure 5.19 shows methods to excite a rectangular waveguide.

Waveguides only support TE and TM modes. Each waveguide mode has a cutoff frequency $f_{c_{mn}}$ given by the following for a rectangular waveguide [3]

$$f_{c_{mn}} = \frac{1}{2\pi\sqrt{\mu\epsilon}} \sqrt{\left(\frac{m\pi}{a}\right)^2 + \left(\frac{n\pi}{b}\right)^2} \quad (5.22)$$

The parameters "m" and "n" represent the number of half wavelengths in the x (length) and y (width) directions, respectively. The WR standard, established by the American National Standards Institute (ANSI) and the Institute of Electrical and Electronics Engineers (IEEE), is used to designate waveguide sizes and performance characteristics. This system uses the letters "WR" followed by a number, which indicates the approximate inside width of the waveguide in hundredths of an inch. For example, a WR-90 waveguide has an inside width of approximately 0.9 inches. Table 5.8 provides a list of commonly used waveguides along with their specifications.

In a rectangular waveguide, the group velocity can be defined as the velocity at which energy propagates through the waveguide. It represents the speed at which the envelope of the wave propagates. The group velocity can be calculated using the following formula [3]

$$v_g = \frac{1}{\sqrt{\mu\epsilon}} \sqrt{1 - \left(f_{c_{mn}}/f\right)^2} \quad (5.23)$$

It is important to note that the group velocity is always less than the speed of light $1/\sqrt{\mu\epsilon}$ for a propagating mode. Furthermore, the group velocity depends on

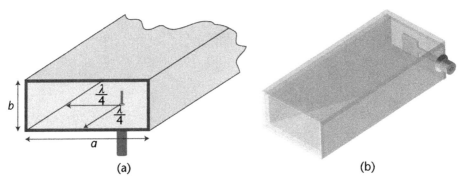

Figure 5.19 Excitation of a waveguide by using (a) a probe and (b) a patch antenna.

the frequency f and increases as the frequency increases for any given mode. So, the frequency dependency of propagation velocity causes dispersion, as mentioned in Chapter 4. Additionally, it is worth mentioning that each mode has a different cutoff frequency $f_{c_{mn}}$. Higher-order modes propagate more slowly than lower-order modes with the same frequency.

To illustrate this, let us calculate the propagation speed for a narrowband signal with a frequency of 10 GHz in a rectangular waveguide with cutoff frequencies of f_{10} = 6.55 GHz and f_{20} = 13.11 GHz. These cutoff frequencies only allow the TE_{10} mode for this signal. We can use (5.23) to determine the speed of propagation, which is approximately 2.26×10^8 m/s. This value corresponds to about 75.5% of the speed of light in free space.

We will later see how waveguide cavities are used as a resonator for qubit readout.

Section 5.3.2 provided a distinction between TEM, TM, and TE waves, emphasizing that transmission lines can be categorized based on the types of waves they support. TEM waves are nondispersive and have no cutoff frequency, while TM and TE waves exhibit dispersion and possess nonzero cutoff frequencies. Several factors must be considered when choosing a transmission line: bandwidth, loss, power-handling capacity, physical size (volume and weight), ease of fabrication (cost), and the potential for integration with other active and passive devices.

5.3.4 Microwave Connectors

Connectors play a vital role as the interface between different transmission lines, enabling their connection and ensuring proper signal transmission. Connectors come in various forms, grades (such as metrology, instrument, and production), frequency ranges, power-handling capabilities, and impedances (such as 50-Ω and 75-Ω), among other characteristics. A comprehensive understanding of the construction and operation of microwave connectors requires extensive study; here, we highlight some practical considerations regarding connectors commonly employed in superconducting quantum hardware.

When choosing connectors, it is crucial to ensure that they meet the specific requirements of the application. Table 5.9 provides an overview of various connector types used in superconducting qubit systems, along with their corresponding frequency ranges and typical applications. It is worth noting that other connector types, such as TNC, MCX, SMB, SMC, and U.FL, find more frequent usage in other types of microwave systems and are not as commonly employed in superconducting qubit systems.

Table 5.8 Dimensions and Cutoff Frequencies of Rectangular Waveguides

Waveguide Name	Recommended Frequency	Cutoff Frequency Lowest Order Mode	Cutoff Frequency Next Mode	Inner Dimensions of the Waveguide Opening	
				A inch [mm]	B inch [mm]
WR2300	0.32–0.45 GHz	0.257 GHz	0.513 GHz	23 [584.2]	11.5 [292.1]
WR137	5.85–8.20 GHz	4.301 GHz	8.603 GHz	1.372 [34.8488]	0.622 [15.7988]
WR90	8.20–12.40 GHz	6.557 GHz	13.114 GHz	0.9 [22.86]	0.4 [10.16]

For qubit applications, SMP and MMPX connectors are widely used for microwave connections on the sample holder. These snap-on connectors allow quick and easy connections and disconnections without threaded mating. This feature makes them suitable for applications requiring frequent cable changes or situations with limited space. They are particularly useful for sample holders in quantum chips due to their space-saving design.

In certain situations, there may be a need to connect connectors of different types, the same type, or even connectors with incompatible genders. In such cases, adapters come to the rescue. Figure 5.20 showcases a variety of microwave adapters, including N-to-SMA, BNC-to-SMA, SMA female–to–SMA female (commonly known as a bullet adapter), and SMA male–to–SMA male (referred to as a barrel adapter). Adapters facilitate connections between different connectors and are also frequently employed to protect metrology-grade connectors, such as 3.5-mm or 2.92-mm, from potential damage. For instance, an N-to-SMA adapter is often utilized at the ports of a network or spectrum analyzer to safeguard the delicate input connectors of these instruments.

Recall that the maximum bandwidth of a cable is determined by its cross-sectional size. As the size decreases, the operating frequency increases. This is why specialized connectors, such as 3.5-mm, 2.92-mm, 2.4-mm, and 1.8-mm, are employed for high-frequency applications, with the 1.8-mm connector capable of operating up to 110 GHz. Metrology-grade connectors, like the 3.5-mm connector used for network or spectrum analyzer ports, offer superior mechanical tolerances, stability, temperature performance, SWR, repeatability, loss, and wear characteristics.

Table 5.9 Connectors with Broad Applications in Superconducting Qubit Systems

Connector Type	Frequency Range	Application in Qubit Experiments
BNC	DC-4 GHz	Room-temperature, low-frequency connections, such as the 10-MHz clock and DC connections, are made to the breakout box.
N-connector	DC-18 GHz	Widely used on measurement instruments, such as spectrum and network analyzers.
SMA	DC-18 GHz	SMA has the widest range of RF and microwave engineering applications, being suitable for almost all room-temperature and cryogenic microwave connections.
SMP	DC-40 GHz	Snap-on connectors allow for quick and easy connections and disconnections without threaded mating. This feature makes the SMP connector suitable for applications requiring frequent cable changes or for limited space. They are particularly useful for sample holders in quantum chips due to their space-saving design.
MMPX	DC-65 GHz	Snap-on connectors allow for quick and easy connections and disconnections without threaded mating. This feature makes the MMPX connector suitable for applications requiring frequent cable changes or for limited space. They are particularly useful for sample holders in quantum chips due to their space-saving design.

5.4 Signal Processing

Always use a torque wrench when connecting or disconnecting them to ensure proper handling of connectors. This helps prevent overtightening, which can lead to connector damage. Additionally, it is essential to inspect connectors visually before use and clean them, when necessary, as part of regular connector care.

5.4 Signal Processing

Until now, our focus has been on generating and transmitting microwave signals. However, they often require some processing before these signals can be effectively transmitted or detected. This processing involves adjusting the amplitude, frequency, and phase of the microwave signal and should not be confused with digital signal processing or similar methods.

Microwave signal-processing techniques can be broadly classified into three categories: amplitude, phase, and frequency. These techniques play a crucial role in the context of superconducting qubits. Sections 5.4.1 to 5.4.3 provide detailed insights into the application and significance of these techniques within a microwave link.

5.4.1 Performance Specifications of Microwave Components

When selecting microwave components, it is crucial to consider their performance specifications to ensure optimal performance. Several key specifications apply to all microwave components, including return loss, insertion loss, operation bandwidth, and power-handling capability. For solid-state components like amplifiers, mixers, and switches, nonlinear parameters such as P1dB and IP3 are also essential considerations. Multiport components like power combiners and couplers require additional parameters such as amplitude and phase imbalance and isolation. These specifications help assess the performance and compatibility of microwave components for specific applications, as we will see in the following sections.

5.4.2 Amplitude Manipulation

Table 5.10 provides a summary of the amplitude-manipulation techniques for microwave signals. Sections 5.4.2.1 to 5.4.2.3 elaborate on these techniques.

Figure 5.20 (a) N-to-SMA adapter; (b) BNC-to-SMA adapter; (c) SMA male-to-SMA male adapter; (d) SMA female-to-SMA female adapter; (e) SMP connector; and (f) MMPX connector.

5.4.2.1 Amplifiers

Amplifiers are crucial in increasing the signal amplitude to surpass the MDS level. Despite introducing additional noise and reducing the overall SNR, as discussed in Chapter 4, amplifiers are necessary to elevate the amplitude above the MDS threshold for successful detection. In specific applications like heating or microwave plasma generation, amplifiers generate specific microwave power levels.

Amplifiers can be categorized based on their usage within the microwave link. Typically, a LNA is employed on the detection side. LNAs feature a low noise figure, moderate gain, and relatively low- to-medium output power. Various types of room-temperature and cryogenic LNAs are utilized in qubit experiments, as explored in Chapter 7.

On the transmit side, a power amplifier with high gain and output power ensures that the signal level from the transmit antenna remains sufficiently high upon reaching the receiver side. However, power amplifiers are not employed in qubit experiments where low power levels are utilized.

Amplifiers differ from passive devices because they consume power and actively deliver microwave energy into the circuit. This is achieved by converting DC bias power into microwave power using an active component, such as a transistor. The key parameter of an amplifier is its gain G, which is defined as the ratio of the output power P_{out} to the input power P_{in}:

$$G = \frac{P_{out}}{P_{in}} \tag{5.24}$$

The gain of an amplifier typically ranges from 15 to 50 dB, depending on its type. Narrowband amplifiers generally exhibit higher gain compared to wideband amplifiers. As the bandwidth of the amplifier increases, its gain tends to decrease. In other words, the gain-bandwidth product remains constant.

Table 5.10 Amplitude-Manipulation Techniques

Operation	Description	Relevant Microwave Components
Amplification	• Increase the signal level to reach the minimum detectable signal level	LNA Power amplifier
Attenuation	• *Tuning:* adjusting the signal level • *Protection:* protecting a device from high power levels	Attenuator RF limiter
Power combining/dividing	• Combine powers • Divide powers	Wilkinson power divider
Power coupling	• Sampling a microwave signal	Directional coupler
Power circulation/isolation	• *Circulation:* Use a single antenna for full duplex (simultaneous transmit and receive) • *Isolation:* allowing the signal to flow in one direction and block it in the other direction	Circulator/isolator
Switching	• Switch between different microwave paths	Microwave switches

The essential performance indicators for amplifiers include gain, noise figure, IP3, and P1dB (1-dB compression point).

A low return loss within the frequency band of operation is crucial for the amplifier to ensure minimal signal reflections. Figure 5.21 illustrates the gain, noise, and S-parameters of a commercial cryogenic LNA operating in the 4–8 GHz frequency range.

Section 4.3.3 discusses the nonlinear behavior exhibited by semiconductor devices like transistors, which makes parameters such as the 1-dB compression point relevant for amplifiers. In Chapter 4, it is observed that the JPA could have an incredibly low 1-dB compression point of about −100 dBm, in stark contrast to a room-temperature LNA with a 1-dB compression point of approximately −30 dBm. Consequently, it is crucial to operate the JPA below its 1-dB compression point to ensure optimal performance.

A chain of amplifiers is employed when a single amplifier cannot provide the desired gain level. As detailed in Chapter 4, according to Friis' formula, the first amplifier's noise figure significantly influences the entire chain's noise figure. Hence, it is essential to have the first amplifier in the chain possess the lowest possible noise figure and high gain. As explained in Chapter 3, the separation error ϵ_{sep} between the two states of the qubit is dependent on the SNR given by $\epsilon_{sep} = \text{erfc}(SNR/2)/2$. Additionally, the qubit readout fidelity F is also determined by the SNR, where $F = (1 + \text{erf}(\sqrt{SNR/2}))/2$ [9]. For example, with an SNR of 10, the gate fidelity is $F = 99.9\%$. Therefore, achieving a high SNR for the detection chain is crucial by reducing the equivalent noise figure of the chain.

In qubit experiments, a chain of amplifiers is utilized at the output of the readout resonator, extending all the way to the room-temperature detection chain, as shown in Chapter 7. The initial amplifier, situated right at the output of the readout resonator, is a superconducting low-inductance undulatory galvanometer (SLUG) or a JPA, which, as discussed in the following section, exhibits lower noise compared to the subsequent best cryogenic amplifier, namely the HEMT amplifier (see Table 5.11). Table 5.11 provides a comparison of various types of LNAs.

In addition to the noise figure and noise temperature, it is common to quantify the number of thermal photons, n_{th}, added by the amplifier with the noise temperature of T_a at each frequency ω. This is expressed as follows

$$n_{th} = \frac{1}{e^{\hbar\omega/k_B T_a} - 1} \tag{5.25}$$

Quantum-Limited Amplification

As Chapter 4 mentioned, nonlinear components generate frequency components in their output that are absent in their input. A single-tone input leads to harmonics, and for two tones, it results in intermodulation products. Section 5.4.3.2 explores how mixers, being nonlinear components, are utilized for up- and downconversion, capitalizing on these nonlinear effects.

Three-wave mixing is a nonlinear process that occurs in nonlinear materials or components, involving three waves: the pump wave with a frequency ω_p, the signal wave with a frequency ω_s, and the idler wave with a frequency ω_i, as shown

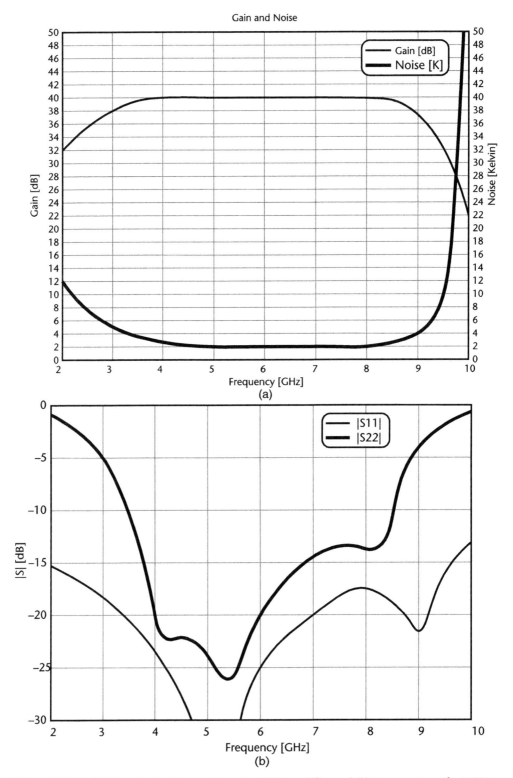

Figure 5.21 (a) Gain and noise temperature of a HEMT amplifier and (b) S-parameters of a HEMT amplifier.

5.4 Signal Processing

Table 5.11 Comparison of Various Amplifiers Used in a Superconducting Qubit System

Technique	Description	Noise Temperature
JPA	Quantum-limited noise performance.	$T_{noise} < 1K$
Traveling-wave amplifier (TWA)	Higher bandwidth and dynamic range for multiplexed readout of multiple qubits.	$T_{noise} < 1K$
SLUG amplifier	Quantum-limited noise performance. Combined amplification and reverse isolation eliminate the need for external circulators.	$T_{noise} < 1K$
Cryogenic HEMT amplifier	Used at the 4-K stage.	$T_{noise} = 2-5K$
Room-temperature LNA	Used in room-temperature detection chain.	$T_{noise} = 75-100K$

in Figure 5.22. The pump wave is typically the strongest among the three, while the signal wave requires amplification or modulation. The idler wave is generated during the process but is usually disregarded and filtered out. Three-wave mixing finds applications in various areas, including signal amplification, generation of new frequencies, and frequency conversion.

As shown in Figure 5.22, the relationship between the pump frequency, signal frequency, and idler frequency in a parametric amplifier is described by the Manley-Rowe relations, expressed as:

$$\omega_p = \omega_s + \omega_i = 2\omega_0$$

Here, ω_0 represents the resonant frequency of the nonlinear component. This equation elucidates how the energy of the pump wave is transferred to the signal and idler waves in the amplifier while conserving the total energy.

In an ideal scenario for a linear, phase-insensitive amplifier, an input state $\langle a_{in} \rangle$ is amplified to an output state $\langle a_{out} \rangle$ with an amplitude gain factor $\sqrt{G}$, where $a_{out} = \sqrt{G} a_{in}$. As explained in Chapter 3, electromagnetic fields can be quantized and thus follow commutation relations $[a_{in}, a^\dagger_{in}] = [a_{out}, a^\dagger_{out}] = 1$. To satisfy these

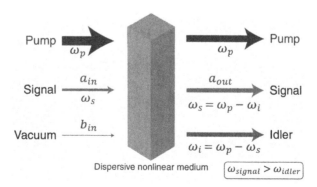

Figure 5.22 Parametric mixing; the signal and pump photons interact via a purely dispersive nonlinear medium.

commutation relations, the gain equation mentioned above needs to be modified to account for the added noise caused by vacuum fluctuations [12]

$$a_{out} = \underbrace{\sqrt{G}a_{in}}_{\text{Amplification}} + \underbrace{\sqrt{G-1}b_{in}^{\dagger}}_{\text{Added idler noise}} \tag{5.26}$$

In this scenario, the "idler" mode b_{in} can be regarded as the vacuum fluctuations of another mode. To ensure that the commutation relation holds, the idler mode is amplified by the gain factor $\sqrt{G-1}$. It has been demonstrated that when dealing with high gain, at least half a noise photon $\hbar\omega/2$ must be added to a signal that is amplified with a gain of $\sqrt{G}$. Now, let us explore how squeezing can decrease the noise level.

Squeezed states are Gaussian states in which the uncertainty of one observable property of a system is reduced but at the cost of increasing the uncertainty in its conjugate observable. This principle can be utilized to decrease the noise in one quadrature of the electromagnetic field while simultaneously increasing the noise in the conjugate quadrature. Therefore, it does not violate Heisenberg's uncertainty relation for the two field quadratures.

By reducing the uncertainty in one quadrature, such as $\Delta\hat{x}$, fewer measurements are needed to achieve the same level of precision when estimating $\langle\hat{x}\rangle$. Parametric amplifiers leverage this concept to achieve noise levels below the standard quantum limit for one field quadrature.

Single-mode squeezing is a unitary transformation parameterized by the squeezing strength r and angle θ. The operator $\hat{p}_\theta$ and its variance $\Delta\hat{p}_\theta$ are both amplified with a gain factor $\sqrt{G} = e^{+r}$, while the conjugate operator $\hat{x}_\theta(r)$ and its variance $\Delta\hat{x}_\theta(r)$ are both compressed by the inverse factor $e^{-r} = 1/\sqrt{G}$, as shown in Figure 5.23(a). This means that measuring $\hat{x}_\theta(0)$ on the squeezed vacuum (a vacuum state that has undergone the squeezing unitary transformation) results in an uncertainty below the standard quantum limit $\Delta\hat{x}_\theta = e^{-r}/\sqrt{2}$.

Squeezing strength can be expressed in decibels as $r_{dB} = 10 \times \log10 (\Delta O_{new}/\Delta O_{old})$, where O is an operator [13]. So, a squeezing strength of 3 dB corresponds to a 50% reduction in uncertainty. The Laser Interferometer Gravitational-Wave Observatory (LIGO) uses a squeezing strength of around 3 dB, while parametric amplifiers can achieve much larger values of around 10 dB.

Adding phase sensitivity to the amplifier can result in squeezing. For this purpose, the idler mode must oscillate at the same frequency as the signal or a multiple thereof, with the capability of tuning its phase $\phi \in [0, 2\pi]$. By substituting the idler mode in (5.26) with $b_{in} = e^{i\phi}a_{in}$, we obtain

$$a_{out} = \underbrace{\sqrt{G}a_{in}}_{\text{Amplification}} + \underbrace{e^{-i\phi}\sqrt{G-1}a_{in}^{\dagger}}_{\text{Phase-dep. noise}} \tag{5.27}$$

These two terms can exhibit phase-dependent constructive or destructive interference, which allows for tuning the amplification orientation, or in other words, the quadrature in which the noise reduction is desired by adjusting the

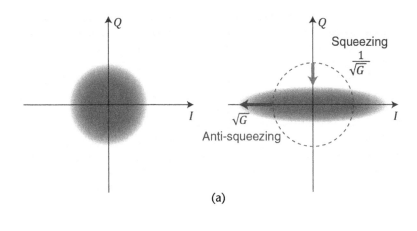

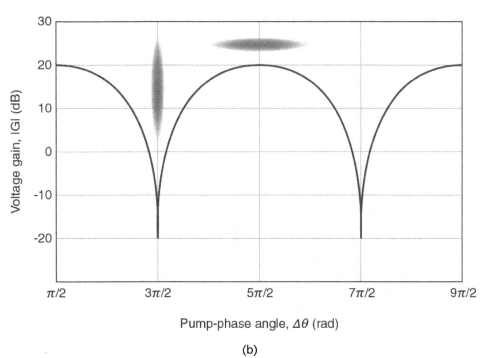

Figure 5.23 (a) Phase-sensitive parametric amplification makes it possible to suppress the noise along one axis resulting in the added noise to the other quadrature. (b) Voltage gain as a function of pump-phase angle. The orientation of amplification can be tuned depending on the phase of the pump.

pump phase, as shown in Figure 5.23(b). As a result of this interference, the noise can be suppressed to levels below the standard quantum limit.

Parametric Amplifiers

When utilizing a low-noise HEMT amplifier with a noise temperature of approximately 2–5K immediately following the qubit's readout resonator, the resulting system noise temperature rises to around 7–10K. This indicates an introduction of approximately 10–20 photons of noise per signal photon at approximately 5 GHz. In practical terms, this noise level is generally deemed excessive for achieving a successful single-shot readout. As a result, there has been considerable interest in developing quantum-limited parametric amplifiers specifically tailored for superconducting qubit readout [9–13].

The JPA can be employed as an alternative to the HEMT amplifier to enhance the SNR. The JPA is well-suited for reading out a small number of qubits due to its narrowband nature, providing a bandwidth of a few tens of megahertz. The traveling-wave parametric amplifier (TWPA) with a broader bandwidth and higher dynamic range is utilized for multiplexed readout of multiple qubits.

Let us provide a quick overview of parametric amplification and explore its application in the context of superconducting circuits. Consider a mechanical pendulum where the oscillation frequency, for small amplitudes, can be approximated by $\omega = \sqrt{g/l}$, with g representing gravity and l denoting the length of the string. By modulating the frequency ω through variations in either l or g, we can increase the amplitude of the oscillations. The g parameter can be modulated by moving the pendulum along a line, as depicted in Figure 5.24(a). This movement pumps energy into the system and results in an amplified oscillation. Note that two frequencies are involved in this process: the natural frequency of the oscillator and the pump frequency, which corresponds to the frequency at which the parameter g is modulated. Typically, the pump frequency is twice the natural frequency.

The behavior of a harmonic oscillator with sinusoidal excitation can be described using a well-known second-order differential equation $\ddot{x} + b\omega_0\dot{x} + \omega_0^2 x = A\sin(\omega_0 t)$. The solution to this equation yields the amplitude of oscillation for the harmonic oscillator $x_{o1} \propto (A/\omega_0^2 b)\sin(\omega_0 t + \phi)$. Now, if the natural frequency ω_0 is modulated with a pump frequency ω_p twice the natural frequency $\omega_p = 2\omega_0$, then the oscillation amplitude is given by $x_{o2} \propto (A/\omega_0^2(b - \delta/2))\sin(\omega_0 t + \phi)$.

Upon comparing the denominators of x_{o1} and x_{o2}, it becomes evident that modulating the natural frequency of the oscillator leads to an increase in the amplitude of vibration. This concept can be applied to parametric amplification in an LC resonator by modulating the inductance value, similar to how a pendulum's "g" parameter is modulated.

In the case of an LC resonator implemented with a SQUID, which consists of two Josephson junctions in parallel, the inductance can be adjusted by applying an external flux that passes through the SQUID's loop, as depicted in Figure 5.24(b). A nearby coil is utilized to couple the magnetic flux Φ and modulate the inductance $L(\Phi)$ at the pump frequency $\omega_p = 2\omega_0$, thereby altering the natural resonant frequency $\omega_0(\Phi)$.

The input signal is introduced into the circuit through a circulator (see Section 5.4.2.3), amplified, and then directed to the output. Figure 5.24(c) visually illustrates the variation in the resonance frequency of the LC resonator as the flux changes. The pump power typically ranges around −100 dBm, and the current flowing through the coil is usually a few milliamps.

Figure 5.24(d) illustrates that increasing the pump power leads to higher gain but simultaneously reduces the bandwidth of the JPA. Figure 5.24(e) depicts the gain versus input power, with the input 1-dB compression point occurring at

5.4 Signal Processing

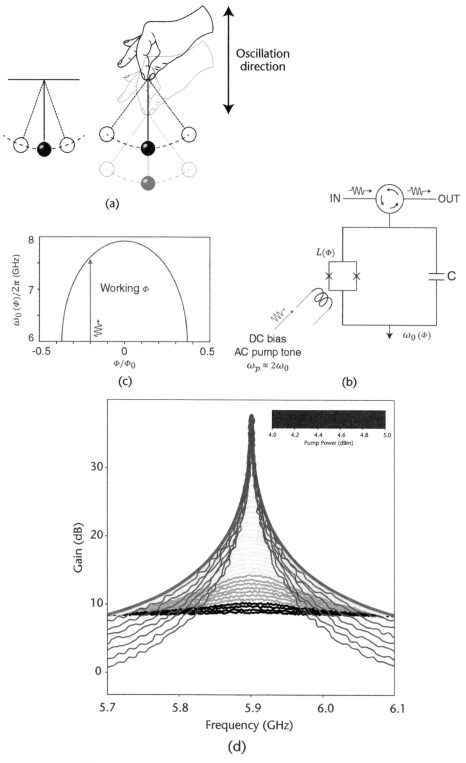

Figure 5.24 (a) A fixed pendulum. A pendulum with modulated g can create parametric amplification and increase the amplitude of oscillations. (b) Schematic of a JPA, (c) the resonance frequency of the LC circuit changing as a function of applied flux, (d) gain variation for various pump powers, (e) 1-dB compression point of a JPA, and (f) TWPA.

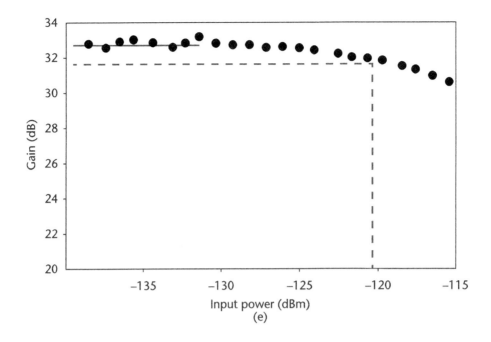

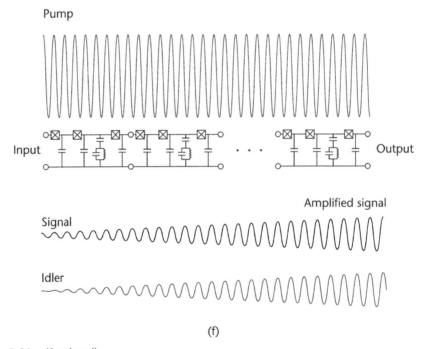

Figure 5.24 *(Continued)*

approximately −120 dBm. Therefore, it is crucial to exercise caution and prevent the JPA from saturating, as saturation can occur rapidly.

While a narrow bandwidth is acceptable for single-qubit readout, applications involving multiplexed multiqubits require a wider bandwidth. To meet this requirement, a TWPA, consisting of cascaded stages of unit cells consisting of

Josephson junctions, inductors, and capacitors, is employed, as shown in Figure 5.24(f). The TWPA can have a bandwidth as wide as a few gigahertz [10–12]. The idler signal depicted in Figure 5.24(f) is an outcome of the parametric amplification process. It represents a third signal generated at a frequency equal to the sum of the pump and signal frequencies but with an opposite phase. The idler signal is not utilized in amplification and is typically filtered out to prevent interference with the desired output signal.

SLUG Amplifier

The following section discusses using circulators to isolate the quantum circuit from downstream measurement stages. Circulators pose challenges in terms of size and scalability. An alternative solution that offers significant advantages is the SLUG amplifier, which combines near quantum-limited noise performance with higher reverse isolation than commercial cryogenic isolators [14, 15].

Figure 5.25(a) illustrates the schematic of a SLUG amplifier, a modified version of the dc SQUID (dc SQUID). Figure 5.25(b) depicts the SLUG flux-to-voltage transfer curve. In typical operation, the input signal is coupled as a current to the node located at the lower left, and the output is obtained from the node at the upper right. The wire-up and operation of the SLUG amplifier are particularly straightforward, requiring only two DC current biases and no microwave pump tones. The pulsed-mode operation of the SLUG amplifier enables the characterization of transmon qubits without the need for cryogenic circulators, and it does not result in any measurable degradation of qubit performance in the measurement chain.

The SLUG gain element demonstrates remarkable robustness and immunity to fluctuations in ambient magnetic fields due to its extremely small magnetic sensing area [14, 15]. Additionally, the SLUG amplifier provides a large instantaneous bandwidth of around 1 GHz and a large dynamic range with saturation powers approaching −90 dBm, enabling multiplexed single-shot qubit readout.

Let us compare the parametric and SLUG amplifiers' noise performance and reverse gain. The Josephson TWPA and the kinetic-inductance traveling-wave (KIT) amplifier exhibit directionality. However, in the ideal case, the reverse gain of these devices is 0 dB, meaning that signals coupled to the output port propagate without attenuation back to the input, which does not provide any reverse isolation. Therefore, it is necessary to use a circulator at the output of the parametric amplifier.

While the noise level of a SLUG amplifier will never match that of an optimized JPA, it is possible to achieve added noise of approximately one quantum at frequencies approaching 10 GHz [14, 15].

Another advantage of the SLUG is that, unlike the JPA, the SLUG amplifier does not require a separate strong microwave pump tone, which means that the qubit does not need to be protected by multiple stages of cryogenic isolation.

5.4.2.2 Attenuators

Sometimes, intentionally reducing the amplitude of a signal becomes necessary for four primary reasons, described in the following. Figure 5.26(a, b) shows fixed and tunable attenuators.

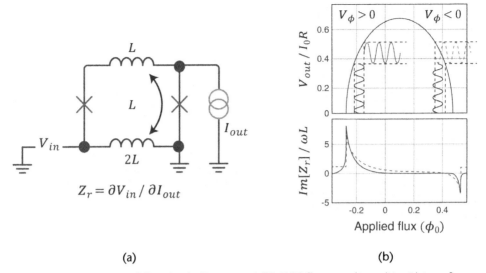

Figure 5.25 (a) SLUG amplifier circuit diagram and (b) SLUG flux-to-voltage ($V - \Phi$) transfer curve. R and I_0 are the resistance and critical current per junction. (*After:* [14].)

- First, it may be necessary for protection purposes, mainly to safeguard sensitive devices. For example, attenuators are often employed at the input of a measurement instrument to prevent damage to the instrument's front end resulting from excessive output power from the DUT.
- Second, attenuation may be used to match circuits with unequal impedances. Figure 5.26(c) demonstrates the insertion of a matched attenuator before a mismatched load. The return loss "seen" at the attenuator's input is improved by an amount equal to twice the attenuation as the signal passes through the attenuator twice. Hence, a 3-dB attenuator enhances the return loss by 6 dB. The notable drawback of this technique is the loss incurred by the attenuator, which comes at the expense of improving the return loss.
- Third, attenuation may be needed for noise suppression in cryogenic environments. Surprisingly, attenuators are employed in qubit experiments to reduce the amount of thermal noise coupled to the dilution fridge. This can only be achieved by thermalizing the attenuators within a cryogenic environment. Chapter 7 further discusses this topic.
- Last, the signal level adjustment may be required for tuning purposes. Digital step attenuators (DSAs) are tunable attenuators utilized to adjust signal levels, such as in a microwave signal generator or within an automatic gain control loop, as depicted in Figure 5.26(d).

Chapter 7 explores how attenuators play a crucial role in a dilution fridge by suppressing room temperature noise. The distribution of attenuation among the various stages of the fridge depends on each stage's cooling power and temperature, as explained in Chapter 7. A 0-dB attenuator, utilized in specific microwave circuits, typically exhibits an insertion loss below 0.5 dB. These attenuators function similarly to wideband 0-Ω resistors and are commonly employed in cryogenic environments to thermalize the central connector of cables. They facilitate better

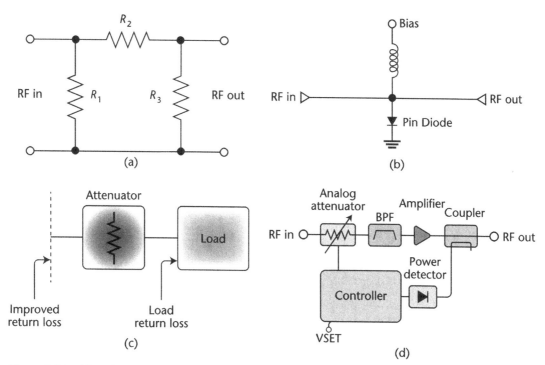

Figure 5.26 (a) A pi-attenuator, where the symmetrical configuration allows for easy matching of the input and output to 50Ω; (b) schematic of a variable attenuator; (c) use of an attenuator to improve matching; and (d) use of an attenuator to adjust the output signal in an automatic gain-control (AGC) feedback loop.

thermal conduction between the center and the body than cables with a dielectric. Furthermore, they can provide mechanical stability by attaching them to a plate.

Some critical operational parameters of solid-state attenuators include attenuation, P1dB, IP3, and maximum power.

5.4.2.3 Signal Addition/Subtraction

In microwave engineering, the separation and combination of microwave signals is frequently required. This section examines several techniques and components commonly employed.

Power Combiner

A power combiner is a component capable of combining or dividing microwave signals. The Wilkinson combiner shown in Figure 5.27(a) is the most well-known type. When employed as a divider, port 1 is the input, while ports 2 and 3 are the outputs, with the signal equally split between them. For signal combining, port 1 is utilized as the output. In scenarios with additional outputs, N-way power combiners can be designed. Moreover, power combiners can be tailored for unequal power division [3].

A quarter-wavelength line section is utilized in the Wilkinson combiner, as observed in Figure 5.27(a). This narrowband design is evident from the S-parameters

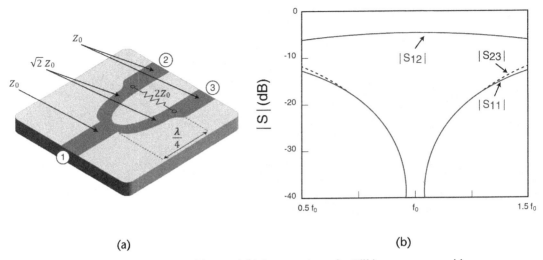

Figure 5.27 (a) Wilkinson power combiner and (b) S-parameters of a Wilkinson power combiner.

in Figure 5.27(b), where the maximum $|S_{12}|$ value of −3 dB (representing equal power division) is valid only within a limited bandwidth. However, the bandwidth of power combiners can be enhanced using a multisection power combiner. The even-odd mode analysis method is employed for analyzing power combiners [3].

Power combiners are not only utilized in test equipment but also find application in quantum hardware setups. For instance, they combine control and readout signals, enabling their transmission over a single coaxial line to the dilution fridge, as depicted in Chapter 7. The critical operational parameters of a power combiner include bandwidth, isolation, amplitude, and phase imbalance.

Directional Coupler

A directional coupler, depicted in Figure 5.28(a), is a four-port microwave component that facilitates the transmission of input power from port 1 to port 2 (known as the through port) with minimal insertion loss. Simultaneously, a portion of the input power, determined by the coupling factor, is coupled to port 3 (the coupled port), while ideally, none reaches port 4 (the isolated port). For a bidirectional coupler, input power at port 2 is coupled to both port 1 and port 4 but not to port 3, resulting in decoupling between ports 1 and 4 and ports 2 and 3.

Directional couplers are commonly employed to sample a fraction of the signal passing through them and redirect it to another circuit, such as a power detector, enabling the monitoring of the output power of an amplifier or a transmitter, as depicted in Figure 5.28(b). They also find applications in microwave tests and measurements, such as network analyzers, for measuring the reflected signal power from a DUT. In a superconducting quantum system, couplers are utilized to inject the pump signal into parametric amplifiers, as illustrated in Figure 5.28(c).

The insertion loss of the forward path L can be expressed as follows $L = 10 \log(P_1/P_2) = -20 \log|S_{12}|$ dB. The fraction of power coupled from port 1 to port 3 is given by C, the coupling coefficient $C = 10 \log(P_1/P_3) = -20 \log\beta$ dB. The leakage of power from port 1 to port 4 is given by I, the isolation, as defined $I = 10 \log(P_1/$

5.4 Signal Processing

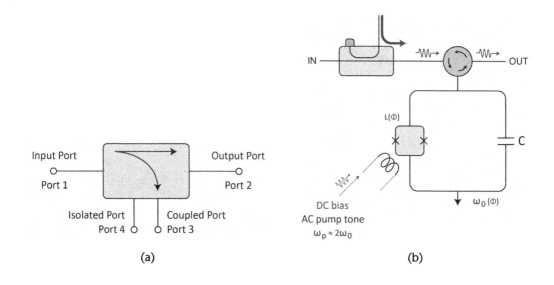

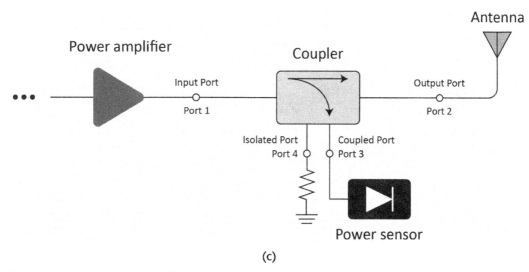

Figure 5.28 (a) Schematic of a directional coupler; (b) a directional coupler used to couple the pump signal to a JPA; and (c) application of a coupler for sampling the output power of a power amplifier.

P_4) = $-20 \log|S_{14}|$ dB. The coupler's ability to isolate forward and backward waves (or the coupled and uncoupled ports) is called directivity $D = 10 \log(P_3/P_4) = 20 \log(\beta/|S_{14}|)$ dB. These quantities are related as follows $I = D + C$. The coupler's important operational parameters include bandwidth, coupling, mainline loss, directivity, coupling flatness, and maximum input power.

Circulator/Isolator

Figure 5.29(a, b) shows that a microwave network is considered reciprocal if the response at one component's port remains unchanged when the ports of excitation and response are interchanged. In the case of a reciprocal component, the scattering matrix exhibits symmetry ($S_{ij} = S_{ji}$), indicating that the performance of the component remains unaffected by the direction of signal flow. Passive components like filters and attenuators fall under the category of reciprocal components, whereas amplifiers do not exhibit reciprocity. Nonreciprocal components, such as circulators, isolators, and gyrators, form an important class of components. These directional devices are constructed using ferrimagnetic materials, which allow us to manipulate their interaction with applied microwave signals by adjusting the strength of a bias field. This phenomenon can also be leveraged to control the behavior of other devices, including phase shifters, switches, and tunable resonators and filters.

Figure 5.29(c) displays the schematic of a circulator, which acts like a unidirectional microwave valve. A circulator is a three-port device designed to permit a signal to exit through the port immediately following the port it entered. When a signal is injected at port 1, it reaches port 2 with minimal attenuation or insertion loss. Conversely, if a signal is injected at port 2, it experiences significant attenuation as it travels to port 1. The signal flow is low-loss from port 2 to port 3 and from port 3 to port 1. This directional property is what gives the component its name: a circulator.

One can create an isolator by terminating the third port of the circulator with a 50-Ω load, as depicted in Figure 5.30(a). In qubit systems, a cryogenic isolator is typically employed at the output of a qubit to block noise and reflections generated by downstream measurement stages from reaching the qubit. The power absorbed by the termination resistor is converted into heat, which needs to be dissipated externally to prevent any increase in the refrigerator's temperature. Therefore, proper thermal anchoring of the isolator's body is crucial to ensure effective heat dissipation.

A single-junction isolator typically exhibits an insertion loss of less than 0.5 dB and an isolation of 20 dB, meaning that it attenuates signals flowing in the reverse direction by 20 dB. Increasing the number of junctions enhances isolation, resulting in greater signal attenuation in the reverse direction and improved noise-blocking performance.

When the output port of the first junction is cascaded to the input of the second junction, it forms a double-junction configuration, as illustrated in Figure 5.30(b). The double-junction cryogenic isolator typically possesses an isolation exceeding 40 dB and an insertion loss of less than 0.5 dB. A double junction's isolation and insertion loss are near twice those of a single-junction component.

Isolator arrays are employed in multiqubit systems to save space. In addition to isolation and insertion loss, the isolation between adjacent ports is also crucial. For instance, 50-dB isolation between adjacent ports means that if a signal is injected at one port, the signal appearing at the adjacent port will be 50 dB lower.

Circulators can be implemented using various technologies. Traditional ferrite circulators are passive devices, meaning that they provide no gain. They rely on ferrite material and a magnetic bias field to break time-reversal symmetry. However, one of the main challenges with this type of circulator is its size and weight. Figure 5.30(c) shows a microstrip circulator with a ferrite slab sandwiched between a microstrip Y-junction and a ground plane, with a permanent magnet supplying the bias field. Magnetic shielding plays a crucial role in the performance of ferrite

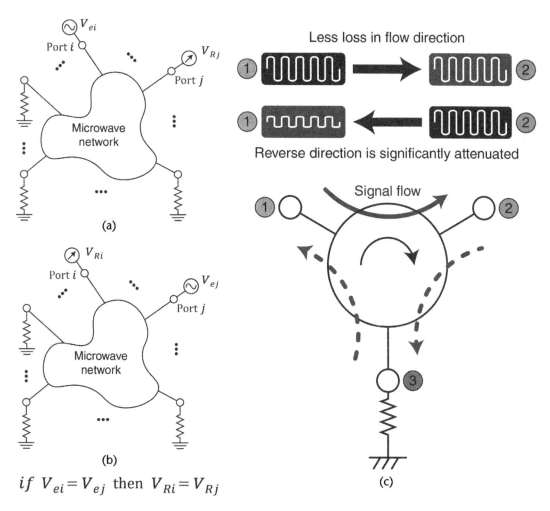

Figure 5.29 Reciprocity theorem: (a) exciting at port i by V_{ei} and measuring at port j (V_{Rj}); (b) swapping the excitation and response if $V_{ei} = V_{ej}$ then $V_{Ri} = .$; and (c) the operation of a circulator.

circulators to prevent any magnetic coupling to the qubit. There are three commonly used shields: the standard steel shield, the double shield, and the mu-metal shield.

Alternatively, nonferrite circulators based on transistors are active and provide gain but require power for biasing the active devices. However, these circulators have poor noise performance and lower power limitations than ferrite ones.

Finally, there are on-chip circulators, which have been the subject of significant research in recent years. Physicists have dedicated considerable effort to developing compact, scalable, magnetic-free, low-noise, and tunable on-chip circulators [16]. These advancements aim to address the scaling problem in quantum computing. These circulators are also frequency-tunable, but their available signal bandwidth is relatively limited, and their fabrication process is complex.

Last but not least, as seen in Section 5.4.2.1, a SLUG amplifier can function as both an amplifier and a circulator, enabling an integrated and scalable solution.

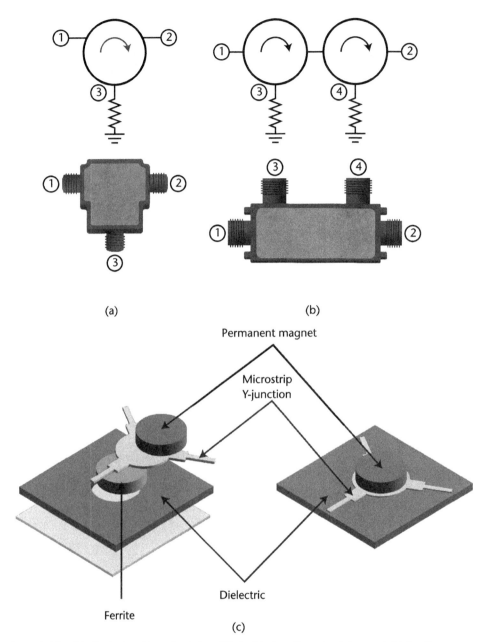

Figure 5.30 (a) Single junction circulator, (b) double junction circulator, and (c) Ferrite circulator.

Microwave Switch

Switches provide flexibility to microwave circuits, enabling signals to be routed without connecting and disconnecting cables. Chapter 7 discusses the advantages of switching in modifying measurement setup configurations to facilitate the reconfiguration of the superconducting qubit hardware setup.

Microwave switches come in various configurations, including single-pole, double-throw (SPDT); double-pole, double-throw (DPDT); and multiplexer (such as SP4T, SP6T, and SP8T), as illustrated in Figure 5.31(a, b). Multiplexer switches are particularly valuable inside a dilution fridge as they enable switching between different samples without requiring the fridge to be warmed up, connections to be changed, and subsequent cooling down for each sample.

Switches can be reflective or absorptive based on their behavior at the open ports, as shown in Figure 5.31(c). Reflective switches reflect the input signal to the source at the open port, leading to high VSWR and poor return loss performance. On the other hand, absorptive switches terminate the open port, resulting in low VSWR and excellent return-loss performance. Absorptive switches are preferred when minimizing reflections back to the RF source is essential, such as preventing standing waves on the transmission line in high-power applications. Generally, reflective switches exhibit higher isolation and lower insertion loss than absorptive switches. However, absorptive switches typically offer better linearity and lower distortion. In most cases, an absorptive switch can be used as a replacement for a reflective switch, but not vice versa.

Microwave switches are commonly controlled using a control signal, often a transistor-to-transistor logic (TTL) signal. In this context, the "high" state corresponds to a voltage between 1.5 and 5V, while the "low" state corresponds to a voltage between 0 and 0.7V.

Switches can also be categorized into electromechanical and solid-state (such as PIN diodes). Electromechanical switches can handle higher power and exhibit lower insertion loss than solid-state switches. However, they have slower switching speeds and are typically utilized in high-power applications or measurement instruments. Additionally, electromechanical switches have a shorter lifespan compared to solid-state switches.

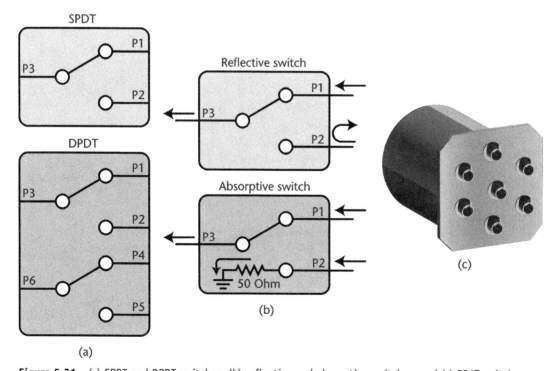

Figure 5.31 (a) SPDT and DPDT switches, (b) reflective and absorptive switches, and (c) SP6T switch.

Critical operational parameters of a switch include isolation, video leakage, IP3, rise time, switching time, P1dB, and repeatability.

Since solid-state switches are semiconductor devices, they can exhibit nonlinearity, making parameters such as P1dB and IP3 relevant. When a switch is driven into compression, harmonics are generated, and insertion loss increases. Therefore, ensuring that the switch operates within its linear region is crucial.

The rise time of a switch plays a crucial role in determining its ability to respond quickly to abrupt changes in the input signal. For example, the bandwidth corresponding to the rise time should be adequate to ensure accurate transmission of square wave input signals up to the fifth or seventh harmonic. The 3-dB bandwidth of the switch can be calculated using the formula $0.35/\tau_R$, where τ_R represents the rise time of the switch module.

The switching time is the speed with which the RF switch operates from OFF to ON state and vice versa. GaAs FET-based RF switches are fastest with less than 1 ns of switching time.

DC Block

As mentioned in Chapter 6, induced currents in the ground loop can negatively impact the qubit performance, microwave equipment, and low-temperature thermometry. In qubit experiments, DC blocks are employed on the microwave lines to mitigate DC ground loops, improve the SNR, and prevent the flow of DC signals into the system by isolating the microwave line ground from the cryostat ground. It is important to note that standard DC blocks only provide DC blocking on the inner conductor, as depicted in Figure 5.32. However, a component with a DC block on both the inner and outer conductors is necessary for our specific requirements. An example of such a component is the Inmet 8039, which offers a blocking effect on both conductors.

Bias-Tee

A diplexer is a passive component that separates or combines two different frequency bands. One type of diplexer is known as a bias-tee, consisting of an RF choke (inductor) and a DC block (capacitor), as illustrated in Figure 5.33(a). When functioning as a combiner, the AC signal passes through the path with the capacitor, while the path with the inductor allows the DC signal to pass through. The combination of these DC and AC signals emerges at the output of the bias-tee.

Bias-tees find widespread usage in microwave circuits for applying a DC bias to active components like transistors or diodes while allowing the AC signal to pass through without any loss. In qubit hardware setups, a bias-tee combines the flux control signals generated by the DC source and the AWG, as depicted in Figure 5.33(b).

Bias-tees can be either passive or active devices. Passive bias-tees do not require an external power supply, whereas active bias-tees rely on a power supply. Active bias-tees may offer additional features like adjustable DC bias voltage and current limiting.

The significant operational parameters of a bias-tee include DC port isolation, maximum applicable DC voltage, and current. A general consideration regarding

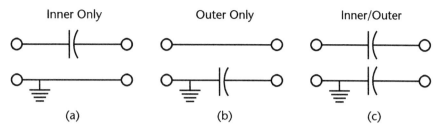

Figure 5.32 (a) DC block on the inner conductor only, (b) DC block on the outer conductor only, and (c) DC block on the inner and outer conductors.

cryogenic components is the use of nonmagnetic materials, as magnetic parts can potentially impact the qubit's performance.

5.4.3 Frequency Manipulation

The ability to modify the frequency content of a signal is crucial to ensure successful signal detection. Table 5.12 provides a summary of the main techniques employed for this purpose. Filtering eliminates unwanted frequency content, including interference, harmonics, and noise. Mixers enable the translation of the signal's frequency content on the spectrum, which is necessary for optimizing the transmission and detection of microwave signals. Resonators are essential for frequency generation and serve as the primary component of cavity QED for qubit readout. Although other techniques and components, such as equalizers, can modify the signal's shape to compensate for distortion, we do not discuss them in this section since they are not widely used in qubit hardware setups.

5.4.3.1 Filters

A filter is a component that selectively attenuates specific frequency components of a signal, effectively removing unwanted frequencies while preserving the desired ones. In a qubit hardware setup, extensive filtering is essential at room and cryogenic temperatures to eliminate noise, interference, and spurs, as explained in Chapter 7.

Filtering can be achieved through two mechanisms: reflective filtering and absorptive filtering. Reflective filtering involves reflecting the unwanted frequencies away from the filter, preventing them from reaching its output. Absorptive filtering, on the other hand, absorbs and dissipates the unwanted frequencies. For instance, a copper powder filter is an example of an absorptive filter (see Chapter 7), while an LC filter can be a reflective filter. When minimizing reflections from the filter is crucial, absorptive filters are preferred, especially when dealing with high-power signals to avoid the formation of standing waves on the lines.

Filters are categorized based on the shape of their frequency response. Figure 5.34(a–d) illustrates four types of filters: low-pass, high-pass, band-pass, and band-stop filters. In a band-pass filter, frequency components within the passband, between the lower and upper cutoff frequencies, pass through the filter with minimal attenuation, as depicted in Figure 5.34(a). Frequency components outside the

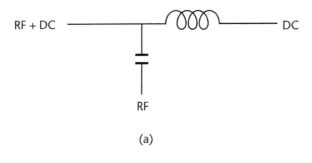

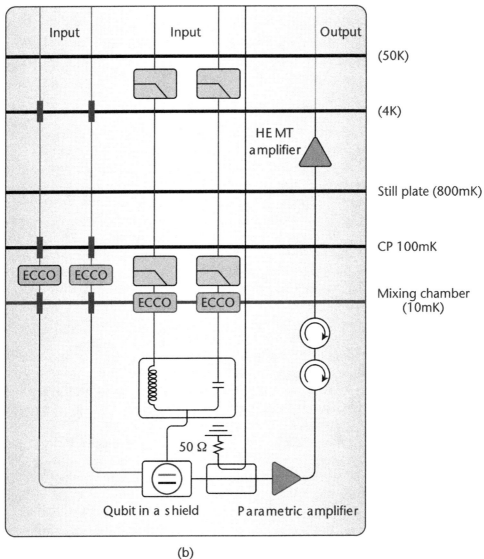

Figure 5.33 (a) Schematic of a bias-tee and (b) application of a bias-tee in a superconducting qubit hardware setup.

5.4 Signal Processing

Table 5.12 Frequency-Manipulation Techniques

Component	Description	Relevant Microwave Components
Filter	Removes unwanted frequency components	Various types of filters (low-pass, bandpass, band-stop(notch), high-pass)
Mixer	Used for up- and downconversion of signals	Scalar mixer IQ mixer
Resonator	Generates microwave oscillations	Planar cavities (microstrip, CPW) Waveguide cavities

passband are significantly attenuated. The same concept applies to low-pass, high-pass, and band-stop filters. Figure 5.34(e, f) provides a visual representation of the effect of a low-pass and a band-pass filter on the frequency spectrum of a signal.

There are several types of filters available, each implemented differently. These include lumped-element filters such as RC and LC filters and distributed-element filters like microstrip, CPW, and waveguide filters. Key filter parameters include the cutoff frequency, out-of-band rejection, and roll-off.

The roll-off represents the steepness of the frequency response during the transition between the passband and the stopband, which is measured in decibels/decade. For instance, a roll-off of 20 dB/decade indicates that the signal decreases by 20 dB for every tenfold increase in frequency. The roll-off value increases with

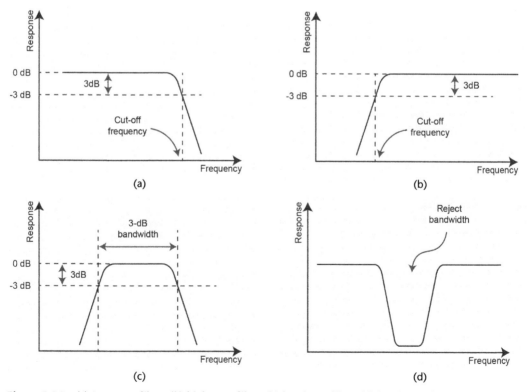

Figure 5.34 (a) Low-pass filter, (b) high-pass filter, (c) bandpass filter, (d) band-stop filter, (e) low-pass filtering of a signal, and (f) bandpass filtering of a signal. Various lumped element low-pass filters: (g) single capacitor filter, (h) Pi filter, and (i) T filter.

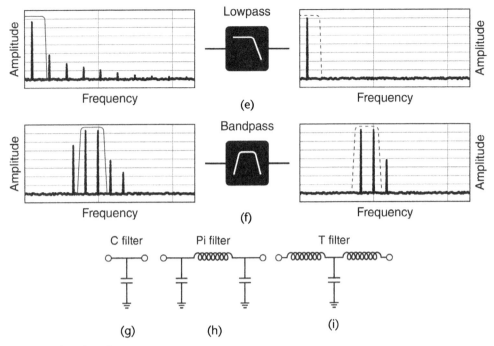

Figure 5.34 *(Continued)*

the filter's order, which relates to the number of energy-storing elements (inductors and capacitors) used. A first-order filter is shown in Figure 5.34(g). A third-order filter, such as a pi filter [see Figure 5.34(i)] with three inductors and capacitors, has a theoretical roll-off of 60 dB/decade, which is higher than that of a simple first-order RC filter with a 20-dB/decade roll-off.

The concept behind low-pass filtering with LC components is as follows: The shunt capacitor in Figure 5.34(h) provides a low-impedance path at high frequencies ($Z = 1/jC\omega$) to divert them from the filter's input into the ground connection. The series inductor allows DC or low-frequency currents to pass through while blocking unwanted high-frequency currents ($Z = jL\omega$). Increasing the number of L and C components in a filter increases the filter's order (e.g., a filter with 5 LC components is a fifth-order filter), improving its selectivity or roll-off. Thus, a pi or T filter has better selectivity than a single capacitor C filter. The same concept can be applied to other filters to understand their operation.

Figure 5.34(h, i) displays the Pi and T filter topologies. Pi filters are suitable for scenarios involving high-impedance sources and loads, whereas T filters are more appropriate for low-impedance sources and loads. This is because capacitors have low impedance at high frequencies. If the load impedance connected in parallel to the capacitor is also low, a significant amount of noise current can flow through the load.

Chapter 7 explores how different filters are employed in a superconducting qubit hardware setup to safeguard the qubits against unwanted signals such as noise and interference. Electromagnetic interference (EMI) filters suppress EMI and are commercially available in various form factors, including PCB mounting and panel-mounting feedthrough.

5.4.3.2 Up- and Downconversion

Let us examine an example from radio-station broadcasting to grasp the concept of frequency translation and its significance in microwave engineering. The voice's frequency content typically falls within the kilohertz range. To achieve efficient radiation, the antenna length must be half the wavelength. Implementing an antenna for a kilohertz signal becomes impractical as it would require lengths in the kilometer range. To address this challenge, the frequency content is translated to a higher frequency, significantly reducing antenna size and enabling practical implementation. While there are other reasons for frequency translation, they are beyond the scope of this book.

In the context of qubit operation discussed in Chapter 3, qubits typically operate within the microwave frequency range of 4–8 GHz. The signal must be translated to a much lower frequency range to ensure reliable sampling of the qubit readout signal for analog-to-digital conversion. This is achieved through downconversion using a mixer. Additionally, mixers are utilized to modulate the microwave tone with the Gaussian pulse generated by the AWG, which controls the qubit's state on the Bloch sphere. This is also called upconversion. Both upconversion and downconversion operations rely on mixers as essential components.

Mixer

A frequency mixer is an electronic circuit with three ports: a LO port, a RF port, and an IF port, as shown in Figure 5.35(a, b). In an ideal mixer, the output signal frequency is the sum or difference of the input frequencies.

$$f_{out} = f_{in1} \pm f_{in2} \tag{5.28}$$

Figure 5.35(a, b) illustrates the downconversion, where RF is the input, and IF is the output, and upconversion, where RF is the output and IF is the input.

In principle, any nonlinear device can be used to create a mixer. A mixer can be as simple as a diode, where nonlinearity generates harmonics, from which the desired upconverted (or downconverted) signals are extracted. GaAs FETs, Schottky diodes, and CMOS transistors are used to build good-quality mixers.

Image Frequency

In an ideal scenario, only one frequency band must be translated during up- and downconversion. However, the presence of frequency components equidistant from the LO frequency, known as image frequency, can pose challenges.

Figure 5.35(b) illustrates that f_{RF1} and its image at f_{RF2} are present in the upconversion case at the RF output port. This type of upconversion is referred to as double-sideband upconversion. The undesired sideband generated at f_{RF2} can introduce interference with other channels. However, as we will soon explore, a clever approach utilizes phasing to suppress one of the sidebands, resulting in single-sideband upconversion. The unwanted sideband in double-sideband upconversion can also present challenges in qubit systems, as it may interfere with other qubits or induce higher-order transitions in a qubit.

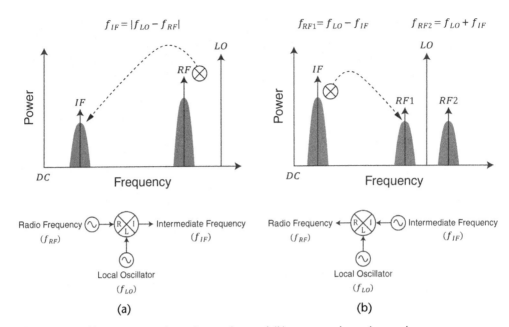

Figure 5.35 (a) Downconversion using a mixer and (b) upconversion using a mixer.

During downconversion, the opposite of the upconversion process occurs. Frequencies mirrored around the LO frequency will downconvert to the same IF, as depicted in Figure 5.36(a, b). This is undesirable since it leads to a mixture of the desired and image signals, thereby corrupting the desired signal.

One way to solve the image problem is to use an image-reject filter, essentially a bandpass filter that eliminates image frequencies. Typically, the image falls well outside the required bandwidth and can be effectively eliminated using filters. However, this approach may not always apply in certain applications where the unwanted or image product is close to the desired signal.

To address the issue of image frequencies in upconversion and downconversion, a solution involves creating two replicas of the input signal, manipulating the signal's phase, and utilizing constructive interference to preserve the desired signal while employing destructive interference to suppress the unwanted signal. The IQ mixer, depicted in Figure 5.36(c), is pivotal in achieving this. It comprises two scalar mixers where the LO is applied with a 90-degree phase difference, generating the input signal's in-phase (I) and quadrature-phase (Q) components. To transform an IQ mixer into a single-sideband (SSB) mixer, the IQ outputs are combined using a 90-degree hybrid. Figure 5.36(c) illustrates how the upper sideband is suppressed. When an SSB mixer is used for downconversion, it is called an image reject (IR) mixer.

As depicted in Figure 5.36(c), the upper sideband is not entirely eliminated, reflecting the practical limitations. Imperfect cancelation of sidebands, LO feedthrough, RF/IF feedthrough, and spurious products can arise due to amplitude or phase imbalances caused by asymmetries in the power combiner, hybrid, and mixer. When discrete components are employed, levels of image rejection of around 20 dB can be achieved, while commercially available units may offer levels of 40

5.4 Signal Processing

dB or higher. Both analog and digital methods can be employed to compensate for amplitude and phase imbalances. These include utilizing a DC offset voltage or employing digital correction methods. In qubit systems that utilize analog methods for upconversion, it is necessary to calibrate the IQ mixer regularly to minimize the level of sidebands. As discussed in Chapter 7, using digital upconversion eliminates IQ mixer calibration.

Figure 5.36(d) illustrates the utilization of the SSB mixer alongside an AWG to generate modulated pulses. These pulses play a crucial role in qubit control, as we saw in Chapter 3.

Now, let us delve into the operational parameters of the mixer. The conversion loss, expressed in decibels, represents the difference (in dBm) between the input RF power level and the output power at the IF frequency $CL = P_{RF} - P_{IF}$.

The conversion loss of a mixer can vary between approximately 4.5 and 9 dB, depending on the specific mixer model. Conversion loss is closely correlated with other metrics, such as isolation and 1-dB compression. Consequently, a single measurement of conversion loss provides a reliable indication of the mixer's overall performance. If the conversion loss falls within a specified range, it generally implies that all other performance measures will meet the required specifications.

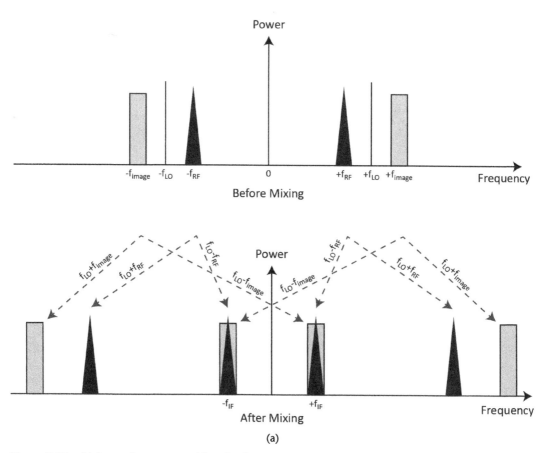

Figure 5.36 (a) Image frequency problem for downconversion, (b) IQ mixer used as a single sideband mixer. Operation of a single sideband mixer to cancel the sidebands, (c) application of a single sideband mixer as a generator [17].

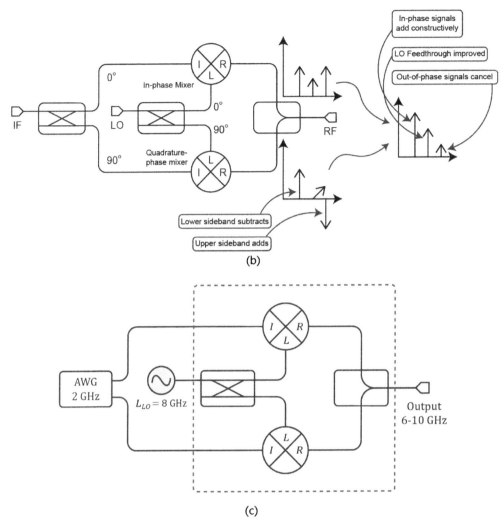

Figure 5.36 *(Continued)*

However, the converse is not always true. For instance, it is possible to have good isolation but poor conversion loss [17].

Isolation quantifies the amount of power that leaks from one port of the mixer to another. It is measured in decibels and defined as the power difference (in dBm) between the input signal and the leaked power at the other ports. In simpler terms, if we apply an input signal at the LO port and measure the power available at the RF port at that LO frequency, the isolation in dB is calculated as follows $P_{ISO(L-R)} = P_{in(@LO)} - P_{out(@RF)}$.

Power leakage from one port of the mixer to another can lead to various issues. Section 5.4.5 examines how the leakage of the LO signal to the RF port in a homodyne receiver results in a DC offset.

Several operational parameters significantly affect mixer performance, including single-tone intermodulation distortion (IMD), multitone IMD, isolation, noise figure, gain, and conversion loss. These parameters collectively contribute to the overall performance and characteristics of the mixer.

5.4.3.3 Microwave Resonators

Microwave resonators are vital in oscillators, filters, and tuned amplifiers. They are also crucial components in qubit experiments, particularly in cavity QED, where they enable the reading of a qubit's state while minimizing interactions. Resonators exist in various domains of physics, each with its examples. For instance, a spring-mass system is a mechanical resonator, an LC circuit is an electrical resonator, and an atomic clock employs an atomic resonator. What unifies these examples is their ability to oscillate through energy exchange between two physical quantities. In a mechanical oscillator, the quantities are potential and kinetic energy, while in an LC oscillator, they are electric and magnetic energy. Figure 5.37(a) illustrates a lossless LC circuit, where the initial charge on the capacitor leads to oscillating current i(t) at the resonance frequency $\omega_r = 1/\sqrt{LC}$. At resonance, the average energy stored in the electric and magnetic fields becomes equal, $W_m = W_e$. The Fourier transform of the sine wave oscillation results in an impulse, as shown in Figure 5.37(a).

In practical applications, every resonator experiences losses. Therefore, modeling these losses and energy decay involves the introduction of a resistor to the LC circuit, as shown in Figure 5.37(b). Due to losses in the resistor, the oscillations gradually diminish over time, as illustrated in Figure 5.37(b). The Fourier transform of an exponentially decaying oscillation is depicted in Figure 5.37(b), where the spectrum becomes broader compared to a pure sine wave oscillation. The width of the spectrum is directly proportional to the decay rate. As the oscillation decays faster, the spectrum becomes wider. Chapter 3 also discussed the concept of spectral broadening within the context of coherence time.

Figure 5.37(b) demonstrates that the bandwidth (BW) of the spectrum is inversely proportional to the quality factor (Q) of the resonator. The quality factor represents the number of oscillations a resonator can sustain before experiencing complete decay due to losses. In essence, it quantifies the oscillator's capability to maintain oscillations. The quality factor is the ratio of the energy stored in the resonator to the energy lost per oscillation.

$$Q = \omega \frac{\text{average energy stored}}{\text{energy loss/second}}$$
$$= \omega \frac{W_m + W_e}{P_{\text{loss}}} = \frac{f_0}{B} \quad (5.29)$$

In the equations provided, W_m and W_e represent the average stored energy in the electric and magnetic fields, respectively. P_{loss} denotes the power loss, f_0 is the resonant frequency, and B represents the bandwidth. A higher quality factor results in a sharper frequency response and narrower linewidth, as Figure 5.37(b) depicts. For example, a tuning fork typically has a quality factor of around 10^3, an atomic clock around 10^{11}, a microstrip resonator on the order of 10^3, and microwave cavities on the order of 10^5.

Microwave resonators exhibit behavior similar to LC circuits near their resonant frequency and can be constructed using any transmission line. A half-wavelength

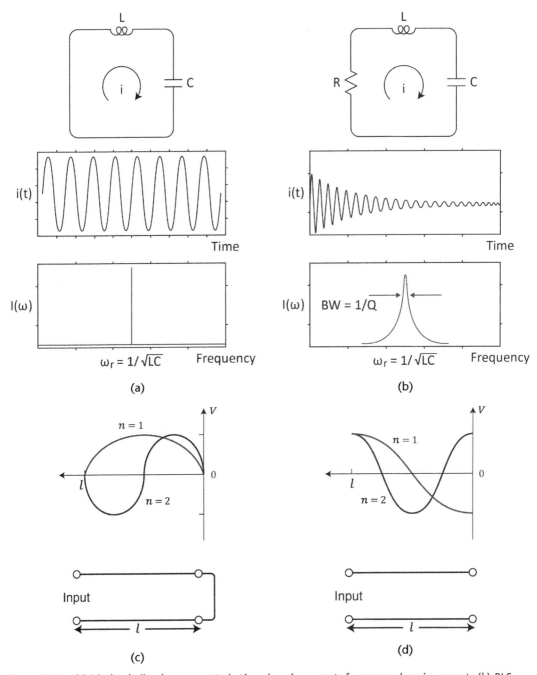

Figure 5.37 (a) LC circuit (lossless resonator); time domain current; frequency domain current; (b) RLC circuit (lossy resonator); time domain current; frequency domain current. Voltage distribution of (c) a short-circuit resonator and (d) an open-circuit resonator.

transmission line with one or both ends open, as shown in Figure 5.37(c, d), can be employed to create a microwave resonator. Determining the voltage distribution is straightforward: At the open end, the voltage reaches its maximum and then changes by half a wavelength over the length of the transmission line. For a shorted end, the voltage is zero and then changes by half a wavelength over the length of the transmission line.

The rate of photon loss, κ, which represents the width of the Lorentzian-shaped transmission power spectrum, is related to the measured quality factor, Q_L, as $Q_L = \omega_r/\kappa$. The photon storage time of the cavity mode under consideration is given as $\tau_n = 1/\kappa$.

Regarding P_{in}, which denotes the probe power applied directly at the resonator input, the resonant transmission peak power, P_r can be calculated as $P_r = IL \cdot P_{in}$, where IL represents the insertion loss. Equation (5.30) gives the Lorentzian-shaped transmission power spectrum:

$$P(\nu) = \frac{P_r}{1 + 4\left(\frac{\omega - \omega_r}{\kappa}\right)^2} \tag{5.30}$$

with $\kappa/2\pi$ the full width at half maximum. The phase shift of the transmitted microwave wave with respect to the incident wave can also be measured and is described by

$$\delta(\nu) = \tan^{-1}\left(\frac{2(\omega - \omega_r)}{\kappa}\right) \tag{5.31}$$

5.4.3.3.1 Resonator Coupling

Resonators typically require coupling to the external world. In Figure 5.38(a), the meander line CPW resonator used in a quantum circuit is capacitively coupled to the input and output lines, allowing the qubit's state to be read out by analyzing the S_{21} parameter.

Figure 5.38(b) illustrates the connection of an external load to a resonant circuit. The coupling between the resonator and external circuits introduces an energy leakage channel that decreases the overall quality factor. To examine the impact of coupling, three quality factors are defined: the unloaded quality factor (Q_u), which describes the resonator without any loading effects from the external circuitry; the external quality factor (Q_e); and the loaded quality factor (Q_L). The following expression defines the relationship between these quantities:

$$\frac{1}{Q_L} = \frac{1}{Q_e} + \frac{1}{Q_u} \tag{5.32}$$

Table 5.13 presents the quality factors of series and parallel resonant circuits. It demonstrates that the external quality factor takes into account the losses caused

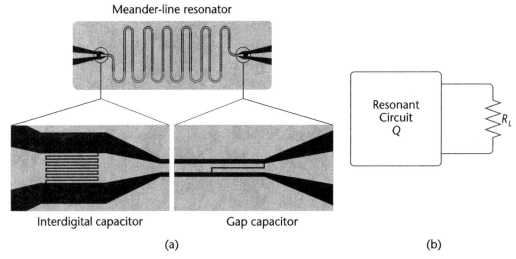

Figure 5.38 (a) Capacitive coupling to a meander line CPW resonator and (b) an external load, R_L, connected to a resonant circuit.

Table 5.13 Unloaded, External, and Loaded Quality Factors for Series and Parallel RLC Circuits

Parameter	Series RLC Resonator $\left(\omega_0 = \frac{1}{\sqrt{LC}}\right)$	Parallel RLC Resonator $\left(\omega_0 = \frac{1}{\sqrt{LC}}\right)$
Unloaded Q	$Q_u = \frac{\omega_0 L}{R} = \frac{1}{\omega_0 RC}$	$Q_u = \omega_0 RC = \frac{R}{\omega_0 L}$
External Q	$Q_e = \frac{\omega_0 L}{R_L}$	$Q_e = \frac{R_L}{\omega_0 L}$
Loaded Q	$Q_e = \frac{\omega_0 L}{R + R_L}$	$Q_e = \frac{R_L \parallel R}{\omega_0 L} = \frac{RR_L}{\omega_0 L (R + R_L)}$

by coupling to the external circuit, denoted as R_L. In contrast, the loaded quality factor combines the resonator's internal losses and the external losses resulting from the coupling.

Transmission-line open-ended resonators are commonly used in quantum circuits. In the case of an open-ended $\lambda/2$ line, the unloaded quality factor is determined by

$$Q_u = \frac{\pi}{2\alpha \ell} = \frac{\beta}{2\alpha} \tag{5.33}$$

Here, ℓ represents the length of the transmission line and the attenuation constant $\alpha = \alpha_c + \alpha_d$ comprises metallic α_c and dielectric losses α_d, provided that radiation losses are negligible. The propagation constant is given by $\beta = 2\pi f/v_p$, where v_p represents the phase velocity.

5.4 Signal Processing

External circuits can be coupled in three regimes: undercoupled, critically coupled, and overcoupled. The coupling coefficient g, defined as follows, can be used to determine the coupling regime of a resonator.

$$g = \frac{Q_u}{Q_e} \tag{5.34}$$

Three coupling regimes can be defined depending on the value of g, as follows:

1. $g < 1$: The resonator is undercoupled to the feedline.
2. $g = 1$: The resonator is critically coupled to the feedline.
3. $g > 1$: The resonator is overcoupled to the feedline.

When a resonator is connected to a transmission line with a characteristic impedance of Z_0, the coupling coefficient for a series resonant circuit is given by ($g = Z_0/R$), while for a parallel resonant circuit is given by ($g = R/Z_0$). Resonators strongly coupled (overcoupled) with low-quality factors (high bandwidth) are well-suited for fast qubit measurements. In qubit measurements, it is typically desirable to have $g_{CPW} \gg 1$. Furthermore, a large bandwidth is advantageous for resetting a qubit to its ground state. This can be achieved by bringing its frequency close to the cavity resonance and utilizing the Purcell-enhanced decay rate. On the other hand, undercoupled resonators with high-quality factors negatively impact the fidelity of qubit readout, as they collect fewer signal photons during the qubit's lifetime. Resonators with high-quality factors can store photons in the cavity for an extended period and potentially be used as quantum memories [19].

The coupling coefficient can be determined by measuring the S_{21} parameter of the cavity at the resonance frequency ω_0. If the resonator appears as a series RLC resonator, then [3]

$$g = \frac{S_{21}(\omega_0)}{1 - S_{21}(\omega_0)} \tag{5.35}$$

Please note that the above S_{21} parameter is expressed in linear units, not in decibels. If the resonator appears as a parallel RLC circuit, it can be demonstrated that the result for g in (5.35) should be inverted. Figure 5.39(a, b) illustrates the reflection measurements of over and undercoupled resonators.

Directly measuring the unloaded Q of a resonator is generally not feasible due to the loading effect of the measurement system. However, it is possible to determine the unloaded Q from measurements of the loaded resonator's frequency response when connected to a transmission line. The loaded quality factor, Q_L, can be directly measured from the S_{21} frequency response. In this case, Q_L is calculated as $Q_L = f_0/BW$, where f_0 is the resonant frequency, and BW is the half-power bandwidth, where the transmission response is 3 dB lower than at resonance. It can be demonstrated that $Q_u = (1 + g)Q_L$, allowing for the determination of the unloaded Q by calculating Q_L and g from the S-parameter. In addition to the

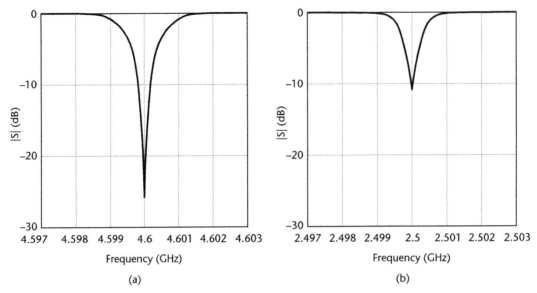

Figure 5.39 (a) Critically coupled resonator with a sharp resonance and a large return loss and (b) overcoupled and undercoupled resonators with small return loss.

two-port (transmission measurement) technique discussed above, the one-port (reflection measurement) technique can also be employed to determine the quality factor of a resonator [18].

5.4.3.3.2 CPW Microwave Resonator

As discussed in Section 5.3.3.3, planar qubits commonly employ CPW transmission lines. Figure 5.40(a) depicts a half-wavelength CPW resonator with the qubit positioned at the center, resulting in a strong coupling due to the maximum voltage at that location. Alternatively, two qubits can be placed at the open ends of the resonator, as shown in Figure 5.40(b). Another option is to add two additional qubits on the opposite side of the line, enabling the coupling of four qubits to the same resonator.

The frequency of the dominant resonant mode f_0 for a CPW line can be expressed as [19]

$$f_0 = \frac{c}{\sqrt{\epsilon_{\text{eff}}}} \frac{1}{2l} = \frac{1}{\sqrt{L_\ell C_\ell}} \frac{1}{2l} \tag{5.36}$$

Where l is the length of the line, which is half the wavelength; C_ℓ and L_ℓ is the capacitance and inductance per unit length of the line, respectively; and ϵ_{eff} is the effective permittivity. The per unit length parameters are given by $L_\ell = (\mu_0/4)(K(k'_0)/K(k_0))$, $C_\ell = 4\epsilon_0\epsilon_{\text{eff}}(K(k_0)/K(k'_0))$. Here, K denotes the complete elliptic integral of the first kind with the arguments $k_0 = w/(w + 2s)$ and $k'_0 = \sqrt{1 - k_0^2}$, where w is the width of the CPW line, and s is the gap dimension between the line and the ground. Superconducting CPW resonators can reach a quality factor on the order of 10^5.

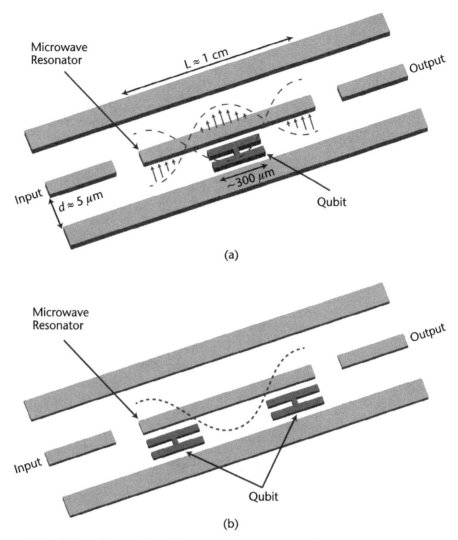

Figure 5.40 (a) Coupling a single qubit to a CPW resonator and (b) coupling two qubits to a CPW resonator.

Waveguide Resonators

Microwave resonators undergo losses caused by metallic, dielectric, and radiation effects. Open structures on dielectrics, such as microstrip or CPW resonators, experience all three types of losses. Radiation losses, in particular, are significant because an open end can act as a slot antenna. Additionally, these structures have high metallic and dielectric losses, which limit their achievable quality factor and operating power.

In contrast, waveguide resonators form closed boxes or cavities [see Figure 5.41(a)] and typically do not incorporate dielectric components. Due to the shielding in all directions, energy leakage as radiation is prevented. Moreover, the larger surface area available for current within waveguide cavities results in lower metallic losses [20, 21]. As a result, waveguide cavities can achieve quality factors up to two orders of magnitude higher than planar resonators. External circuits can be coupled to the cavity using a small aperture, probe, or loop.

For the dominant cavity mode, TE_{101}, the electric field exhibits maximum intensity in the middle of the cavity, as illustrated in Figure 5.41(b). Therefore, as shown in Figure 5.41(c), placing the qubit at this location maximizes the coupling [20, 21]. The advantages of 3D cavities for qubit applications are discussed in Chapter 3.

The resonant frequency of the TE_{mnl} or TM_{mnl} mode of a lossless rectangular cavity can be calculated as follows [3]

$$f_{mn\ell} = \frac{c}{2\pi\sqrt{\mu_r \epsilon_r}} \sqrt{\left(\frac{m\pi}{a}\right)^2 + \left(\frac{n\pi}{b}\right)^2 + \left(\frac{\ell\pi}{d}\right)^2} \quad (5.37)$$

Where a, b, and d are the waveguide dimensions in Figure 5.41(a). If $b < a < d$, the dominant resonant mode (lowest resonant frequency) will be the TM_{101} mode. The dominant TM resonant mode is the TM_{110} mode. The cavity's resonant modes, TE_{mnl} or TM_{mnl}, indicate the number of variations in the standing wave pattern in the x, y, and z directions, respectively.

The unloaded Q of the cavity with lossy conducting walls but lossless dielectric can be found as $Q_c = (2\omega_0 W_e)/P_c$ where P_c is the dissipated power due to metallic losses. The unloaded Q of the cavity with a lossy dielectric filling and perfectly conducting walls is $Q_d = (2\omega_0 W_e)/P_d = \epsilon'/\epsilon'' = 1/\tan\delta$ where P_d is the dissipated power in the dielectric. The total unloaded Q becomes

$$Q_u = \left(\frac{1}{Q_c} + \frac{1}{Q_d}\right)^{-1} \quad (5.38)$$

Waveguide cavities with other shapes, such as cylindrical ones, are utilized in certain applications including some qubit architecture.

5.4.4 Phase Manipulation

A phase shifter is a two-port device that adjusts a microwave signal's phase. It can be implemented using microwave diodes, field-effect transistors (FETs), or ferrite materials such as YIG or iron oxides. Ferrite phase shifters utilize a bias field to control the amount of phase shift. These phase shifters find wide applications in various systems, including steerable communication links like phased-array antennas, where the radiation beam can be electronically steered by manipulating the phase of each antenna element in the array. Phase shifters are also used in other applications, such as cancelation loops, to enhance the linearity of amplifiers.

The phase of an analog tunable phase shifter can be adjusted using a control voltage. In contrast, a digital phase shifter utilizes a series of control bits. Analog-tunable phase shifters typically offer lower insertion loss than their digital counterparts. However, it is essential to note that the phase shift is not a linear function of the control voltage, as demonstrated in Figure 5.42(a). Therefore, it is crucial to accurately characterize the phase shifter and establish the correct association

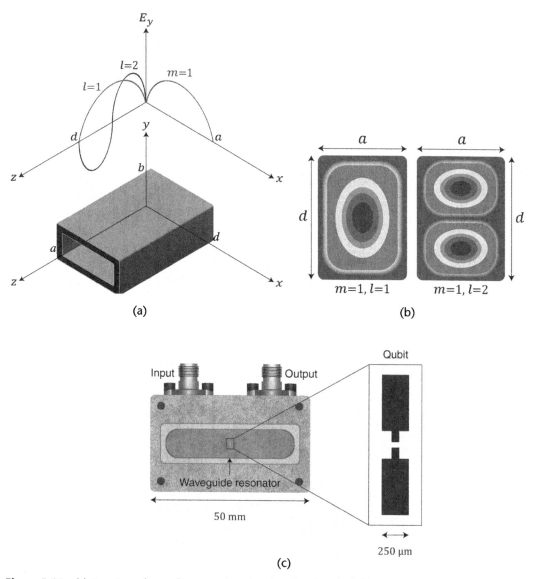

Figure 5.41 (a) A rectangular cavity resonator, showing the electric field variations for the TE_{101} and TE_{102} resonant modes; (b) top view of the field distribution for the first two modes of a rectangular cavity; and (c) a transmon qubit coupled to a rectangular cavity, where the large pads act as capacitors, and the Josephson junction is placed between the two pads.

between each voltage and its corresponding phase shift. The control function of a phase shifter refers to the relationship between the voltage or digital control value and the resulting phase shift. Ideally, this relationship would be linear.

Digital phase shifters, commonly employed in microwave systems, have limited resolution determined by the number of bits they possess. For example, an 8-bit phase shifter offers a resolution of approximately 1.4 degrees ($360°/2^8$). Another significant aspect of phase shifters is their insertion loss variation with phase control, indicating the extent to which the insertion loss changes as the phase shifts. In an ideal scenario with a perfect phase shifter, there would be no variation in insertion loss. This characteristic is crucial for both phased-array antennas and cancelation applications, as a constant insertion loss versus phase eliminates the need to adjust gain as the phase is modified. However, some amplitude variation will always exist in practical circuits.

The variation of IP3 and the insertion loss with phase control are two critical operational parameters of a phase shifter.

Tunable phase shifters are critical in implementing photonic quantum gates in qubit systems [22]. They also facilitate the on-chip integration of control and measurement systems for superconducting qubits, enabling the adjustment of the phase of an on-chip coherent microwave source to access a universal set of single-qubit gates with a single off-chip microwave source [22]. One approach to

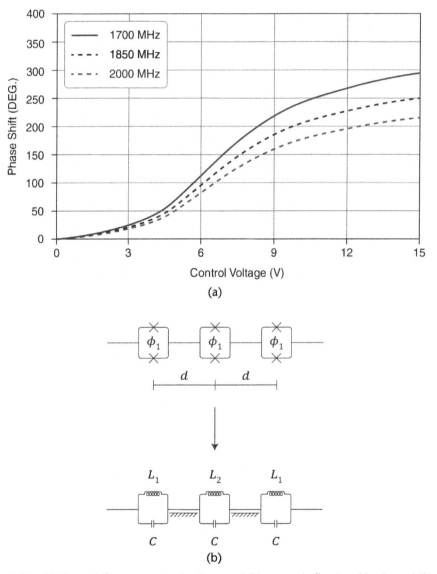

Figure 5.42 (a) Phase shift versus control voltage and (b) magnetic-flux-tunable phase shifter.

implementing an on-chip magnetic-flux-tunable phase shifter is utilizing three equidistant SQUIDs on a transmission line, as illustrated in Figure 5.42(b).

5.5 Signal Detection

Signal detection is the final crucial step in a microwave link, aimed at successfully detecting the received signal. The definition of successful detection can vary depending on the system requirements and the type of receiver employed. Multiple receiver architectures are available, including heterodyne, direct conversion, image reject, and low-IF receivers, each with its advantages, disadvantages, and trade-offs concerning performance, complexity, and cost. In this context, we now explore two widely used architectures in superconducting quantum computers: the homodyne architecture and the superheterodyne architecture.

Figure 5.43 presents a general overview of the signal-detection process, followed by nearly all architectures. However, each architecture modifies certain parts of the detection chain to achieve particular objectives, such as addressing image frequency problems or enhancing selectivity. The fundamental principle of detecting the readout signal of a qubit is similar to that of detecting communication signals, as discussed in Section 7.3.2.2.

The various components of a receiver, as depicted in Figure 5.43, are described as follows:

- *Amplification:* Chapter 4 discussed the concept of the MDS and showed that signal detection fails if the signal level falls below the MDS level. For instance, when utilizing a diode envelope detector, the signal must be sufficiently strong to overcome the diode's forward voltage drop. Therefore, amplifiers increase the signal level above the MDS level. On the receiving end, LNAs maintain a low noise level to facilitate signal detection. However, the nonlinear behavior of the amplifier can affect the detection performance. Chapter 4 examined how intermodulation products generated by the amplifier or mixer nonlinearities can leak into the desired channel and corrupt the signal. Hence, when operating in tightly spaced channels and high interference

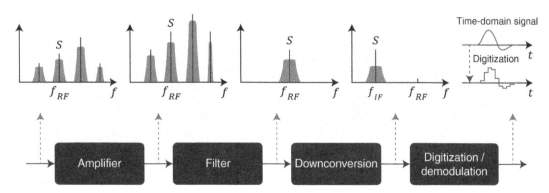

Figure 5.43 Signal flow in a general receiver architecture.

environments, it is crucial to consider not only the gain and noise figure but also the amplifier's P1dB and IP3 to ensure optimal performance.

- *Filtering:* Filtering is critical in microwave signal processing, serving two primary objectives: suppressing strong out-of-band signals and eliminating image frequencies and spurs. A bandpass image reject filter eliminates out-of-band signals including images, as shown in Figure 5.44(a). Subsequently, a channel select filter is used to choose the desired channel, commonly used in a heterodyne architecture (refer to Section 5.4.3.2). On the other hand, a homodyne architecture (refer to Section 5.4.3.2) utilizes a single low-pass filter to eliminate out-of-band signals, as shown in Figure 5.44(b).

Figure 5.44(a) illustrates the image problem, where signals might exist at the image frequency. Recall that two signals with the same distance from the LO frequency would be downconverted to the same frequency, resulting in the corruption of the desired signal. To address this issue, a filter, as depicted in Figure 5.44(a), is employed to eliminate the image frequency.

The placement order of the filter and LNA may raise some questions. Typically, the LNA is positioned immediately after the readout resonator in qubit experiments. As discussed in Section 4.3.2.4, the first element's NF

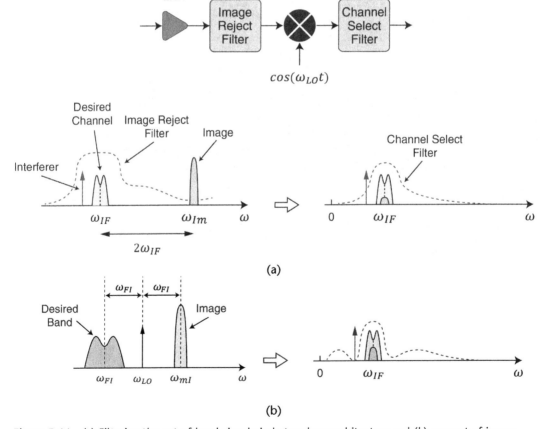

Figure 5.44 (a) Filtering the out-of-band signals in heterodyne architecture and (b) concept of image frequency.

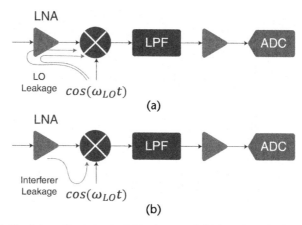

Figure 5.45 (a) Self-mixing effect due to LO leakage and (b) interferer leakage.

determines the entire chain's overall NF. Hence, a low NF and high-gain LNA are used as the first element to minimize the NF. However, this configuration has a drawback: The LNA can quickly saturate due to strong out-of-band signals, and image frequencies can corrupt the detection process. Nonetheless, this is not a concern in qubit experiments since highly shielded qubit environments typically do not have strong out-of-band signals.

If we swap the filter and the LNA, the overall NF is significantly increased since the filter acts as a lossy component with a relatively high NF. However, in high-interference environments, placing the filter before the LNA may be necessary to prevent the LNA from entering compression if a large out-of-band signal overloads it or if image frequency issues exist.

- *Downconversion:* Downconversion is a crucial step in microwave signal processing, as it brings the signal's frequency content to a range suitable for digitization or demodulation. A mixer and an oscillator are utilized to achieve this, as depicted in Figure 5.45. Depending on the receiver architecture, the downconversion process can occur in single or multiple stages. In homodyne detection, the signal is directly downconverted to the baseband, while double-conversion superheterodyne architectures involve two downconversions.

Regarding IQ mixers, as discussed in Section 5.4.3.2 amplitude and phase imbalances can introduce errors that cause the constellation points on the IQ diagram to deviate from their ideal locations. Therefore, mixer calibration is essential to minimize imbalance and improve the mixer's spur performance.

The LO is typically a variable frequency oscillator that tunes the receiver to different channels. Chapter 4 mentions that the LO must have low-phase noise and deliver a clean output. Otherwise, issues such as leakage to neighboring channels and reciprocal mixing may arise. Furthermore, the absence of a clean LO spectrum can lead to mixing its harmonics with spurious input signals, which can be transmitted to the IF.

Another issue to consider is the LO leakage, as depicted in Figure 5.45(a). When the LO signal leaks to the input of the mixer and reflects back, it mixes

with itself and generates a DC offset, which is called the self-mixing effect. Additionally, the LO leakage can reach the receive antenna and radiate, potentially affecting the sensitivity of other receivers operating in the same frequency band. Also, the interferer can leak as shown in Figure 5.45(b).
- *Digitization/demodulation:* The downconverted signal is commonly sampled using a digitizer, also called an ADC. The obtained digital samples can be employed for demodulation or subjected to additional processing using digital signal processing (DSP) techniques.

While not employed in all types of receivers, an equalizer can compensate for nonlinear effects, thereby mitigating distortions and enhancing the overall fidelity of the received signal.

5.5.1 Homodyne Detection

The term "homodyne" originates from the combination of "homo" (same) and "dyne" (mixing), referring to the same frequency of the LO and the received signal. Figure 5.46(a) shows that this receiver brings the received signal directly to the baseband centered at DC. An IQ mixer is used after the LNA to downconvert the signal. Subsequently, a low-pass filter is employed to suppress harmonics and spurs generated in the previous stages. This architecture is simple and cost-effective due to its few components. However, it also has its drawbacks.

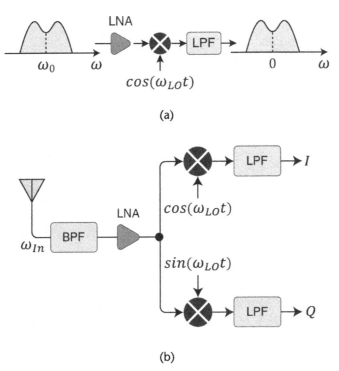

Figure 5.46 (a) Simple homodyne receiver and (b) homodyne receiver with quadrature downconversion.

The homodyne receiver is also utilized for the room-temperature detection of a qubit's readout signal. Figure 5.46(b) illustrates a homodyne receiver with quadrature downconversion. The LO frequency is adjusted to downconvert the readout signal to a frequency suitable for digitization. Subsequently, the I and Q components are fed into a digitizer, and the resulting output is transmitted to the data acquisition software on a PC.

One of the main drawbacks of the homodyne detection architecture is the presence of DC offset caused by the previously mentioned self-mixing issue, which some design tweaks can minimize. As depicted in Figure 5.46(b), many modern receivers employ homodyne detection with quadrature detection followed by digital signal processing.

Another drawback of the homodyne detection architecture is its susceptibility to noise and interference, as it does not reject the image signal in the opposite sideband. However, a method developed by Campbell in 1993 [8] allows for eliminating the other sideband by modifying the receiver's architecture.

5.5.2 Superheterodyne Detection

Figure 5.47(a) showcases a typical single-conversion superheterodyne receiver. Figure 5.47(b) further illustrates the signal flow as it progresses through the superheterodyne detection chain. The initial RF filter offers selectivity, effectively filtering out the out-of-band and image signals. When operating at a lower frequency, IF filters can provide narrower passbands at the same Q-factor compared to an equivalent RF filter. Consequently, the first RF filter possesses a wider bandwidth and lower selectivity. After the downconversion process, a high-Q channel selection IF filter, such as a SAW filter, is employed to extract the desired channel.

The IF selection involves a trade-off between image rejection and channel selectivity. The separation between the received signal and the image frequency is twice the IF frequency, as illustrated in Figure 5.48(a). A higher IF provides significant rejection of the image frequency but suffers from a lower Q of the channel selectivity filter. Conversely, as depicted in Figure 5.48(b), a lower IF is more effective in suppressing in-band interferers and benefits from a higher Q of the channel selectivity filter. So, a trade-off exists between achieving high image frequency suppression and ensuring optimal channel selectivity.

A double-conversion architecture can achieve both high suppression of the image frequency and high channel selectivity. Figure 5.48(c) demonstrates the utilization of two IFs. In this architecture, the first stage converts the input frequency to a high IF, ensuring a low image response. Subsequently, the second stage converts this frequency to a low IF, resulting in excellent selectivity within the second IF filter. Commercial qubit receivers, like the SHFQC from Zurich Instruments, employ a double-conversion superheterodyne architecture, as illustrated in Figure 5.48(d) [23].

5.5.3 Direct RF Sampling

Recall that digital signal generation can reduce the size and cost per channel, thus opening new opportunities for qubit scaling. Similarly, digital detection techniques

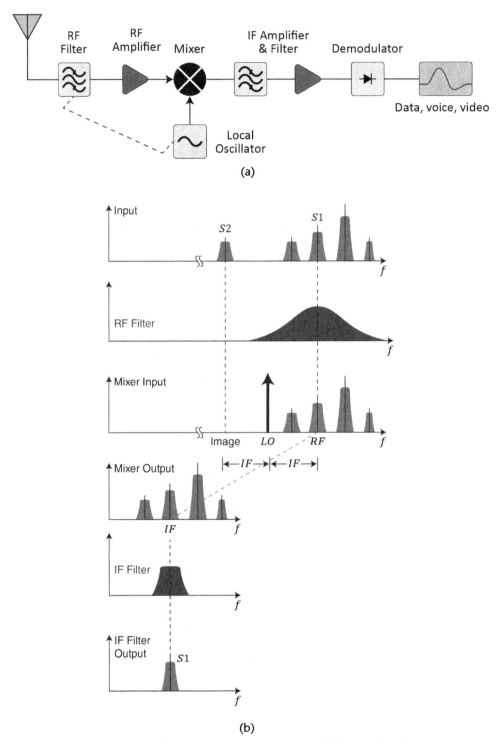

Figure 5.47 (a) Single-conversion superheterodyne receiver and (b) signal flow in a superheterodyne detection chain.

can play the same role for each receive channel. This approach employs fast, high-resolution, and accurate analog-to-digital conversion of the received signal to eliminate the downconversion stage, as depicted in Figure 5.49. Consequently, there is no need for a LO and mixer at the IF stage. This simplified architecture eliminates potential noise sources, images, and other errors, including LO leakage and quadrature impairments. As a result, the tedious IQ mixer calibration process is no longer necessary, and synchronization becomes simpler.

The continuous advancement in ADCs has made it feasible to sample microwave signals directly. RF system-on-chip (RFSoC) modules bypass the entire room-temperature downconversion chain, allowing for direct sampling and digitization of the microwave signal from the fridge. Direct sampling is also utilized in the atom and ion qubits detection chain.

While direct RF sampling offers several advantages, several drawbacks must be considered. First, high-speed ADCs capable of sampling microwave frequencies are still quite expensive. It is another challenge to achieve a high dynamic range to accommodate the wide range of signal levels, as it necessitates using ADCs with a high dynamic range. This can be technically challenging and costly, especially when aiming for high sampling rates.

Moreover, direct RF sampling systems require advanced digital signal processing techniques and calibration methods, adding to the overall complexity.

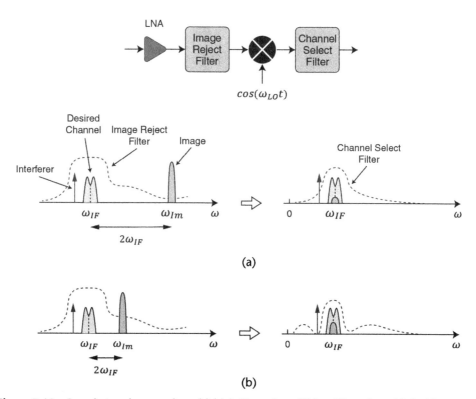

Figure 5.48 Superheterodyne receiver: (a) high-IF receiver; (b) low-IF receiver; (c) double-conversion receiver; and (d) double-conversion receiver used for qubit readout signal detection.

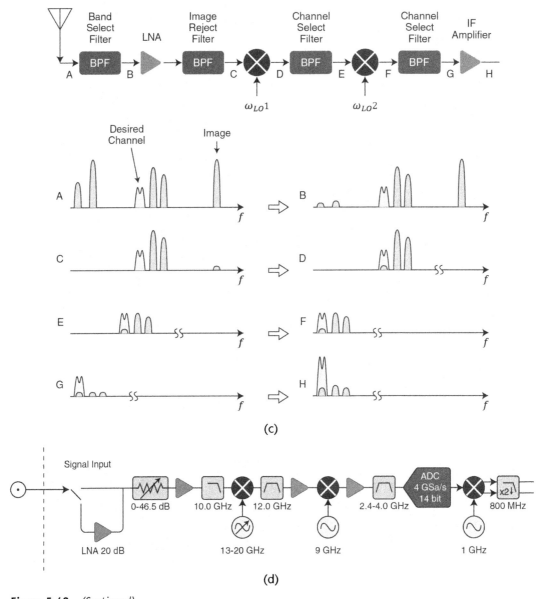

Figure 5.48 *(Continued)*

Additionally, these systems are susceptible to phase noise in the ADC clock and other system components, impacting their performance.

Last, direct RF-sampling systems often require more power than traditional heterodyne systems due to the higher sampling rates and demanding digital signal processing requirements. In superconducting qubit systems, direct RF sampling through an ADC is employed to digitize signals. This occurs after bringing the signal to the appropriate frequency range via downconversion.

5.5 Signal Detection Signal Processing

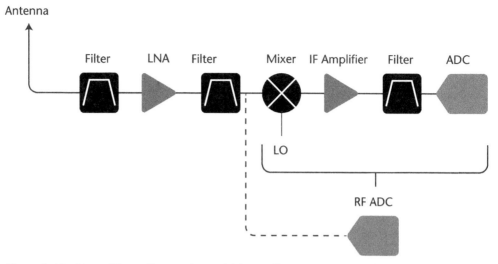

Figure 5.49 Direct RF sampling receiver; a high sampling rate ADC can replace the entire downconversion chain in this configuration.

References

[1] Balanis, C., *Advanced Engineering Electromagnetics*, Wiley, 2012.
[2] Collins, R. E., *Foundations of Microwave Engineering*, Wiley-IEEE Press, 2001.
[3] Pozar, D. M., *Microwave Engineering*, Wiley, 2011.
[4] Jackson, J. D., *Classical Electrodynamics*, Wiley, 1998.
[5] Harrington, E., *Time-Harmonic Electromagnetic Fields* (Second Edition), Wiley, 2001.
[6] Sadiku, M. N. O., *Numerical Techniques in Electromagnetics with Matlab* (Third Edition), Taylor & Francis, 2009.
[7] Peng, K., et al., "X-parameter Based Design and Simulation of Josephson Traveling-Wave Parametric Amplifiers for Quantum Computing Applications," *2022 IEEE International Conference on Quantum Computing and Engineering (QCE)*, Broomfield, CO, 2022, pp. 331–340.
[8] Razavi, B., *RF Microelectronics*, Prentice Hall Communications Engineering and Emerging Technologies, 2011.
[9] Boissonneault, M., et al., "Improved Superconducting Qubit Readout by Qubit-Induced Nonlinearities," *Physical Review Letters*, Vol. 105, No. 10, 2010, p. 100504.
[10] White, T., et al., "Traveling Wave Parametric Amplifier with Josephson Junctions Using Minimal Resonator Phase Matching," *Applied Physics Letters*, Vol. 106, No. 24, 2015.
[11] Remm, A., et al., "Intermodulation Distortion in a Josephson Traveling-Wave Parametric Amplifier," *Physical Review Applied*, Vol. 20, No. 3, 2022.
[12] Krantz, P., et al., "A Quantum Engineer's Guide to Superconducting Qubits," *Applied Physics Reviews*, Vol. 6, No. 2, 2019.
[13] Ripoll, J. J. G., *Quantum Information and Quantum Optics with Superconducting Circuits*, Cambridge University Press, 2022.
[14] Thorbeck, T., et al., "Reverse Isolation and Backaction of the SLUG Microwave Amplifier," *Physical Review Applied*, Vol. 8, No. 5, 2017, p. 054007.
[15] Hover, D., et al., "High Fidelity Qubit Readout with the Superconducting Low-Inductance Undulatory Galvanometer Microwave Amplifier," *Appl. Phys. Lett.*, Vol. 104, No. 15, 2014.

[16] Barzanjeh, S., et al., "Mechanical On-Chip Microwave Circulator," *Nature Communications*, Vol. 8, No. 1, 2017, p. 953.

[17] Marki, F., and C. Marki, "Mixer Basics Primer A Tutorial for RF and Microwave Mixers," *Microwave Journal*, 2018, https://www.microwavejournal.com/articles/print/32203-microwave-mixers-tutorial.

[18] Kamigaito, O., "Circuit-Model Formulas for External-Q Factor of Resonant Cavities with Capacitive and Inductive Coupling," 2020, https://arxiv.org/pdf/2005.05843.pdf.

[19] Göppl, M., et al., "Coplanar Waveguide Resonators for Circuit Quantum Electrodynamics," *Journal of Applied Physics*, Vol. 104, No. 11, 2008.

[20] Reshitnyk, Y., et al., "3D Microwave Cavity with Magnetic Flux Control and Enhanced Quality Factor," *EPJ Quantum Technology*, Vol. 3, No. 1, 2016, pp. 1–6.

[21] Kumar, N. P., et al., "Bound States in Microwave QED: Crossover from Waveguide to Cavity Regime," 2022, *arXiv:2208.00558*.

[22] Kokkoniemi, R., et al. "Flux-Tunable Phase Shifter for Microwaves.," *Scientific Reports*, Vol. 7, No. 1, 2017.

[23] SHFQC User Manual, Zurich Instruments, https://docs.zhinst.com/shfqc_user_manual/overview.html.

CHAPTER 6
Principles of Electromagnetic Compatibility

> A new idea must not be judged by its immediate results.
>
> —*Nikola Tesla*

This chapter introduces the fundamental concepts of electromagnetic capability (EMC) necessary to operate a superconducting qubit hardware setup. The chapter begins by exploring the crucial role of signal integrity and EMC in a qubit hardware setup. Then it delves into critical techniques for ensuring superconducting qubits' reliable and efficient operation, including shielding, filtering, and grounding. Subsequently, Chapter 7 applies the concepts and techniques learned in this chapter to the qubit hardware setup.

6.1 Signal Integrity

Chapter 3 explained that microwave pulses are crucial in controlling and reading out the quantum states of superconducting qubits. The pulse parameters, such as the shape, amplitude, and width of these pulses applied to a qubit, determine the qubit's final state on the Bloch sphere. Therefore, any changes to the pulse parameters used to apply a quantum gate can potentially harm quantum-gate fidelity.

Chapter 4 further discussed how attenuation, distortion, noise, and interference can negatively impact the signal's shape. Furthermore, noise and interference can also cause qubit dephasing and relaxation, as discussed in Chapter 3. Figure 6.1 shows the effect of distortion on a Gaussian pulse used for qubit control as it travels over a microwave transmission line. Changes in the pulse parameters, such as the width and amplitude, can lead to deviations in the final state from its ideal point, affecting the quantum gate fidelity. Chapter 3 also described how pulse calibration helps to improve gate fidelity and minimize the effect of signal impairments.

Signal-integrity analysis studies how the shape of a signal changes as it propagates in a system and allows us to evaluate whether the signal can be detected successfully. Analytical and simulation tools and measurement techniques are employed at the board and system levels to ensure good signal integrity. This chapter focuses on techniques that mitigate interference, which is the subject of EMC. Next, Chapter 7 discusses various noise-suppression techniques.

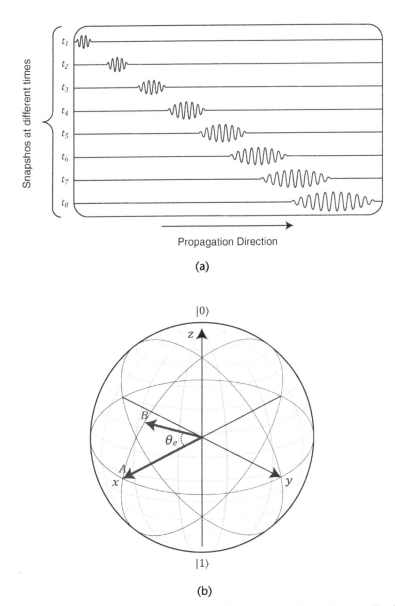

Figure 6.1 (a) Pulse broadening due to dispersion effect. The Gaussian pulse's amplitude and width affect the qubit's final state on the Bloch sphere. (b) The goal is to bring the qubit state to a superposition state at point A on the Bloch sphere. However, due to the broadening of the Gaussian pulse, the final state deviates from point A and ends at point B.

6.2 EMC

Chapter 4 covered the concept of interference. Dealing with interference is the subject of EMC. Three conditions, listed as follows, must be satisfied to meet the EMC requirements for a system [1–3]:

- The system should not interfere with other systems.
- The system should not interfere with itself.

- The system should be resistant to emissions and interference from other systems.

In electronic systems that use narrow pulses, the spectral content spans a wide range of frequencies. This broad frequency range increases the likelihood of interference in electronic systems. Therefore, designers must incorporate EMC considerations into their electronic-system designs to ensure proper operation in the presence of external interference and prevent the creation of interference that could disrupt other systems. These considerations become even more crucial in today's RF environment, where numerous radio signals are present.[1] As covered in Chapter 7, a considerable amount of effort is put into minimizing the interference and noise coupled to a semiconductor qubit system, as these unwanted signals cause the loss of quantum information.

EMC can be approached at two primary levels: board-level design and system-level design. Board-level EMC design involves techniques and considerations at the printed-circuit-board (PCB) level to ensure good signal integrity. This consists of considerations on the board layout, component placement, and managing power, analog, and digital signals. On the other hand, system-level design considerations focus on meeting EMC requirements at the system level using techniques such as filtering, shielding, and grounding. This chapter discusses the system-level EMC design considerations that are directly applied to the qubit hardware setup in Chapter 7.

6.2.1 Interaction of an Electronic System with the Environment

An electronic system's interactions with its surroundings comprise three primary components: the generator, coupling path, and receptor. The generator's signal is coupled to the receptor through the coupling path. The coupling path can be either a conducted path, such as cables and wires, or a radiated path, such as electromagnetic radiation. Figure 6.2 shows that thermal noise at room temperature (generator) can couple to the qubit (receptor) through either the cables that run inside the fridge (conducted coupling path) or radiation (radiated coupling path).

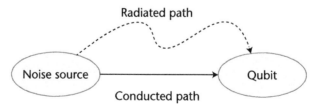

Figure 6.2 Noise coupled to the qubit through the conducted or radiated paths.

---

1. Electronic products need to meet certain regulatory requirements set by the Federal Communications Commission (FCC) in the United States or the Conformité Européene (CE) in the European Union. These requirements essentially ensure that the product does not generate an excessive amount of electromagnetic radiation that could interfere with the function of other electronic components.

The EMC terminology for these two coupling cases is *conducted emission* and *conducted susceptibility* for the case of direct coupling and *radiated emission* and *radiated susceptibility* for the case of indirect coupling. The term *emission* is used for the generator, while *susceptibility* is used for the receptor. In EMC terminology, the source is called the aggressor, whereas the system affected by the source is called the victim.

We can manipulate the generator, the coupling path, or the receptor to protect the system from unwanted signals. For instance, we can identify and eliminate the source of interference or noise. However, this may not always be feasible, so we must use filtering and shielding techniques to block the coupling paths. We can also implement measures and design considerations to make the receptor less vulnerable to interference, such as onboard filtering, shielding, and layout optimization.

6.2.2 Interference Sources

The primary sources of interference in an electronic system are the far-field or antenna coupling and the near-field or crosstalk, as illustrated in Figure 6.3. Various techniques are utilized to suppress coupling effects depending on the operating frequency and nature of the source (far-field versus near-field and electric versus magnetic).

Near-field sources are classified as electric and magnetic. As shown in Figure 6.3, a near-field magnetic source, such as a magnetic dipole, has a low impedance in the near field, where the impedance Z is defined as the ratio of the electric E and magnetic fields H (i.e., $Z = E/H$). As we move away from the source, the magnetic field decays more rapidly than the electric field. Motors and transformers are examples of near-field magnetic sources.

In contrast, as shown in Figure 6.3, an electric source, such as the near field of an electric dipole, has a high impedance near the source, and the electric field decays faster than the magnetic field as we move away from the source. Near-field

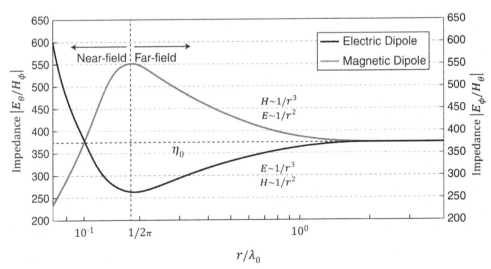

Figure 6.3 Variation of an electric dipole and magnetic dipole impedance with distance.

electric and magnetic probes can be used with a spectrum analyzer to locate near-field interference sources.

As shown in Figure 6.3, in the far field, the ratio of the electric and magnetic fields, called the free-space characteristic impedance $Z = E/H = \eta_0 = \sqrt{\mu_0/\varepsilon_0} = 377\Omega$ is constant for both sources. TV stations and cell phones are examples of far-field sources.

6.2.3 Crosstalk

Crosstalk refers to near-field interference within a system where the source and receptor of electromagnetic emission are located within the same system. An example of crosstalk is the unintended coupling between PCB traces in close proximity. The electric and magnetic fields in a PCB extend to the vicinity, and when the density of circuit traces or cables is high, crosstalk can become a severe issue. As shown in Figure 6.4, there are two types of crosstalk: capacitive and inductive. Various techniques minimize crosstalk, including shields and specific design considerations, such as signal-routing techniques.

In a multiqubit system, crosstalk can occur due to spurious inter-qubit couplings [4–6]. Techniques such as modifying the qubit design and using control pulses to decouple qubits, called dynamical decoupling, are used to suppress the effect of crosstalk [5, 6]. Additionally, crosstalk can be an issue when using a high density of cables going into the fridge, as discussed in Chapter 2 [7].

6.3 Electromagnetic Shielding

As discussed in Chapter 4, noise and interference can be coupled to a device through the radiated path. Electromagnetic shielding can block the radiated path and protect the device from noise and interference. In addition, a shield can prevent radiation generated by a device from escaping outside the shield, thereby protecting other devices from the generated noise and interference. Figure 6.5 illustrates these two functions of the shield.

While the interference coupled through the conducted path is tackled with filtering, shielding blocks the radiated path. Therefore, shielding the qubit and all

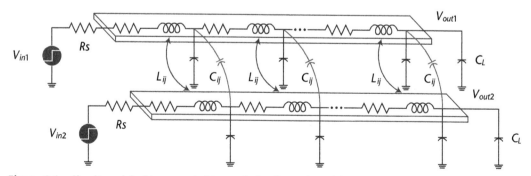

Figure 6.4 Circuit model of two coupled transmission lines. Capacitive and inductive coupling happens between the traces.

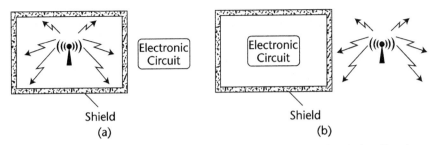

Figure 6.5 A shield being used to (a) contain radiated emissions and (b) block radiated emissions.

the instruments and components in the setup is crucial to blocking the radiated coupling path to the qubit. Sections 6.3.1 to 6.3.4 discuss how to design an effective shield and avoid mistakes that can reduce its effectiveness, such as leaving holes in the enclosure or poor grounding.

6.3.1 Shielding Effectiveness

A shield is typically a metallic enclosure that isolates a device from its surroundings. Figure 6.6 shows the incident, reflected, and transmitted signals when a shielding material is used. Various mechanisms, such as reflection, attenuation, and multiple reflections, contribute to the effectiveness of a shield.

The effectiveness of a shield is defined by a parameter called shielding effectiveness SE, which is the ratio between the incident electric field on the shield $\hat{E}_i$ and the transmitted electric field on the other side of the shield $\hat{E}_t$ (see Figure 6.6) [1, 2]. The shielding effectiveness is given by

$$SE = 20\log_{10}\left|\frac{\hat{E}_i}{\hat{E}_t}\right| \tag{6.1}$$

A 120-dB shielding effectiveness indicates that the transmitted field magnitude is reduced by a factor of 10^6. The shielding effectiveness for magnetic fields is defined as

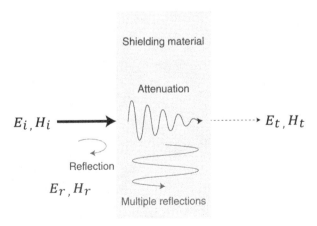

Figure 6.6 Reflection loss, absorption loss, and the multiple reflections in a shield.

6.3 Electromagnetic Shielding

$$SE = 20\log_{10}\left|\frac{\hat{H}_i}{\hat{H}_t}\right| \tag{6.2}$$

The intrinsic impedance of the medium defines the relationship between electric and magnetic fields for a uniform plane wave. If the incident field is a uniform plane wave and the media on both sides of the shield are the same, the two definitions of shielding effectiveness are equivalent. However, they are not identical for near fields or different media on each side of the shield.

Three mechanisms contribute to shielding effectiveness, as shown in Figure 6.6: reflection loss R, absorption loss A, and multiple reflections M, which are expressed by the following equation:

$$SE_{dB} = R_{dB} + A_{dB} + M_{dB} \tag{6.3}$$

The first term, reflection loss, is associated with the portion of the reflected incident field at the interfaces of the shield. According to Table 6.1, reflection loss is the predominant mechanism for shielding low-frequency near-field and far-field electric sources. For this purpose, thin metal sheets are sufficient, as most of the electric field is reflected at the first interface.

The second term in (6.3) describes the absorption loss of the wave as it propagates through the shield. This is caused by the exponential attenuation of the wave amplitude with the factor $e^{-\alpha z}$, where α is the attenuation constant of the material and z is the distance from the shield's first interface. In good conductors, the attenuation constant α is the inverse of the skin depth δ (i.e., $\alpha = 1/\delta$). The skin depth is given by $\delta = 1/\sqrt{\pi f \mu \sigma}$, where f is the frequency, μ is the permeability, and σ is the conductivity of the shield. The shield must be sufficiently thick for effective

Table 6.1 Shielding Mechanisms for Different Types of Sources

	Far field		Near field			
			Electric Source (High-Impedance)		Magnetic Source (Low-Impedance)	
Source Frequency	Low frequency	High frequency	Low frequency	High frequency	Low frequency	High frequency
Predominant shielding mechanism	Reflection	Absorption	Reflection	Absorption	Absorption	Absorption
Type of shielding	Good conductor	Thick good conductor	Good conductor	Thick good conductor	High permeability material (Mu-metal)	Thick good conductor (several skin depths)
Reflection and absorption formulas	$R_{dB} = 168 + 10\log_{10}\left(\frac{\sigma_r}{\mu_r f}\right)$ $A_{dB} = 131.4t\sqrt{f\mu_r\sigma_r}$ $= \frac{8.686t}{\delta}$ (t in meters)		$R_{e,dB} = 322 + 10\log_{10}\left(\frac{\sigma_r}{\mu_r f^3 r^2}\right)$ $A_{dB} = 131.4t\sqrt{f\mu_r\sigma_r}$ $= \frac{8.686t}{\delta}$ (t in meters)		$R_{m,dB} = 14.57 + 10\log_{10}\left(\frac{fr^2\sigma_r}{\mu_r}\right)$ $A_{dB} = 131.4t\sqrt{f\mu_r\sigma_r}$ $= \frac{8.686t}{\delta}$ (t in meters)	

absorption loss, especially at high frequencies, where the skin effect becomes the dominant shielding mechanism. Table 6.1 recommends using thick conductors for shielding high-frequency electric and magnetic sources.

Low-frequency magnetic sources can be shielded using high-permeability materials such as ferromagnetic or Mu-metals. The shielding effect by a high-permeability material is shown in Figure 6.7(a), where the material acts as a low-resistance path for magnetic fields, diverting them from the victim.

The following three factors may degrade the effectiveness of high-permeability magnetic shields:

- The permeability of ferromagnetic materials decreases with increasing frequency.

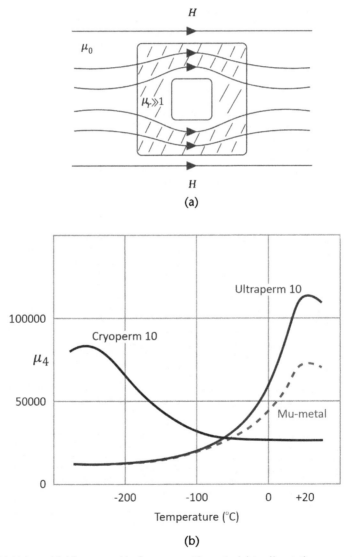

Figure 6.7 (a) Using a highly permeable ferromagnetic material to divert the magnetic field and (b) the cryoperm, ultraperm, and mu-metal permeability versus temperature [8].

- The permeability of ferromagnetic materials decreases with increasing magnetic field strength, as a large magnetic field can saturate the material.
- The permeability of conventional ferromagnetic materials decreases with decreasing temperature.

Therefore, high-permeability materials are ineffective at high frequencies, low temperatures, and large magnetic fields.

Using two shields can mitigate the saturation effect caused by large magnetic fields. The outer shield has low permeability and low susceptibility to saturation, which reduces the intensity of the incident magnetic field to prevent saturation of the inner shield.

An additional consideration for qubit experiments is that the permeability of conventional ferromagnetic materials reduces with temperature, as shown in Figure 6.7(b). Therefore, special ferromagnetic materials, such as cryoperm, must be used for qubit experiments at millikelvin temperatures [8].

The third term, M in (6.3), represents the contribution of multiple reflections and transmissions, which occur as waves bounce back and forth between the shield's interfaces. These reflections can create fields that add to the initial field transmitted across the right-hand side interface, thus reducing the shielding effectiveness. Unlike the reflection and absorption loss, the multiple-reflection factor M is generally negative. However, for most practical shielding scenarios, the effect of multiple reflections is insignificant and can be neglected.

Figure 6.8 compares the shielding effectiveness of copper and aluminum shields, both 0.508-mm-thick. It can be observed from Figure 6.8 that reflection loss is the primary shielding mechanism at low frequencies, while absorption loss becomes dominant as the frequency increases.

Note that the shielding effectiveness can be reduced under the following conditions:

- Any penetration into the shield, such as holes, cables, slots, or seams, can drastically reduce its effectiveness. Therefore, penetrations into the shield must be treated appropriately.
- A shield with no ground is not as effective as expected. Accordingly, shields must be grounded.

6.3.2 Effect of Penetration in the Shield

Unless appropriately treated, any penetration in the shield can drastically reduce its shielding effectiveness. Seams, cables, holes, or slots are necessary for cooling or when something comes in or out of the enclosure. The electrical size, which is the ratio of the physical dimension to the wavelength, of the holes or slots in the shield determines how much energy can leak through the enclosure. If the opening is larger than half the wavelength, the power leaks easily in or out of the enclosure. Therefore, it is essential to avoid large holes or slots.

A typical penetration is a hole or slot in the shield wall. Babinet's principle can be used to explain the slot effect in the enclosure. According to Babinet's principle, fields inside the shield can radiate outside through the slot opening, reducing its

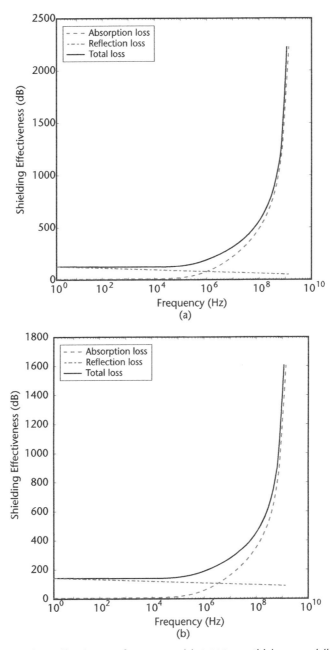

Figure 6.8 (a) Shielding effectiveness for copper with 0.508-mm thickness and (b) shielding effectiveness for aluminum with 0.508-mm thickness.

effectiveness. Moreover, fields outside the shield can be picked up inside through the slot opening. In both cases, the slot acts as an antenna.

Figure 6.9 shows a slot and the complementary structure. If the slot in the left of Figure 6.9 has a linear dimension that is half the wavelength, then the radiation properties of this slot will be the same as a dipole antenna of the same length, with E and H components interchanged, as shown in the right of Figure 6.9. Note that the slot width does not significantly affect the radiation.

6.3 Electromagnetic Shielding

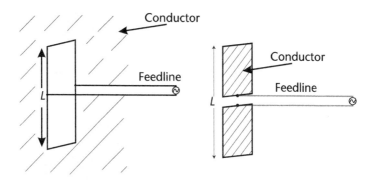

Figure 6.9 Babinet's principle. A slot is equivalent to a complementary radiating structure.

Babinet's principle expresses that the far fields radiated by the slot ($E_{\theta s}, E_{\phi s}, H_{\theta s}, H_{\phi s}$) and those radiated by the complementary structure ($E_{\theta c}, E_{\phi c}, H_{\theta c}, H_{\phi c}$) shown in Figure 6.9 are related by the following relationship: $E_{\theta s} = H_{\theta c}$, $E_{\phi s} = H_{\phi c}$, $H_{\theta s} = -E_{\theta c}/\eta_0^2$, and $H_{\phi s} = -E_{\phi c}/\eta_0^2$ (where η_0 is the free-space characteristic impedance). In light of this, it can be seen that apertures can function as effective antennas with the same conductor dimensions as the aperture. The radiation is generated since the slot largely perturbs the current path. So, if an opening needs to be created in the enclosure for ventilation purposes, avoiding slots and using holes is necessary, as holes do not change the current path significantly, as shown in Figure 6.10(a, b).

The next consideration concerns the lids that provide access to the interior of the shielded enclosure. Sometimes the lids have slots around them, as shown in Figure 6.10(c). Hence these gaps are treated by placing conductive gasket material in the gap or closely spaced screws to "short out" the slot antennas [2].

A component on a PCB can be shielded (see Section 6.2), the PCB itself can be shielded, and even the entire room can be shielded. Rooms used for high-accuracy electromagnetic testing or highly secured environments must be shielded. However, these rooms also need to have ventilation. To achieve this, a honeycomb array of waveguides below the cutoff frequency is used in shielded room walls. This design allows airflow into the room while preventing the propagation of frequencies below the waveguide's cutoff frequency into the enclosure [2].

6.3.3 Effect of Grounding on the Shield

Grounding the shield serves two essential purposes. First, it ensures safety by preventing high-voltage shocks from touching the shield. The shield of any instrument or component powered by a wall outlet must be connected to the outlet's safety ground. This ground connection ensures that the enclosure's potential is not too high, which could otherwise give electric shocks to anyone touching it.

The second reason for grounding a shield is to prevent unwanted coupling between circuits. As illustrated in Figure 6.11(a), an ungrounded shield can facilitate coupling between circuits inside or between cables going into the enclosure and the circuits [1, 2]. This can lead to crosstalk within the system and reduce the shielding effectiveness of the enclosure. On the other hand, a grounded shield does

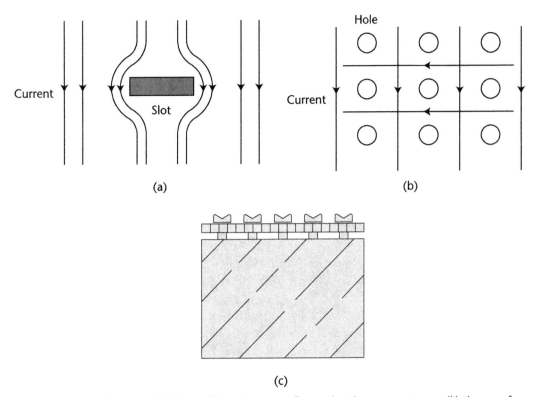

Figure 6.10 (a) A slot on a shield perturbing the current flow and acting as an antenna; (b) the use of many small holes, which offers the same ventilation effect as a large slot and does not perturb the current much as a large slot would; and (c) slots radiating according to Babinet's principle with closely spaced screws used to break up the slot length. (*After:* [2].)

not create crosstalk within the system, as shown in Figure 6.11(b). Therefore, it is essential to ground all shields to the circuit neutral to prevent unwanted coupling and ensure the shield's proper functioning.

6.3.4 Shielding Techniques for Qubits

For a qubit or any other microwave system, it is crucial to identify the noise sources and keep them away from the system if possible. For example, sources such as motors and loop antennas that produce predominantly magnetic fields at high-current or low voltage levels, including pumps and compressors of the dilution fridge, should be kept away from the qubit hardware setup. However, it is also essential to have sufficient shielding for the qubit, to suppress the noise and interference caused by various near-field and far-field sources.

For room-temperature electronics, a shield made of a good conductor, such as aluminum that is a few millimeters thick, is used to provide shielding. The PCB version of a shield called a shield can is shown in Figure 6.12(a). A shield can is usually used to protect sensitive components such as detectors or receivers or contain the radiation of components such as amplifiers. It is crucial to be cautious when selecting the shield size, as a shield with a large volume can excite cavity modes that may couple to the circuits and components on the PCB, creating unwanted effects such as instabilities (see microwave cavities in Section 5.4.3.3). Hence, minimizing the shield's volume is vital to suppress cavity modes. In some cases, microwave absorbers inside the shield help suppress cavity modes, particularly when the resonance source cannot be identified, or the shield itself cannot be modified.

The following two types of shields are used to protect the qubits:

- A relatively thick oxygen-free high-conductivity (OFHC) copper enclosure, as shown in Figure 6.12(b), shields the quantum chip mounted on the sample holder. Before placing the larger enclosure (on the far left of Figure 6.12(a)), a smaller intermediate cavity with a higher resonant frequency is utilized to prevent energy coupling to the resonant mode of the larger enclosure's cavity. OFHC copper is used due to its high thermal conductivity, which is essential in cryogenic environments to conduct the heat away from the experiment. This shielding is effective against far-field and near-field electric and magnetic sources at high frequencies.

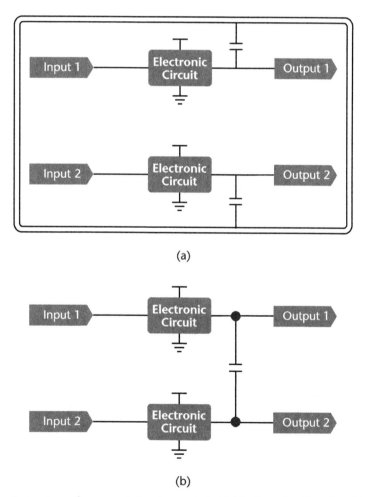

Figure 6.11 Comparison of ungrounded and grounded shields: (a) a circuit enclosed in an ungrounded shield and (b) its equivalent circuit, and (c) a circuit enclosed in a grounded shield and (d) its equivalent circuit.

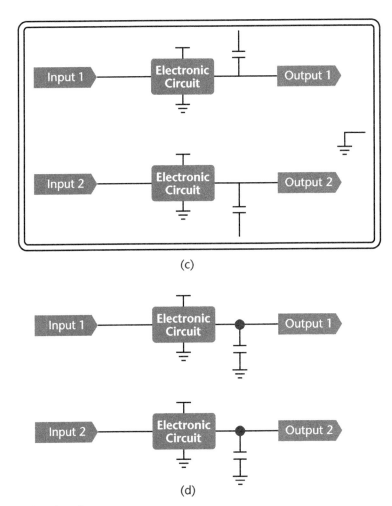

Figure 6.11 *(Continued)*

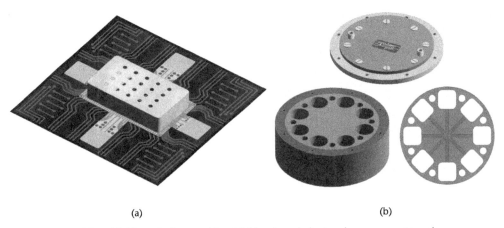

Figure 6.12 (a) A shield can being used to shield onboard electronic components and (b) electromagnetic shielding of the qubit.

- A cryogenic mu-metal shield with high permeability is used to shield against low-frequency magnetic sources. For example, a double-walled magnetic shield made from tempered cryoperm 10 can be utilized. Chapter 7 provides further details on shielding.

6.4 Filtering

Chapter 5 discussed various types of filters; Chapter 7 demonstrates that various low-pass filters, such as copper powder and infrared filters, filter out high-frequency noise and interference.

Filters are used to remove undesired frequency components, like harmonics and spurs, across different stages of the generation, processing, and detection chain. The 1/f noise from electronic instrumentation and 50/60 Hz frequency from the power grid can negatively impact coherence times if they reach the qubit. Figure 6.13(a) shows a breakout box used for room-temperature filtering of the signals that go into the fridge. Typically, BNC connectors are used on the front panel, and low-pass filters are inside the box. A shielded cable at the breakout box's rear panel is connected to the fridge's input. Figure 6.13(b) shows the interior of a typical

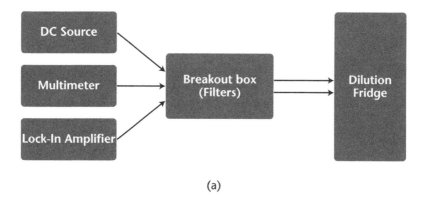

(a)

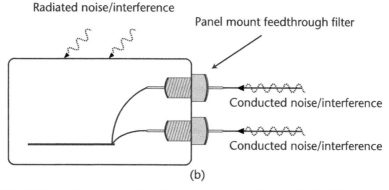

(b)

Figure 6.13 (a) Room-temperature filtering using a breakout box, with the filtered output of the breakout box going into the dilution fridge and (b) the interior of a typical breakout box, with panel-mount filtering blocks conducted interference and shield blocking the radiated interference.

breakout box. The shield blocks radiated noise, and panel-mount feedthrough filters are used to minimize the effect of noise coupled through the conducted path. Chapter 7 provides a comprehensive overview of room-temperature and cryogenic filtering in superconducting qubit hardware setup.

6.5 Grounding

Grounding is crucial to an electronic system's functionality, particularly at high frequencies; however, it is often overlooked. This section discusses the significant impact of grounding on a circuit's performance, particularly at high frequencies.

There are two types of grounds: safety and signal. A safety ground, called chassis ground, is typically needed to protect against shock hazards [1, 2]. In addition to providing shock protection, the safety ground plays a crucial role in draining charge and rerouting ESD currents away from delicate electronics. When designing a component or an instrument in an enclosure, we must always consider providing safety ground and proper enclosure grounding by connecting the circuit's ground to the enclosure.

The signal ground is the path for the current to return to its source. However, it is crucial to note that there is no guarantee that a signal will return to its source along the designated paths, even if this was the designer's intention. The signal return path must be determined and purposefully designed for EMC design to succeed. The reason for this is explained as follows.

A ground plane, such as the ground plane of a PCB, is ideally an equipotential plane with zero impedance $Z = R + jL\omega = 0$, where R is the resistance, L is the inductance, and ω is the radian frequency. This idea may be appropriate at dc or low frequencies but is never accurate at higher frequencies. At high frequencies, conductors have significant impedance due to the inductive part of Z, where the conductor's resistance, including the skin effect, is negligible compared to its inductance. Because of this impedance, any currents that flow through the assumed "ground" will result in different potentials at various spots on the ground's surface. Therefore, the assumed ground cannot be considered an equipotential surface. The voltage difference between the two points of the ground, $V_G = V_{G1} - V_{G2}$, acts like a voltage source, which drives common-mode currents I_{C1} and I_{C2} through the signal and return wires, as shown in Figure 6.14(a). These common-mode currents are similar to antenna currents and can cause radiation.

It is important to note that currents return to their source via paths of the lowest impedance. As the ground's impedance, Z, varies with frequency, the lowest impedance path for each frequency component of a signal may differ. Therefore, different frequency components may return via different paths, which can distort the waveform (see Chapter 4). Therefore, poor grounding can cause distortion and radiation, among other effects.

A well-designed grounding strategy is crucial for maintaining signal integrity and minimizing noise in a hardware setup. While there are various techniques and considerations for designing a good high-frequency ground, we focus on one crucial technique used in semiconductor qubit hardware setup: breaking ground loops. Modifying the grounding of a fridge has been proven to improve dephasing

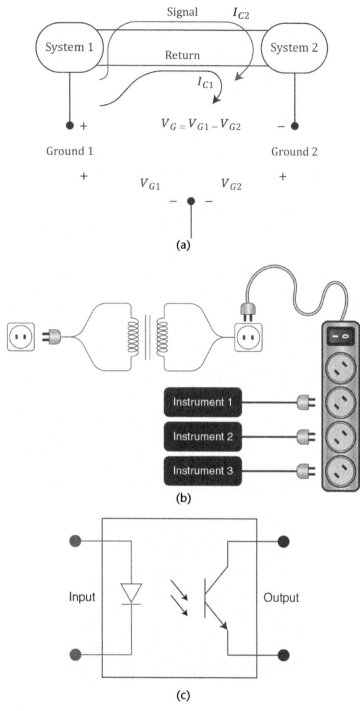

Figure 6.14 (a) Common mode currents being generated on interconnect cables due to ground voltage differences, (b) breaking ground loop using isolating transformer, and (c) an optocoupler.

times [9, 10]. The potential problems for qubit experiments due to ground loops include generating unwanted noise signals through common-mode currents on interconnect cables. Moreover, nearby magnetic fields passing through the ground loops induce noise signals. These unwanted signals can couple into the fridge and consequently to the qubit, resulting in dephasing and corruption of quantum information.

It is necessary to ensure that there is no electrical contact between the circuits to break the ground loop between them. This process of isolating two circuits without electric contact is called galvanic isolation. The transmission of an electric signal between two circuits without any electric contact can be achieved through the following methods:

- *Isolating transformer* (1:1 transformer): An inductive coupling, such as with a transformer, as shown in Figure 6.14(b).
- *Optocoupler:* Alternatively, the electric signal can be converted into an electromagnetic or optical wave, which travels the medium between the two circuits and is converted back into an electric signal on the other end. This can be accomplished through optocouplers with different bandwidths for analog and digital signals, as shown in Figure 6.14(c).[2]

The first step to improve the grounding in the setup is to isolate the power source for the qubit hardware setup from the electricity network using an isolating transformer called a 1:1 transformer. For GPIB connections, an isolator such as the GPIB-120A Bus Expander/Isolator from National Instruments can be used. This isolator uses an optocoupler to isolate the GPIB port on the instrument's side and the port on the PC's side. Similarly, isolators can be used for USB and Ethernet connections.

Chapter 8 shows that an AC bridge is used to precisely measure a dilution fridge's internal temperature. When connecting to the fridge, there is likely to be a significant difference between the ground of the instrument chassis and the ground of the fridge. If the two grounds are connected, 50/60-Hz noise currents can be introduced into the sensor wiring [11]. Therefore, it is essential to break ground loops using galvanic isolation in this case as well.

6.5.1 Grounding of Wires

Protection against the magnetic field is achieved by minimizing the loop area between a signal and its return path (e.g., using twisted-pair or neighbor conductors in flat cables). Using a twisted-pair cable is also effective in reducing electric-field coupling if we have balanced terminations at both ends with respect to the reference conductor [2]. Shielded cables or twisted pairs are recommended for the room-temperature qubit hardware setup to minimize crosstalk and external electromagnetic interference. Further, proper grounding is necessary for the effective functioning of the cable's shielding property.

---

2. Common mode choke is another technique that can break the ground loop, but it is not usually used in the qubit setup.

Enclosing cable wires with a shield may not reduce the cable's radiated emissions. A cable shield must be attached to the ground to reduce radiated emissions effectively. Note that effective cable shielding reduces the electric field coupling (capacitive coupling) so long as either end of the shield is grounded. The shield must be grounded at both ends to mitigate magnetic-field coupling (inductive coupling). This allows the current to flow back along the shield, producing a magnetic flux that cancels the flux due to the current induced by the noise generator.

Many undesirable issues can be caused by ungrounded or poorly grounded cables. For example, connecting the cable shield peripherally to the enclosure may cause internal noise currents to flow out of the enclosure along the shield's exterior, leading to radiation. Therefore, removing an overall shield may sometimes decrease a cable's radiated emissions. The cable shield may also become a monopole antenna if the voltage of the point where the shield "pigtail" is attached varies, such as the logic ground of an electronics PCB [2]. If the length of the cable shield is around a quarter wavelength, it becomes an effective radiator. Resonances in the radiated emissions are often observed due to common-mode currents generated on cables with ungrounded or poorly grounded shields. Eliminating these resonances and their enhanced radiated emissions can often be achieved when the cable is disconnected.

References

[1] Ott, H. W., *Electromagnetic Compatibility Engineering* (Second Edition), New York, NY: Wiley, 2009.

[2] Paul, C. R., *Introduction to Electromagnetic Compatibility* (Third Edition), Hoboken, NJ: Wiley, 2006.

[3] McLaren, P. D., *EMI Troubleshooting Cookbook for Product Designers*, New York, NY: Springer, 2014.

[4] Sarovar, M., "Detecting Crosstalk Errors in Quantum Information Processors," *Quantum*, Vol. 4, September 2020, arXiv:1908.09855.

[5] Da Silva, M. P., A. Melville, and A. Megrant, "Reducing Crosstalk in Superconducting Qubits," *Physical Review Applied*, Vol. 16, 2021.

[6] Nuerbolati, W., et al., "Canceling Microwave Crosstalk With Fixed-Frequency Qubits," *Appl. Phys. Lett.*, Vol. 120, No. 17, April 25, 2022, https://doi.org/10.1063/5.0088094.

[7] Reilly, D. J., "Challenges in Scaling-up the Control Interface of a Quantum Computer," *2019 IEEE International Electron Devices Meeting (IEDM)*, San Francisco, CA, 2019, pp. 31.7.1–31.7.6, doi: 10.1109/IEDM19573.2019.8993497.

[8] Cryoperm, https://www.mushield.com/material-sales/cryoperm/.

[9] Finnegan, et al., "Vibration-Induced Electrical Noise in a Cryogen-Free Dilution Refrigerator: Characterization, Mitigation, And Impact on Qubit Coherence," *Applied Physics Letters*, Vol. 114, No. 24.

[10] Birenbaum, J. S., *The C-shunt Flux Qubit: A New Generation of Superconducting Flux Qubit* (PhD thesis), University of California, Berkeley, CA, 2014.

[11] Lakeshore Cryotronics, "Model 372 AC Resistance Bridge Installation Best Practices for Measurement in a Dilution Refrigerator."

CHAPTER 7
Control Hardware for Superconducting Qubits

If science is to progress, what we need is the ability to experiment.
—*Richard Feynman*

This chapter brings together all that we have learned so far to gain a comprehensive understanding of the control hardware for superconducting qubits. Since there are significant similarities between the control hardware of superconducting, spin, and topological qubits, the knowledge gained in this chapter can be widely applied to semiconductor qubits.

We begin by examining the room-temperature setup, where signal generation and detection occur. Next, we explore the cryogenic setup and various techniques, including noise reduction and low-noise amplification.

We conclude the chapter by exploring control and measurement automation, providing insights into how these processes can be automated to enhance efficiency and accuracy.

7.1 High-Level Description of the Setup

Just like today's computers, quantum computers are expected to follow the path of integrating discrete components into chips, resulting in size and cost reduction while improving performance. However, the fundamental principles of microwave-signal generation, processing, transmission, and detection in superconducting qubit systems remain unchanged. It is crucial to comprehend these principles to advance and enhance quantum computing technology. Chapters 1–6 established a strong foundation for qubit operation and microwave engineering, including discussions of qubit control and readout and the concept of a superconducting quantum computer functioning as a microwave link that incorporates techniques from all four areas of microwave engineering (generation, transmission, processing, and detection) to ensure proper operation.

This chapter applies these ideas to the cryogenic and room-temperature hardware setup for superconducting qubits [1–4]. To better understand the setup, let us begin by providing a high-level description, mapping each part of the setup to the concepts of qubit control and readout we learned in Chapter 3. This section aims to examine how the hardware functions as a whole.

Subsequently, the following sections delve into a more detailed, low-level analysis of the building blocks comprising the setup, thoroughly studying each component.

The setup shown in Figure 7.1 represents a typical configuration for superconducting qubits, which can also be adapted with modifications for topological qubits like the top-transmon and spin qubits. The input involves signal sources that generate pulses and microwave signals to control and read the qubits. These room-temperature signals enter a dilution fridge and interact with the qubits in the quantum-processing unit (QPU). The primary purpose of the fridge is to create a controlled, low-noise environment that allows the qubit to maintain coherence for an extended period. Once the qubits are interacted with, the readout signals are amplified, downconverted, and digitized for postprocessing in the host computer, as shown in Figure 7.1.

Let us now focus on the signal-generation block shown in Figure 7.1. Chapter 3 explains that microwave pulses control the qubit's state on a Bloch sphere. Moreover, continuous and pulsed spectroscopy techniques enable the readout of the qubit's state using microwave signals. In addition, Chapter 3 demonstrated that magnetic fields are coupled to the qubit's SQUID using DC and pulse signals for global and local tuning of the qubits. Chapter 5 discussed the need for a pump signal to operate a cryogenic parametric amplifier, which is generated by a microwave source. These are the signals necessary for qubit control, tuning, and readout.

The setup at cryogenic temperatures serves two main functions. First, it aims to minimize noise coupled to the qubit by implementing various noise-suppression techniques such as attenuators, low-conductivity microwave cables, filters, and thermal and electromagnetic shielding. Second, it requires ultralow noise amplification of the readout resonator's signal to ensure successful detection at room temperature by maximizing the SNR.

Room-temperature signal detection involves several steps. It begins with low-noise amplification at room temperature. Downconversion is necessary to shift

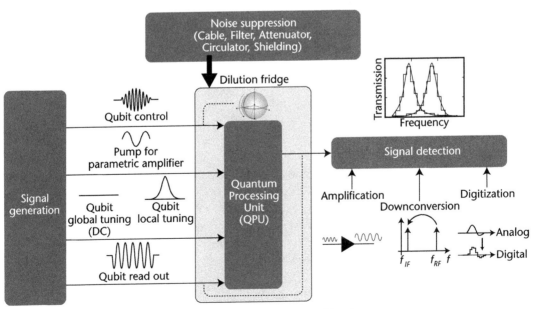

Figure 7.1 High-level description of the superconducting control hardware.

the signal to a frequency range suitable for digitization. Finally, a digitizer is used to sample the analog signal for subsequent postprocessing in the host computer.

7.2 Low-Level Description of the Setup

Figure 7.2(a) provides detailed information about each block diagram presented in Figure 7.1. Additionally, Figure 7.2(b) illustrates the real-world setup within a dilution fridge. To offer additional clarity, Table 7.1 explains each section of the setup.

Currently, almost all labs researching superconducting quantum computers employ a setup featuring control and readout chains operating at room temperature. However, a notable trend is emerging, wherein there is a shift toward integrating and miniaturizing the room temperature setup and relocating it into the cryogenic environment of the fridge. This approach not only results in a notable reduction in the amount of noise coupled from room temperature to the fridge but also facilitates qubit scaling. Section 7.5.2 discusses cryogenic CMOS chips to implement a cryogenic control and detection scheme.

Sections 7.3 and 7.4 conduct a more detailed investigation into each building block depicted in Figure 7.2.

7.3 Room-Temperature Setup

This section discusses signal generation, processing, and detection at room temperature. Although the industry is moving toward fully digital signal generation and detection, studying both the analog and digital techniques is essential. The ideas and principles of analog techniques are highly insightful and contribute to a deeper understanding of a system. Additionally, many labs continue to utilize analog techniques due to their simplicity, relatively low cost, and ease of implementation, especially when working with a small number of qubits.

7.3.1 Signal Generation

In qubit experiments, three types of signal sources are utilized: the AWG, microwave-signal generator, and DC voltage source, as depicted in Figure 7.2(a). As we'll soon discover, all signal generation can be accomplished digitally, offering numerous advantages, including reduced component count and a more cost-effective solution per channel.

7.3.1.1 Analog-Signal Generation

Let us begin by discussing qubit control. As explained in Chapter 2, any single-qubit gate can be represented as rotations on the Bloch sphere. At the qubit's transition frequency, microwave pulses enable arbitrary rotations around the x- and y-axes of the Bloch sphere, where the pulse amplitudes and phases define the rotation angle and axis orientation, respectively. By combining rotations around the axes of the Bloch sphere, as detailed in Chapter 3, rotations around any arbitrary axis of the Bloch sphere can be achieved. For example, Section 3.3.1.5 discussed using pulse-shaping techniques, such as Gaussian and DRAG pulses, to prevent the excitation of transitions to higher qubit levels.

Figure 7.3 illustrates the pulse-generation process using I and Q components of an AWG. The microwave-signal generator applied to the LO port allows for the upconversion of the Gaussian pulse to the qubit's transition frequency. Software packages like Python and MATLAB are employed to design these pulses. The generated codes are then imported into the AWG and stored in memory. (See Chapter 5 to learn about AWG operation.)

Now, let us discuss the qubit readout. In Figure 7.2(a), the same input line is used for both qubit control and readout, with a power combiner employed to combine these two signals. However, having separate drive and readout lines is advantageous as it allows complete control of the on-chip couplings. Moreover, having separate lines enables faster gates since driving does not happen through the resonator [1].

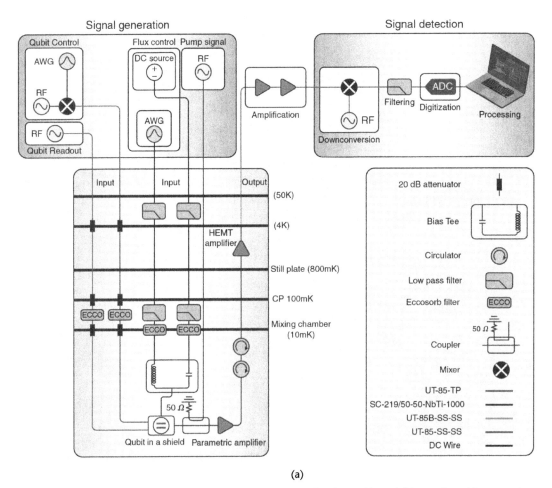

(a)

Figure 7.2 (a) Schematic of a hardware setup for a superconducting qubit and (b) a real-world cryogenic hardware setup.

7.3 Room-Temperature Setup

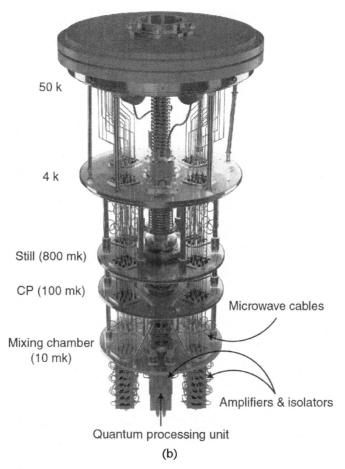

Figure 7.2 *(Continued)*

Section 3.3.3 discussed two schemes for qubit readout. The first scheme involves continuously driving the qubit and readout resonator simultaneously using a microwave generator with a CW output. However, this continuous-readout approach leads to an ac-Stark shift of the qubit frequencies during the qubit drive due to the presence of photons in the readout resonator.

The second method utilizes a pulse for readout. In this scheme, the readout resonator is driven after the qubit drive is switched off. By doing so, the ac-Stark shift issue is circumvented since no photons are in the resonator while the qubit is being driven. The AWG generates the readout pulse, which is then upconverted using a mixer and a microwave generator, as illustrated in Figure 7.2(a).

A microwave-signal generator is required for the pump signal of the cryogenic parametric amplifier. As shown in Figure 7.2(a), the pump signal generated by a room-temperature microwave source is applied to the coupled port of the directional coupler, which is combined with the readout signal coming from the qubit.

Furthermore, as depicted in Figure 7.2(a), a microwave signal generator is needed as the LO for the downconverter in the room-temperature detection chain. To maintain phase coherence, it is advantageous to split the LO employing a power splitter and use it for both the readout up and downconversion.

Table 7.1 Description of the Room-Temperature and Cryogenic Hardware Setup Shown in Figure 7.2(a)

Operation	Instruments/Components

Signal generation (room temperature)	**Qubit control**
	Analog techniques
	• Microwave signal generator
	• Arbitrary waveform generator
	• Mixer
	Digital techniques
	• Direct digital synthesis
	Qubit frequency tuning
	• DC source for global qubit tuning
	• AWG for fast control pulses for individual qubit tuning
	Qubit readout
	Analog techniques:
	• Continuous readout
	  • Microwave signal generator
	• Pulsed readout
	  • Microwave signal generator
	  • Arbitrary waveform generator
	  • Mixer
	Digital techniques:
	• Direct RF sampling

Qubit in the fridge (cryogenic temperature)	**Noise suppression** • Low–thermal conductivity cables • Attenuators • Filter • Circulator **Signal amplification** • Directional coupler • TWPA • HEMT amplifier
Signal detection (room temperature)	**Signal amplification** • LNA **Downconversion** Analog techniques • Mixer • Microwave-signal generator Digital technique • Direct RF sampling **Analog-to-digital conversion** Digitizer

An additional channel of the AWG is utilized to generate pulses for quick and local tuning of the qubit's transition frequency. For this purpose, the AWG produces short pulses applied to the on-chip flux bias lines (FBLs), enabling rapid and localized shifts in the qubit transition frequency.

Figure 7.2(a) illustrates the utilization of a DC voltage source to tune the qubit's transition frequency globally through the external superconducting coil. The coil holder shown in Figure 7.4(a) is positioned at the rear of the sample holder, as depicted in Figure 7.4(b). An ultraclean battery-powered voltage source such as SRS SIM928 can be employed to bias the coil. Alternatively, an AC-powered DC voltage source, such as Yokogawa DC 7651, can be utilized along with a low-pass filter to eliminate the 60-Hz noise originating from the Yokogawa source. The low-pass filter and its frequency response are shown in Figure 7.4(a, b).

The impact of the DC source and filtering on the qubits' phase coherence can be evaluated by measuring the Ramsey oscillations. The results obtained using no filter and a battery are comparable to those obtained using an AC-powered voltage source with a filter. Not implementing a filter or utilizing a filter on only one of the lines can have detrimental effects on qubit coherence. Thus, it is recommended to employ differential filtering to maximize noise rejection. Figure 7.4(c) showcases an example of an RC filter used in a DC line with a cutoff frequency of $1/2\pi RC$. The low-pass filter employed in Figure 7.4(c) in conjunction with the superconducting coil is a T filter. A T filter is necessary since the coil exhibits low impedance at low frequencies (see Chapter 5).

In Figure 7.4(c), R_1 is a bias resistor, providing the required current to the superconducting coil. Moreover, the combination of R_1 and C acts as a low-pass filter on both leads, and R_2 ensures a high-impedance environment at the filter output. Our objective is to effectively suppress 60-Hz noise while maintaining a reasonable RC time constant (τ) on the order of 100 ms. The R_1 and R_2 ratio selection is made to optimize the filter cutoff, and the total resistance $2R_1 + 2R_2$ is adjusted to accommodate the full range of the voltage source. Figure 7.2(a) depicts a bias-tee combining the DC and the AWG signal.

7.3.1.2 Digital-Signal Generation

We have observed that to control a single qubit, two AWG channels (I and Q), a microwave-signal generator, and an IQ mixer are required, as illustrated in Figure

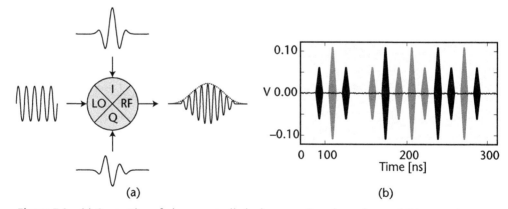

Figure 7.3 (a) Generation of phase-controlled microwave Gaussian pulses and (b) a sample sequence of concatenated Gaussian pulses. (Black (gray) pulses correspond to x(y) gates of either π or $\pi/2$ rotations.)

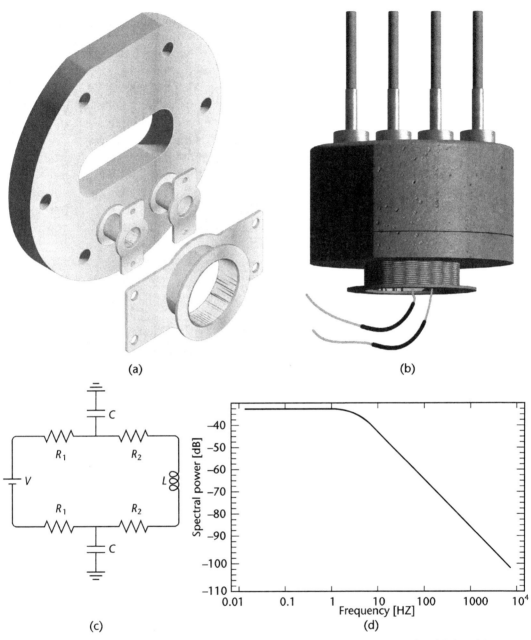

Figure 7.4 (a) Coil holder. (b) The coil holder, which is used for the global tuning of multiple qubits, is installed on the back of the sample holder. (c) An RC low-pass filter is used to voltage bias the superconducting. (d) The frequency response of the filter with $C = 6.8\ \mu F$, $R_1 = 4.7\ k\Omega$, $R_2 = 3\ k\Omega$. The L represents the inductance of the coil.

7.5(a). A pulse is commonly employed for readout, necessitating its own AWG channel, mixer, microwave-signal generator, and an additional power combiner since the readout pulses sometimes share the same line as the control line [see Figure 7.5(a)]. In addition to the number of components, this type of setup requires calibration to achieve high gate fidelities. One important calibration step involves calibrating the IQ mixer offsets and gains to minimize the impact of carrier leakage and spurs, which can drive undesirable transitions.

As demonstrated in Chapter 5, the sampling rate of digital sources has reached significantly high levels, resulting in a broad bandwidth that can efficiently produce microwave signals up to a few tens of gigahertz. This advancement obviates the need for upconversion, using a technique called broad-bandwidth DDS. With DDS, a single channel can generate control and readout pulses at different frequencies, offering significant advantages in operating single and multiple qubits. Figure 7.5(b) illustrates how this capability simplifies the setup by reducing the number of components and eliminating the need for mixer calibration. This simplification is crucial for scaling quantum hardware, reducing potential sources of infidelity, and decreasing hardware costs per qubit. Furthermore, this technique enables precise phase relations between signals at disparate frequencies and across multiple channels, facilitating multiqubit gates [5, 6].

One drawback of digital-signal generation is its inferior phase noise compared to analog sources, as discussed in Chapter 5. However, randomized benchmarking schemes have demonstrated that digital-generation techniques can achieve high-fidelity qubit control pulses with an average gate error rate of less than 5×10^{-4} [5].

As mentioned in Chapter 5, modern RF DACs can directly synthesize tones above their sampling rates using the second Nyquist zone mode. This feature enables the system clock rate to remain compatible with digital-logic systems while still allowing the generation of high-frequency pulses with arbitrary profiles [6].

7.3.2 Signal Processing

Signal processing at room temperature serves two primary purposes: filtering and signal detection. Sections 7.3.2.1 and 7.3.2.2 delve into these two functions and explore their significance.

7.3.2.1 Room-Temperature Filtering

Recall the discussion of different types of filters and their corresponding operational parameters from Section 5.4.3.1. Table 7.2 highlights using a low-pass filter

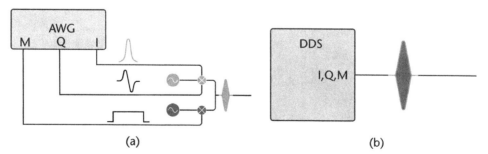

Figure 7.5 (a) The baseband control pulse and the qubit readout pulse (M) need three AWG channels. Two microwave signal generators and two mixers are needed to upconvert the baseband pulses. A power combiner is needed to send the control and readout pulses over a single line [5]. (b) Using a direct digital synthesis (DDS) generator with a high sampling rate makes it possible to generate control and readout microwave pulses without an external upconverter.

in the DC control line and after the mixer to eliminate harmonics and spurious signals in the detection chain. Now, let us discuss implementing these filters in a qubit setup.

It is necessary to filter the DC and low-frequency signals before they enter the fridge to ensure that the coherence times are not affected by the 60-Hz frequency from the power grid and 1/f noise from electronic instrumentation. The filtering process is illustrated in Figure 7.6(a) within a breakout box. The front panel of the breakout box typically features BNC connectors [Figure 7.6(b)], while the box contains the low-pass filters. A shielded cable connected to the input of the fridge is located on the rear panel, as depicted in Figure 7.6(c).

When designing or selecting a low-pass filter for a qubit experiment, careful consideration should be given to the cutoff frequency. In order to rapidly tune the qubit transition frequency, nanosecond flux pulses are applied to the FBLs. To transmit a 10-ns pulse without significant shape degradation and attenuation, a bandwidth of over 100 MHz is required. To prevent pulse degradation, frequency

Table 7.2 Low-Pass Filters Used at Room and Cryogenic Temperatures

Stage	Where	Description
Room temperature	DC control line for the superconducting coil	Suppress 60-Hz and 1/f noise as well as high-frequency noise
	Detection chain	Placed after the mixer to remove the harmonics and spurs in the detection chain

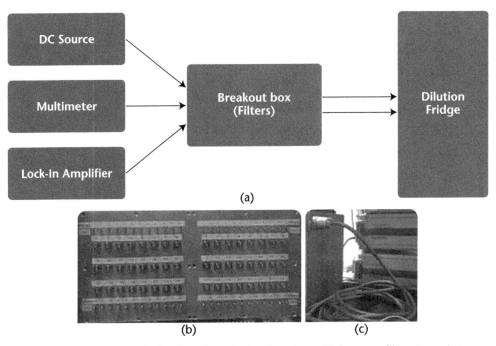

Figure 7.6 (a) Schematic of a breakout box; the breakout box with low-pass filters is used at room temperature to filter the DC and low-frequency noise that enter the fridge; (b) the front panel of a breakout box; and (c) the back panel of a breakout box with a shielded cable that goes to the fridge DC input.

components up to the fourth or fifth harmonic must be preserved. Therefore, the cutoff frequency of the low-pass filter should fall within the range of 400–500 MHz, where the smooth roll-off characteristic helps prevent pulse shape degradation.

In some cases, a low-pass filter is also employed after the mixer in the detection chain to mitigate the impact of spurs generated by the mixer's nonlinearities, as depicted in Figure 7.2(a).

7.3.2.2 Signal Detection

Recall the signal-detection principles explained in Section 5.4, including utilizing homodyne and superheterodyne architectures. While the homodyne detection scheme is simpler, it performs less efficiently than the more complex superheterodyne scheme employed in commercial qubit detection systems.

According to Figure 7.2(b), the homodyne detection chain involves the following steps:

1. Amplification of the readout signal begins inside the fridge, using cryogenic parametric and HEMT amplifiers. The signal is further amplified at room temperature using LNAs.
2. The signal is downconverted using an IQ mixer and a microwave signal generator.
3. Filtering may be applied, with a low-pass filter added after the mixer to eliminate frequency-doubled and other spurious signals.
4. The down-converted signal is digitized or sampled using a digitizer.

To prepare the signal for detection, amplification is necessary. The readout resonator's signal is first amplified inside the fridge using a JPA or TWA, followed by a HEMT amplifier. As discussed in Chapter 4, the noise figure of the first element in the chain has the most significant impact on the overall noise figure. Therefore, it is crucial to use an amplifier with the lowest possible noise figure right after the resonator output. (Refer to Table 4.1 for a comparison of noise figures of various amplifiers.) Thus, parametric amplifiers are preferred as the first element in the detection chain. The signal is usually further amplified using a cryogenic HEMT amplifier commonly installed at the 4K stage.

After being amplified inside the fridge, the signal at room temperature requires additional amplification of 40–60 dB to bring the signal level above the detection threshold for successful detection (see the discussion of minimum detectable signal in Chapter 4). Typically, a room temperature LNA has a noise figure of around 1–3 dB, with a gain of approximately 20–30 dB.

The next step is to downconvert the amplified signal, bringing the frequency down to a suitable range for sampling. A digitizer, such as the 1 GS/s Acqiris AP240 acquisition board, can be used to sample the signal and prepare it for post-processing on the host computer.

The amplified signal is directed to an IQ mixer for downconversion, with the LO microwave tone provided by the microwave-signal generator intentionally detuned by 1 ~ 10 MHz from the cavity signal. The frequency of this signal needs adjustment so that the downconverted signal falls within the operational bandwidth of the digitizer. Note that the IQ mixer must undergo calibration to achieve optimized performance. Section 5.4.3.2 explains that the phase and amplitude imbalances of the IQ mixer must be addressed through calibration to remove the DC offset and sidebands.

The digital-homodyne technique also allows for digitally reconstructing the I and Q signals. In this method, a single output of the IQ mixer, $S_{IF}(t)$, is utilized, where the sine and cosine signals at the IF are digitally multiplied by the input signal. The resulting outputs are the digital I and Q, which can be combined to produce [7].

$$I = \frac{2}{T}\int_T S_{IF}(t)\cos(\omega_{IF}t)\,dt, \quad Q = \frac{2}{T}\int_T S_{IF}(t)\sin(\omega_{IF}t)\,dt \qquad (7.1)$$

$$S_{IF}(t) = A\cos(\omega_{IF}t + \phi) = I\cos(\omega_{IF}t) - Q\sin(\omega_{IF}t) \qquad (7.2)$$

So far, our discussion has been based on analog downconversion techniques. The industry, however, has shifted toward the direct RF-sampling techniques discussed in Chapter 5. For sufficiently high sampling rates, this architecture reduces the number of components required (such as the mixer and LO); eliminates the need for downconversion and, therefore, the IQ mixer and its calibration; and reduces the cost per qubit.

Figure 7.7 depicts a fully digital quantum-control solution encompassing generation and detection modules. This comprehensive solution addresses qubit control and readout requirements at room temperature.

7.3.3 Further Considerations at Room Temperature

In addition to the control, readout, tuning, and detection components that operate at room temperature, numerous other details play a crucial role in the successful

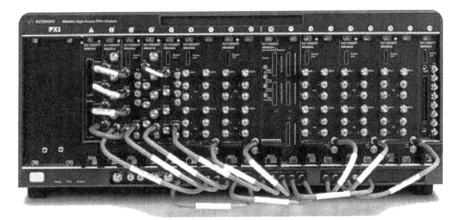

Figure 7.7 The world's first fully digital quantum control solution. (Image Courtesy of Keysight Technologies.)

operation of a superconducting qubit system. Sections 7.3.3.1 and 7.3.3.2 explore some of these aspects.

7.3.3.1 Synchronization

When addressing the phase degree of freedom in a quantum superposition, maintaining coherence extends beyond the internal coherence of the qubit. It also encompasses the synchronization between the qubit and the master clock that governs its operations [8]. Factors such as temperature variations and prolonged acquisitions can lead to significant phase drift.

To ensure long-term phase stability and synchronization, it is crucial to phase-lock all signal generators, measurement instruments such as the network analyzer, AWGs, and acquisition cards using an ultralow–phase noise 10-MHz Rubidium-frequency standard. This frequency standard, also called the reference signal, provides stable 10-MHz outputs that can be connected to instruments like signal generators and data-acquisition cards. A Rubidium clock typically exhibits phase noise around -130 dBc/Hz at a 10-Hz offset.

Instabilities in the clock not only impact the master clock but also affect the phase coherence of the qubit [3]. As discussed in Chapter 4, phase noise results in a broadened spectrum. Assume the two following scenarios: one where the LO interacts with a stable qubit but an unstable LO, and another where the qubit exhibits noise with fluctuating transition frequency, but the LO remains clean and phase stable. These two scenarios have equivalent dephasing effects, highlighting the contribution of the LO's phase instability to dephasing [4].

7.3.3.2 Grounding

Chapter 6 discussed grounding and identified potential issues affecting qubit experiments due to ground loops. These issues include the following:

- Unwanted noise signals generated by common mode currents on interconnect cables due to ground voltage differences;
- Coupling of noise voltage induced by strong magnetic fields into the ground loop and subsequently affecting the qubit and low-temperature thermometry.

Chapter 6 recommended employing an isolating transformer or optocoupler to break the ground loop between two circuits by eliminating electrical contact. It is well-known that modifying the fridge grounding can improve dephasing times [4]. Hence, the initial step involves isolating the experiment's power source from the electricity network using an isolating transformer called a 1:1 transformer.

All cryostat connections, including vacuum lines, are DC-isolated to prevent ground loops, except for a single, well-defined grounding cable. Standard SMA vacuum feed-throughs, such as the 34_SMA-50-0-3/111_NE from Huber and Sühner, are employed at the 300-K plate of the dilution refrigerator to provide microwave access into the fridge.

DC blocks on the microwave lines isolate the microwave ground from the cryostat ground. Hence, a DC block with capacitors on both conductors, such as the Inmet 8039, should be utilized for this purpose.

For GPIB connections, an optocoupler-based isolator like the GPIB-120A Bus Expander/Isolator from National Instruments can be utilized. Isolators are also available for USB and Ethernet connections [5].

7.4 Cryogenic Setup

Recall that the cryogenic setup serves two primary functions: suppressing the noise coupled into the fridge and amplifying the qubit readout signal. Sections explore these two functions and provide a more comprehensive discussion.

7.4.1 DC Wiring

Recall that DC lines adjust the qubit transition frequencies through an external flux created by the superconducting coil, as depicted in Figure 7.1(a). Typically, a superconducting coil can generate a magnetic field within the range of a few hundred Gauss per ampere. To enable tuning of each qubit across a wide range of Φ_0 values, a bias resistor with a DC voltage source is employed. Additionally, an on-chip FBL is utilized, eliminating the need for an external magnetic field. Each qubit has its own FBL, connected to a voltage source that drives an attenuator at room temperature. Each FBL is linked to a channel of an AWG for rapid tuning of the qubits.

Recall that these DC signals are generated by room-temperature sources and transferred through a breakout box to the fridge's input. Owing to their high thermal conductivity, copper or phosphor-bronze (PhBr) twisted-pair looms with wire diameters AWG35 and AWG36 are employed only from room temperature to the 4-K stage. From 4K to the mixing chamber (MXC), superconducting NbTi twisted pairs are used, as they possess low electrical and thermal conductivity, bypassing the limitations imposed by the Wiedemann-Franz law. All wires are thermally anchored at each temperature stage by being wrapped around a copper post and secured with GE varnish. Micro sub-D connectors are typically utilized for connections inside the cryostat, while resistive connectors are avoided in the low-temperature section of the coil wiring to prevent heating.

As discussed in Section 7.2, filters are implemented in the DC lines to reduce out-of-band noise.

7.4.2 Heat Loads

The heat load of the dilution refrigerator can be categorized into the following three major components, as well as a smaller contribution [4]:

1. *Passive load:* This refers to the heat transfer from higher to lower temperature stages. It includes heat conducted through installed cables and heat flowing via posts that separate different plates of the dilution refrigerator.
2. *Active load:* This pertains to the dissipated heat, known as Joule heating in attenuators and microwave cables due to applied microwave signals, the

dissipated DC signals used for biasing HEMT amplifiers at 4K, and flux biasing the qubits at the MXC.
3. *Radiative load:* This involves the transfer of black-body radiation from stages and shields at higher temperatures to stages and shields at lower temperatures.
4. *Residual gas load:* Although typically negligible during regular operation (as the pressure is maintained below 10^{-5} mbar), residual gas contributes to the overall heat load.

As we will explore in Section 7.4.3, minimizing the passive load by utilizing cables with low thermal conductivity is possible. However, according to the Wiedemann-Franz law, materials that are poor thermal conductors are also poor electrical conductors, leading to increased dissipation when microwave signals are applied. This challenge can be addressed by effectively thermalizing the cables, attenuators, and microwave components through thermal anchoring, which is discussed in Section 7.4.3.1.

Table 7.3 illustrates the different stages of a Bluefors XLD400 dilution fridge and the corresponding cooling power available at each stage. It is important to note that the cooling power decreases as we move to lower temperature stages. An excessive heat load at any stage can significantly affect the fridge's cooling performance and potentially lead to operational issues. Therefore, it is crucial to ensure that the heat load on each stage remains below the available cooling power.

7.4.3 Noise-Suppression Techniques

As discussed in Chapter 3, the interaction between the qubit and its environment leads to qubit relaxation and dephasing, leading to decoherence. Furthermore, Chapter 6 explained that noise can couple from the noise source through the mediator to the receptor. Various measures can be implemented to mitigate this issue on all three components (source, mediator, and receptor) to reduce or eliminate noise. For instance, identifying and eliminating the noise source or minimizing its impact through shielding and filtering techniques are possible approaches. Additionally, enhancing the qubit's noise resilience can be achieved by incorporating shielding, considering design aspects of the QPU board, exploring alternative platforms like 3D qubits, or even adopting different technologies such as topological qubits. The

Table 7.3 Temperatures and Available Cooling Powers on the Indicated Stages of a Bluefors XLD400 Dilution Fridge

Stage Name	Temperature (Kelvin)	Cooling Power (Watts)	Cable Length (Millimeters)
50K	35	30 (at 45K)	200
4K	2.85	1.5 (at 4.2K)	290
Still	882×10^{-3}	40×10^{-3} (at 1.2K)	250
CP	82×10^{-3}	200×10^{-6} (at 140 mK)	170
MXC	6×10^{-3}	19×10^{-6} (at 20 mK)	140

primary objective of this section is to explore techniques that effectively block the pathways through which noise couples to the qubit.

As depicted in Figure 7.8, the coupling paths to the qubit can be categorized as follows:

- *Conducted path:* In this path, the noise source is directly connected to the qubit. For instance, thermal noise at room temperature is conducted to the qubit through the qubit control and readout lines.
- *Radiated path:* In this path, the noise source emits radiation that affects the qubit. An example of such noise is EM and infrared radiation.

To mitigate the noise coupled through the conducted and radiated paths, the various techniques highlighted in Table 7.4 can be employed. Each technique is extensively explained in the subsequent sections.

7.4.3.1 Noise Suppression for Conducted Path

This section covers the techniques to suppress the noise mediated by the conducted path.

Microwave Cabling

As mentioned, the noise generated at room temperature can couple to the qubit through the conductive paths utilized for qubit control and readout. The next two sections examine room-temperature and cryogenic microwave cabling.

Room-Temperature Microwave Cabling

Table 7.5 overviews the three types of microwave cables required for room-temperature applications. First, low-loss and phase-stable cables are utilized with microwave instruments such as network analyzers and microwave-signal generators. Note that bending the cable alters its physical length at the bend point and can cause loosening or constriction of the dielectric, shielding, and braiding around the center conductor, leading to variations in electrical length and subsequent phase alterations. Therefore, conventional microwave cables can introduce phase changes when subjected to bending. Second, hand-formable microwave cables are employed to establish connections between room-temperature components of the detection chain, including amplifiers, mixers, and filters, as shown in Figure 7.9(a).

Third, the UT-85-TP cable is employed to connect the room-temperature components to the input and output ports of the fridge.

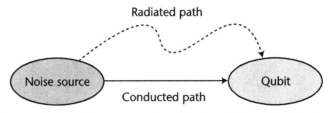

Figure 7.8 Noise couples from the source to the qubit through the radiated and conducted paths.

7.4 Cryogenic Setup

Table 7.4 Summary of Techniques Used for Noise Suppression in the Fridge

	Technique	Where?	Tool	Description
Conducted path	Attenuation	The input of the RF line of the fridge	Low thermal conductivity cables Attenuators	Reduces and attenuates thermal noise at the input of the fridge
	Isolation	Output RF line of the fridge	Circulators/isolators	Blocks the thermal noise to reach the qubit
	Filtering	Input of the fridge Detection chain	Pi filter/ RC filter/ Copper powder filter/ Purcell filter	Low-pass filter High-pass filter Bandpass filter
	Ground loop breaking	Room temperature	1:1 transformer Optocoupler DC block	Break the ground loop by providing disconnections at DC.
	Vibrational damping and decoupling	Inside fridge	Modification to cabling	Flattening a semirigid cable Jacketing a flexible cable in order to restrict movement within the cable
Radiated path	Shielding	Qubit in the fridge	Superconducting shield Magnetic shield Thermal shield	Shields that protect the qubit from electromagnetic radiation and magnetic fields

Cryogenic Microwave Cabling

To mitigate noise coupling into the fridge, microwave cables with low thermal conductivity, such as stainless-steel cables, are employed. This helps minimize the amount of heat transferred to the fridge. However, using cables with low thermal conductivity comes at the expense of increased signal losses. According to Wiedemann-Franz's law, materials with low thermal conductivity also exhibit low electrical conductivity. Table 7.5 demonstrates that a stainless-steel cable, operating at 10 GHz, experiences an insertion loss per unit length almost six times larger than a typical microwave cable.

In the absence of a microwave signal applied to the stainless-steel cable, the conducted thermal noise from room temperature into the fridge is lower than that of a typical microwave cable. However, when a microwave signal is applied, the heat generated in the fridge from the stainless-steel cable surpasses that of a typical cable. To mitigate this, the generated heat is effectively expelled from the fridge by thermally anchoring the cable at each fridge stage, as illustrated in Figure 7.9(b). The cable's outer conductor is clamped and heat-sunk using oxygen-free high-thermal conductivity (OFHC) copper braces. The cable's inner conductor is indirectly thermalized by the attenuator (see the next section). Since the cable's dielectric has very low thermal conductivity, the heat from the center pin cannot be efficiently transferred to the outer conductor, where it is thermalized. However, the heat transfer from the center pin of the attenuator to its body is more efficient. Therefore, the only point where the cable's center pin can be effectively thermalized is where it is connected to the attenuators. Like the cables, an OFHC clamp can thermalize the attenuators, as shown in Figure 7.9(b).

Table 7.5 Types of Cables Used in a Superconducting Qubit Hardware Setup

Cable Type	Where?	Ground/Center Conductor	Loss @ 0.5, 10, 20 GHz (dB/m)	Why?
Standard microwave coax cable	Microwave test equipment		0.7, 1.6, 2	
Hand-formable microwave cable	Microwave front-end			Used for connections between different components of the microwave detection chain
UT-85-TP	Room temperature	Silver- or tin-plated copper wire / tin-plated Copper	0.45, 2.21, 3.29	Used for microwave connection from fridge output at room temperature to the input of microwave detection chain
UT-85-SS-SS	At the input side	Stainless steel / stainless steel	2.92, 9.32, 18.91	Low thermal conductivity
SC-219/50-NbTi-NbTi-1000	From the sample to the first amplifier inside the fridge		0.3 @<4K @300K 3, 13.6, 19.4	
UT-85B-SS	Inside the fridge on the signal output side from 2.1K to room temperature	Stainless steel / SP BeCu	1.02, 3.33, 6.94	Minimum signal loss on the output side

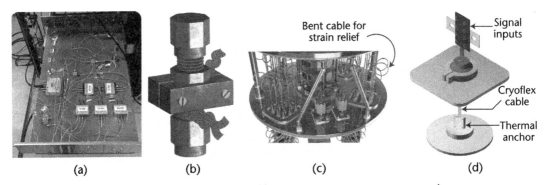

Figure 7.9 (a) Using hand-formable microwave cables to connect room-temperature microwave components; (b) thermalization of an attenuator's outer conductor using a copper clamp and an OFHC braid (the same concept can be used for cables); (c) microwave cables are bent for strain relief; and (d) cryoflex cabling.

Figure 7.9(c) demonstrates bends in the cryogenic microwave cables with a bending radius of approximately 1 cm. This bending configuration helps alleviate strain on the connectors during the cooling-down process.

Using stainless steel cables to reduce the noise coupling on the output side of the fridge, where the qubit readout signal is transmitted for detection, is impractical.

This is due to the extremely low level of the qubit readout signal, which makes it nearly impossible to detect in the presence of large signal losses. Therefore, low-loss cables are employed on the output side of the fridge.

Superconductors do not adhere to the Wiedemann-Franz law, which means that they can simultaneously possess both low thermal conductivity and low losses. Table 7.5 illustrates that the superconducting SC-219/50-NbTi-NbTi-1000 cable exhibits a very low insertion loss of 0.3 dB/m below 4K when in the superconducting state. The insertion loss increases significantly as the cable transitions out of the superconducting phase above 4K. Superconducting cables are utilized from the qubit's output to the fridge stage with a temperature below 4K, as depicted in Figure 7.2(a). For temperatures above 4K, UT-85B-SS cryogenic cables with low thermal conductivity and low losses are employed.

Regarding passive heat load among the microwave cables, the UT-85-SS cable on the output side of the fridge has the highest heat load, followed by the stainless-steel cable UT-85-SS-SS and the superconducting cable UT-85-NbTi [4].

Cabling poses a significant bottleneck for scaling quantum computers due to the challenges associated with increased interconnection density, including fridge size, heat load, noise coupling, signal integrity, and calibration issues. In order to address this scaling challenge, there has been a recent focus on flexible cryogenic microwave cables or cryoflex cables. These cables offer several advantages compared to traditional microwave coaxial cables, such as lower thermal conductivity, high-density interconnections, a smaller form factor, and easy installation. They can be conveniently thermally anchored at each level within the fridge using OFHC copper clamps, as depicted in Figure 7.9(d). Furthermore, these cables can integrate filtering and signal conditioning components, providing additional functionality.

Techniques such as crossbar addressing and multiplexing on a single line have been developed to drive single-qubit operations on multiple qubits, thereby reducing the number of required microwave lines. This approach allows for the reuse of each cable for multiple functions. However, while this approach is viable for a limited number of qubits, it is not a practical solution for scaling to thousands or millions of qubits due to issues such as crosstalk and limited bandwidth for frequency multiplexing.

Microwave Connectors

Recall our discussion of microwave connectors and adapters in Chapter 5. Let us now quickly review the microwave connectors used inside the fridge. A specific length is provided when purchasing cables, and the SMA connectors need to be crimped or soldered onto the cables. Soldering the superconducting cable can be challenging, so in such cases, crimping is often employed to connect SMA connectors. Soldering the SMA connector to the cable for other types of cryogenic cables is typically done. Alternatively, cable vendors can be requested to provide fully assembled cable sets, although this option is usually more expensive.

As mentioned in Chapter 5, snap-on connectors like MMPX or SMP offer the advantage of quick and easy connections and disconnections without the need for threaded mating. This feature makes them well-suited for applications where space for connectors is limited and frequent cable changes are required. Thanks to their space-saving design, they are particularly useful for microwave connections on sample holders for quantum chips. To accommodate high-density connections within the fridge, ganged connectors are employed.

Noise Suppression Using Attenuators

In addition to utilizing low thermal conductivity cables, microwave attenuators are incorporated into the input microwave lines of the fridge to mitigate thermal noise. This section comprehensively explains the necessary attenuation levels and their allocation across various temperature stages to minimize thermal noise effectively.

The idea behind employing attenuators is to attenuate the noise level. The heat generated during this process is pumped out of the fridge by thermalizing the attenuators. This is achieved by connecting their body to each stage using OFHC clamps, as depicted in Figure 7.2. Without proper thermalization, the attenuator's losses contribute to a higher noise level, and heat accumulation leads to thermal radiation that impacts the temperature at the lower fridge stages. For instance, a 20-dB attenuator reduces the signal level by a factor of 100, dissipating 99% of the signal within the attenuator, thereby emphasizing the importance of effective thermalization.

Recall from Section 4.3.2.4 how the thermal motion of charge carriers in a resistor R generates a random voltage with a zero average value but a nonzero RMS value, as described by Planck's black-body radiation law.

At microwave frequencies and for high temperatures where $\hbar\omega \ll k_B T$, the result can be simplified to $V_n = \sqrt{4kTBR}$ with the maximum available noise power of $P_n = kTB$.

The above linear relation approximates Planck's black-body radiation formula and is only valid for temperatures above a few kelvins. Below 4K, the relation between noise power and temperature is not linear anymore; therefore, (4.5) must be used for noise calculations below 4K. As seen in Figure 7.10(a), a factor of 10 change in the temperature results in a 10-dB change in the thermal noise power at high temperatures. At very low temperatures, the noise power becomes a quadratic function of temperature (i.e., a factor of 10 change in the temperature results in a 20-dB change in the thermal noise power). The plot in Figure 7.13(a) shows that the noise power changes by 60 dB from room temperature at 300K to 10mK. Therefore, 60-dB attenuation must be distributed among various stages of the fridge to diminish the thermal noise at the mixing chamber effectively.

There are two reasons for distributing the attenuators among various stages of the fridge: the limited achievable noise floor at each stage and the limited cooling power.

As depicted in Figure 7.10(a), the noise originating from room temperature (300K) can be only attenuated by 20 dB as it reaches the 2K stage. Therefore, a 20-dB attenuation is sufficient, as further attenuation would not reduce the noise level below the thermal noise floor of the 2-K stage.

Let us now delve into the distribution of attenuators based on the limited cooling power at each stage. One might wonder why we distribute the attenuation across the stages instead of applying 60 dB of attenuation solely at the last stage. The answer lies in the available cooling power at each stage. The cooling power

7.4 Cryogenic Setup

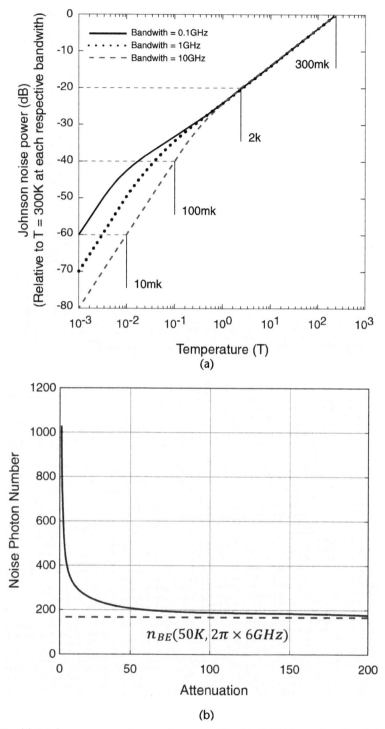

Figure 7.10 (a) Total power versus temperature normalized to 300K for various bandwidths and (b) noise photon number versus attenuation at 6 GHz and 50K.

diminishes as we go down to lower temperature stages (refer to Table 7.3). Moreover, an excessive heat load at each stage can substantially impact the fridge's cooling performance and potentially disrupt its proper operation. Therefore, if we were to dissipate 60 dB of attenuation in the MXC, the cooling power available would be insufficient to remove the heat generated by such high attenuation effectively. Consequently, we cannot add significantly more than 20 dB of attenuation at the MXC stage. Hence, it is crucial to distribute the total attenuation across the different temperature stages to ensure that the active load on the lower stages remains well below the available cooling powers at those stages.

It has been shown that an optimal choice, striking a balance between a low number of noise photons at MXC and a low heat load on the MXC, is a total attenuation of 60 dB consisting of 20-dB attenuators thermally anchored at the 4-K, cold plate (100-mK), and MXC (10-mK) stages [4].

Conventional room-temperature attenuators are unsuitable for the cryogenic environment due to several factors. First, the material used in cryogenic attenuators must be nonmagnetic and specifically designed to withstand the extreme temperatures encountered in cryogenic systems. Additionally, cryogenic attenuators require stable attenuation characteristics across a wide temperature range to ensure consistent performance. Finally, minimizing self-heating in cryogenic attenuators is essential for efficiently achieving the desired low temperatures.

It is also insightful to examine the noise suppression achieved by attenuators using an alternative approach that considers the concept of photon occupation number. According to the Bose-Einstein distribution, the photon occupation number can be expressed as:

$$n_{BE}(T,\omega) = \frac{1}{\exp\left(\frac{\hbar\omega}{k_B T}\right) - 1} \tag{7.3}$$

The photon occupation number $n_{BE}(T,\omega)$ is dimensionless and can be thought of as photon flux spectral density (i.e., as a number of photons per hertz frequency interval per second [4]). As an example, at room temperature T = 300K and a frequency of 6 GHz, n_{BE} (300K, $2\pi \times$ 6 GHz) $\approx$ 1,043. For superconducting qubits, the thermal photon occupation number must be in the range of 10^{-3} photons at the sample.

The noise photon occupation number n_i at stage i with attenuation A_i is given by [4]

$$n_i(\omega) = \frac{n_{i-1}(\omega)}{A_i} + n_{BE}(T_{i,\text{att}},\omega) - \frac{1}{A_i} n_{BE}(T_{i,\text{att}},\omega) \tag{7.4}$$

The number of noise photons at stage i consists of three components. The first component is the noise from the previous stage, which is attenuated by a factor A_i. The second component is the noise floor of stage i itself, and the third component is the noise floor reduced by the attenuation factor A_i. Equation (8.6) illustrates that attenuation effectively reduces the number of noisy photons originating from the previous stage. Simultaneously, it has a negligible impact on the noise floor at the stage where it is installed, particularly for high attenuation values. Eventually, the photon noise at each stage is primarily determined by the thermal noise floor at that particular stage. Therefore, using attenuation values larger than necessary to thermalize the incoming radiation fields onto that stage is unnecessary, as demonstrated in Figure 7.10(b).

The required attenuation on stage i can be estimated as $A_{i,\text{ref}} = (n_{\text{BE}}(T_{i-1},\omega))/(n_{\text{BE}}(T_i\omega))$. By neglecting the noise floor photon number, we can determine a lower bound for the total attenuation needed to achieve $n_{MXC} = 10^{-3}$, where $n_{\text{BE}}(T = 300K, \omega)/10^{-3} = 60$ dB.

This demonstrates a lower bound in cases where the black-body radiation emitted by attenuators at all other temperature stages is disregarded. These findings validate the previously presented results derived from Planck's black-body radiation law.

Cryogenic Filtering and Shielding

So far, we have delved into techniques for reducing noise using low-thermal conductivity cables and attenuators. Now, let us look at how filtering and shielding methods can further enhance the effectiveness of noise suppression.

Cryogenic Filtering

Figure 7.2(a) and Table 7.6 illustrate the utilization of filters in both DC and microwave lines within the fridge. In DC lines, utilizing the attenuation technique is not feasible due to potential heating issues. Therefore, low-pass filters are employed to suppress thermal noise. Recall that it is crucial to transfer the frequency components of a signal up to the fifth harmonic to prevent pulse degradation. Hence, for nanosecond pulses, the cutoff frequency of the low-pass filter should be set between 400 and 500 MHz, ensuring a smooth roll-off to preserve the pulse shape.

In qubit control and measurement, different filter requirements arise. Filters with high cutoff frequencies are needed for XY control and readout, typically in the gigahertz range. On the other hand, filters with low cutoff frequencies are necessary for Z control. As Figure 7.1(a) depicts, microwave control lines employ filters with high cutoff frequencies. In contrast, the FBL incorporates a filter with a low cutoff frequency of a few hundred megahertz.

Copper Powder Filters

Let us discuss an intriguing phenomenon related to the behavior of lumped-element and transmission line filters at high frequencies. An interesting pattern emerges if we plot a microstrip low-pass filter's transmission response ($|S_{21}|$) over a wide frequency range. Initially, the transmission (attenuation) typically decreases (increases) in a monotonic fashion, as depicted in Figure 7.11(a). However, beyond 3.2 GHz, the transmission (attenuation) increases (decreases) due to re-entrant modes. Consequently, the low-pass filter effectively transforms into a high-pass filter beyond a specific frequency as shown in Figure 7.11(b). This behavior may not pose significant issues if the system's operational bandwidth is limited.

Table 7.6 Low-Pass Filters Used at Cryogenic Temperatures

Stage	Where	Property	Type of Filter
Cryogenic	DC and flux bias line	Low cutoff frequency	Copper powder filter RC and LC filters
	Microwave line	High cutoff frequency	IR filter Microwave filters

A similar phenomenon to re-entrant modes in microstrip filters can be observed in lumped-element filters with capacitors and inductors. In real-world scenarios, capacitors and inductors are not purely capacitive or inductive; they possess parasitic components. The parasitic model of a capacitor and inductor is depicted in Figure 7.11(c, d). A capacitor is not solely a capacitor but a combination of a capacitor, inductor, and resistor. As the frequency increases, the capacitor behaves less like a capacitor and more like an inductor, as demonstrated in Figure 7.11(c). The same holds for the inductor, as shown in Figure 7.11(d). Consequently, at sufficiently high frequencies, beyond the self-resonant frequency, the capacitor transforms into an inductor, and the inductor becomes a capacitor. Therefore, a low-pass filter exhibits characteristics of a high-pass filter at sufficiently large frequencies, as shown in Figure 7.11(b).

To circumvent the re-entrant mode problem, Martinis et al. devised the copper powder filter [8], which exhibits a low-pass frequency response across a wide bandwidth. This filter achieves significant attenuation at high frequencies, as Figure 7.12(a) depicts.

As illustrated in Figure 7.12(b), the copper powder filter comprises an insulated spiral coil wrapped around a metal rod filled with metal powder grains, typically in the range of a few tens of microns. When an AC current passes through the coil, it generates an AC magnetic field that induces eddy currents in the metal grains. The skin effect causes the current to flow predominantly near the surface, resulting in increased effective resistance and dissipation of energy. Each grain is naturally coated with an oxide layer, providing insulation from neighboring grains and creating a substantial effective surface area. This surface area contributes to significant attenuation at high frequencies through skin-effect damping.

Metal powder is typically mixed with epoxy, such as Stycast 2850, to improve thermal conductivity and dissipate heat from the metal grains. The filter's performance depends on various parameters, including wire diameter, wire length, powder material, and grain size. Longer wires offer improved attenuation but also increase insertion loss. Stainless steel powder exhibits superior attenuation compared to copper or bronze due to its higher resistivity. However, nonmagnetic powders like bronze and copper are preferred for experiments sensitive to magnetic fields, even if they provide slightly less attenuation. The coil is wound half-clockwise and half-counterclockwise to minimize the magnetic field produced by the current flowing through the coil.

Two powder filters are arranged in series, one at the 4.2-K stage and the other at the 10-mK stage, as depicted in Figure 7.2(a). The purpose of the 4.2-K filter is to eliminate thermal noise, spurious RF signals, and microwave radiation. However,

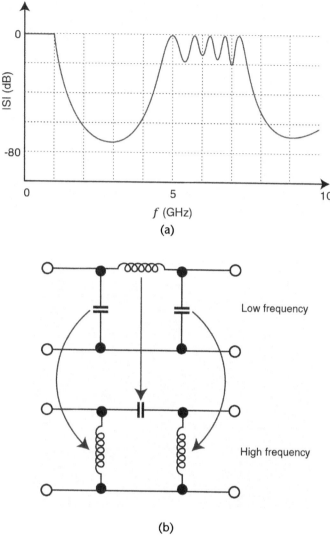

Figure 7.11 (a) The transmission frequency response of microstrip low-pass filters over a large frequency range. The re-entrant modes in a low-pass microstrip filter cause the filter to act as a high-pass filter with a passband starting around 5 GHz. (b) A low-pass lumped-element filter acts as a low-pass filter well below the self-resonance frequency of the capacitor and inductor. Above self-resonance frequency, the capacitor becomes the inductor, and the inductor becomes the capacitor. This results in a high-pass filter above the self-resonance frequency. (c) Equivalent circuit model of a capacitor and its impedance frequency response. (d) Equivalent circuit model of an inductor and its impedance frequency response.

an additional filter is necessary at the qubit temperature to mitigate thermal noise originating from the 4.2-K filters.

Despite their effectiveness, powder filters have certain drawbacks, such as their size, weight, and the labor-intensive process involved in their fabrication. To simplify the fabrication process, researchers have developed PCB-based copper powder filters [9]. Nevertheless, powder filters can exhibit undesired parasitic

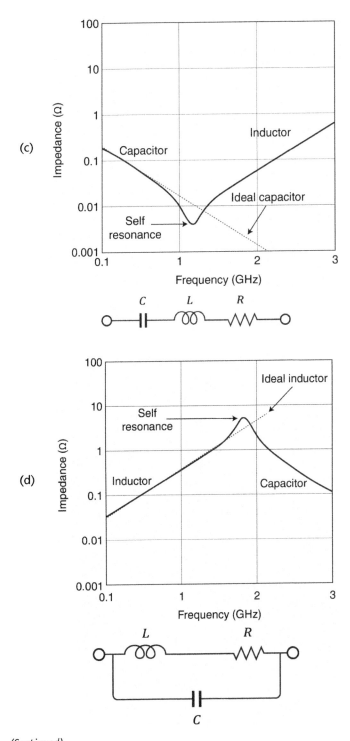

Figure 7.11 *(Continued)*

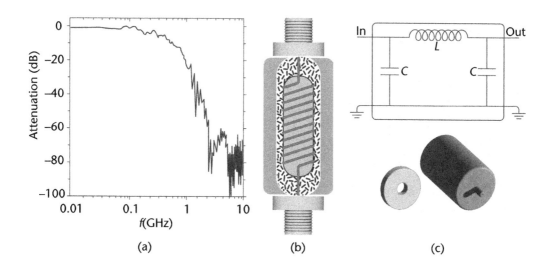

Figure 7.12 (a) Frequency response and (b) structure of a copper powder filter. (c) A pi filter can be embedded by using discoidal capacitors to lower the cutoff frequency of the copper powder filter.

resonances, resulting in a nonlinear frequency response, as depicted in Figure 7.12(a). Additionally, they possess a DC resistance of approximately 10Ω, which leads to a power dissipation of 10 μW when subjected to a flux bias within the range of 1 mA. This exceeds the cooling power available at the lowest temperature stage, around 5 μW. One solution to address this issue is to incorporate superconducting niobium wires.

Copper powder filters typically have a relatively high cutoff frequency. To lower the cutoff frequency, a combination of a filter with a lower cutoff frequency, such as a pi filter, can be employed in conjunction with the powder filter. Since the coil within the copper powder filter functions as an inductor, adding two capacitors at both ends of the filter is sufficient. To accomplish this, discoidal capacitors are soldered onto the plugs at each end of the filter, as demonstrated in Figure 7.12(c). Discoidal capacitors possess lower parasitic inductance than surface-mount device (SMD) chip capacitors, making them suitable for this application.

Infrared Filter

Infrared radiation results in quasi-particle excitations and affects the coherence time of qubits. To safeguard devices operating below 1K from the detrimental effects of infrared radiation, an infrared (IR) filter is employed, typically just before the attenuator at the MXC. One approach to producing such a filter involves using Eccosorb absorbers [4].

The IR filter is a low-pass absorptive filter, dissipating high-frequency noise above the measurement frequency band. Its passband typically ranges from DC to 6 GHz. Like other cryogenic system components, the IR filter must be thermally anchored to the fridge. Using a thermally anchored IR filter also improves the thermalization of the center conductor in a coaxial line. The housing of IR filters is constructed using OFHC copper to provide optimal thermal conductivity.

7.4.3.2 Noise Suppression for Radiated Path

As mentioned in Section 7.4.3.1, noise can couple to the qubit through the radiated path, and shielding techniques are employed to block this radiation path. This discussion covers electromagnetic and thermal shielding to demonstrate how these techniques protect the qubit against noise coupled through the radiated path.

Electromagnetic Shielding

Recall the principles of electromagnetic shielding discussed Chapter 6. In the qubit hardware setup context, we now explore how these principles effectively shield the qubit from external electromagnetic interference. The qubit readout, control, and tuning circuitry are fabricated on a silicon chip and connected to the microwave-PCB using bond wires, as shown in Figure 7.13(a). The microwave PCB connects to the microwave cables inside the fridge using SMP or MMPX microwave connectors, with a smaller form factor than SMA connectors.

Figure 7.13(b) shows the implementation of an electromagnetic shield to suppress stray electromagnetic fields and reflections caused by impedance mismatches in circuit connections. As shown in Figure 7.13(b), the microwave PCB is positioned on a sample holder at the far right, while a milled part is employed in the middle to cover the silicon chip and transmission lines. Prior to installing the larger enclosure on the far left, this smaller intermediate cavity with a higher resonant frequency is utilized to prevent energy coupling to the resonant mode of the larger enclosure's cavity. An infrared-absorbing epoxy is sometimes applied to the enclosure lid's surface facing the sample to improve the shielding effectiveness.

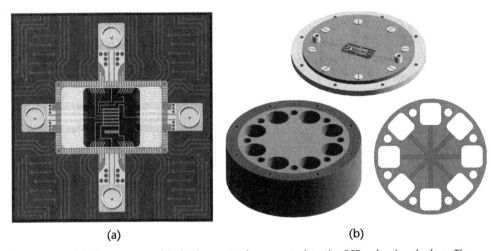

Figure 7.13 (a) The quantum chip in the center is connected to the PCB using bond wires. The SMP or MMPX connectors connect the PCB to the microwave lines. (b) Electromagnetic shielding of the qubit.

After placing the qubit inside the enclosure, additional shielding layers are employed to achieve the desired level of total shielding effectiveness, as depicted in Figure 7.14(a, b). As discussed in Chapter 6, materials with high permeability can be used for shielding magnetic fields. However, the permeability of these materials typically decreases as temperature decreases, as shown in Figure 7.14(c). Special materials such as cryoperm are utilized in a dilution fridge to address this issue, as its permeability remains high at low temperatures. The cryoperm serves as the outermost shield in the fridge at 2.7K, as illustrated in Figure 7.14(a).

Figure 7.14(a, b) illustrates using two gold-plated copper cans at 500 mK and 100 mK, respectively. Both cans feature a thin layer of a superconducting magnetic shield made of tin (Sn), which is electroplated onto the copper surface before gold plating. The innermost copper can at 100 mK is additionally coated with a 2-mm layer of infrared absorbing epoxy. IR filters absorb infrared photons to thermally stabilize the input and output lines, as shown in Figure 7.14(a).

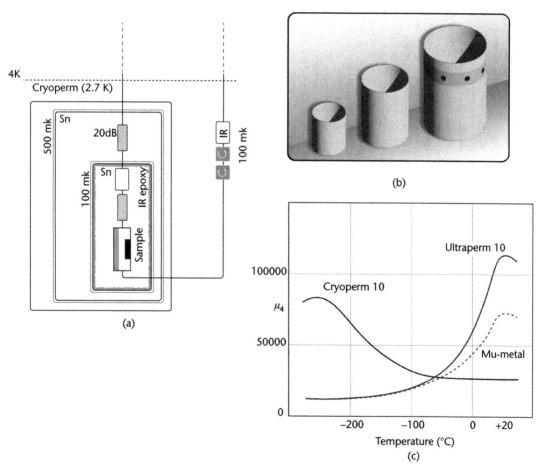

Figure 7.14 (a) Cryogenic shielding consisting of magnetic and thermal shields. Cu cans at 100 mK and 500 mK serve as infrared radiation shielding. Sn plating on the Cu cans and cryoperm 10 at 2.7K provides magnetic shielding. The 100 mK can also have infrared absorbing epoxy. (b) Shields for 100 mK, 500 mK, and 2.7K. (c) The cryoperm, ultraperm, and mu-metal permeability versus temperature.

Thermal Shielding

Thermal radiation, the primary contributor to the thermal budgets for cryostats, requires special consideration. When an object's surface receives electromagnetic radiation, it absorbs a fraction α, reflects a fraction β, and transmits a fraction τ. This follows the condition $\alpha + \beta + \tau = 1$ due to energy conservation. In most cryostat applications, the body is opaque when $\tau = 0$. This means that all energy is either absorbed or reflected. If $\beta = 0$ ($\alpha = 1$ from energy conservation), the body is known as black and absorbs all energy.

In cryostats, an orifice in a thermal shield can behave like a black body, where internal surfaces gradually absorb radiation through multiple reflections. Therefore, it is crucial to avoid gaps and slots when designing a thermal shield. However, in cases where gaps are necessary for thermal contraction compensation, multilayer insulation (MLI) pads can be used to cover them. In ultrahigh vacuum (UHV) applications where MLI is not viable, high-absorptivity material traps can surround the gaps and absorb light on the thermal shield wall, reducing multipath reflection on the inside.

The total energy radiated per unit surface area across all wavelengths per unit time for a black body is given by:

$$E_b(T) = \sigma \cdot T^4 \qquad (7.5)$$

where σ is Stefan-Boltzmann's constant and T is the temperature. Introducing the total hemispheric emissivity $0 < \varepsilon < 1$, which defines the fraction of the emitted radiation with respect to that of a black body, is given by $E(T) = \varepsilon \cdot \sigma \cdot T^4$.

The emissivity value depends on the material, surface finish, and cleanliness. For example, clean and well-polished metallic surfaces have small emissivity, whereas nonmetallic surfaces have higher emissivity. Therefore, cryostat thermal shields are typically made of metallic surfaces (copper or aluminum) with a clean and reflective surface finish, such as gold. Table 7.7 shows hemispherical emissivity values for a few materials.

Emissivity is a function of temperature, and for metals at cryogenic temperatures, emissivity reduces almost proportionally with temperature. This is desirable since it enhances the low-emissivity properties of cryogenic-cooled thermal shields in cryostats. As mentioned earlier, gold-plated copper cans coated with infrared-absorbing epoxy are used as a thermal shield in a dilution fridge.

Output Noise Suppression Using a Circulator

Blocking the noise generated by downstream measurement stages from reaching the qubit is crucial. The cryogenic HEMT amplifier typically exhibits a noise temperature ranging from 2 to 5K, making it the primary noise source on the output line.

Recall that employing attenuation techniques on the output line is impractical as it contradicts the purpose of the amplifier. Hence, it becomes imperative to block the noise generated by the amplifier from reaching the qubit. This can be achieved by incorporating a circulator or isolator at the output of the parametric amplifier, as depicted in Figure 7.2(a).

7.4 Cryogenic Setup

Table 7.7 Hemispheric Emissivity for Some Materials

Material/Temperature	4K	80K	300K
Copper mechanically polished	0.02	0.06	0.1
Gold	0.02	0.02	0.02
Aluminum electropolished	0.04	0.08	0.15

As discussed in Chapter 4, a circulator serves as a unidirectional microwave valve that allows the signal from the parametric amplifier to pass through to the HEMT amplifier with minimal attenuation. It effectively diverts the reflected noise from the HEMT amplifier and redirects it to a 50-Ω termination instead of allowing it to reach the sample, as explained in Chapter 5. It is crucial to ensure that the circulator or isolator is properly thermalized. As discussed in Section 5.4.2.3, double-junction circulators provide improved isolation at the expense of higher insertion loss.

For multiqubit experiments, isolator arrays are utilized to save space. Ensuring adjacent port-to-port isolation is crucial for isolator arrays and considering the figures of merit mentioned in Section 5.4.2.3.

Traditional circulators tend to be bulky and heavy, often utilizing permanent magnets composed of ferrite. Therefore, it is advisable to avoid placing them in close proximity to the qubits, as the magnetic field can adversely affect their performance. To address the scaling challenge in quantum computing, physicists have invested considerable effort in developing compact, scalable, magnetic-free, low-noise, and tunable on-chip circulators. Although the signal bandwidth remains relatively limited, these novel circulators also offer frequency tunability. However, the fabrication process for these circulators is intricate. Alternatively, a SLUG microwave amplifier can be used for qubit readout. The SLUG amplifier combines low-noise amplification and reverse isolation, which is better than a commercial cryogenic isolator.

7.4.4 Signal Amplification

Two key concepts that we need to keep in mind are described as follows:

- First, as shown in Section 3.3.2.2, the qubit readout fidelity F is determined by the SNR, which it can be calculated using the formula $F = (1 + \text{erf}(\sqrt{SNR/2}))/2$ [10]. For example, with SNR = 10, the corresponding readout fidelity is $F = 99.99\%$. Although time averaging can enhance the SNR by a factor of $\sqrt{N}$, where N is the number of signals averaged, achieving a tenfold improvement in SNR could potentially reduce the required average times by a factor of 100. This leads to the next point.
- Second, as discussed in Section 4.3.2.4, the initial amplifier in the chain significantly impacts the overall noise figure NF and, consequently, the output SNR, where $SNR_{out} = SNR_{in}/NF$. Therefore, it is crucial to utilize an ultralow noise amplifier like the parametric amplifier right after the readout resonator to maximize the output SNR. This allows for fast single-shot readout.

Table 7.8 Amplification in a Superconducting Qubit Experiment

Technique	Description	Noise Temperature
JPA	Quantum-limited noise performance	$T_{noise} < 1K$
TWA	Higher bandwidth and dynamic range for multiplexed readout of multiple qubits	$T_{noise} < 1K$
SLUG amplifier	Quantum-limited noise performance; Combined amplification and reverse isolation eliminating the need for external circulators	$T_{noise} < 1K$
Cryogenic HEMT amplifier	Used at 4K stage	$T_{noise} = 2$–$5K$
Room-temperature LNA	Used in room-temperature detection chain	$T_{noise} = 75$–$100K$

The amplifiers used at cryogenic temperatures are listed in Table 7.8. The noise temperature of the amplifier, denoted as T_a, helps us calculate the number of thermal photons, n_{th}, added by the amplifier at each frequency ω, which is expressed as $n_{th} = 1/(e^{\hbar\omega/k_B T_a} - 1)$.

For conventional HEMT-based readout operating at frequencies between 6 and 7 GHz, the added system noise is typically around twenty quanta, $n_{th} \sim 20$. With typical parameters, this results in a single-shot qubit measurement fidelity of approximately 50% within a 500-ns timeframe [11]. To achieve a fast and high-fidelity single-shot readout, it is necessary to employ an amplifier with noise performance approaching the standard quantum limit, $n_{th} = 1/2$. This limit represents the minimum achievable noise level by a phase-insensitive linear amplifier [11].

This section discusses the utilization of various cryogenic amplifiers, including the JPA, the SLUG amplifier, and the cryogenic HEMT amplifier. For a comprehensive review of these amplifiers, please refer to Section 5.4.2.1.

7.4.4.1 Parametric Amplifiers

If a HEMT amplifier with a noise temperature of approximately $T_N \approx 2$–$4K$ is utilized as the first amplifier at the output of the qubit's readout resonator, the resulting system noise temperature will be around 7–10K. This corresponds to an addition of approximately 10–20 noise photons per signal photon at around 5 GHz. In practical terms, this noise level is generally considered too high to achieve a single-shot readout with a satisfactory SNR. The limitations in SNR have generated interest in developing quantum-limited parametric amplifiers, which add noise close to the minimum amount allowed by quantum mechanics.

From Chapter 5's discussion of the fundamentals of parametric amplifiers, we learn that averaging N times enhances the SNR by a factor of $\sqrt{N}$. Thus, achieving a 10-dB improvement in SNR can reduce the required number of averages by a factor of 100. Parametric amplifiers based on Josephson junctions offer improved noise temperature, as depicted in Figure 7.15. Consequently, a parametric amplifier can eliminate the need for signal averaging and enable a high-fidelity single-shot readout of qubits.

As mentioned in Chapter 5, a TWPA offers a higher bandwidth than a JPA, making it suitable for the multiplexed readout of multiple qubits. Figure 7.15 illustrates

7.4 Cryogenic Setup

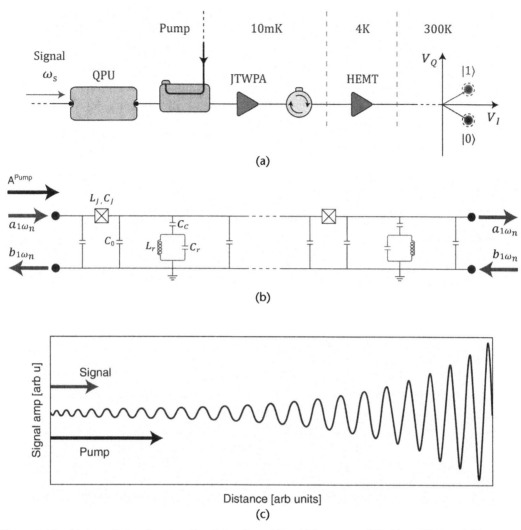

Figure 7.15 (a) A weak signal probes the state of a qubit, which gets amplified by a TWPA and (b) schematic of JTWPA consisting of a lumped-element transmission line made from unit cells of Josephson junctions, shunt capacitors, and phase-matching resonators. (In reality, a device has several Josephson junctions between each phase-matching resonator.) (c) The amplified signal along the length of the JTWPA.

the utilization of a TWPA in a qubit setup. The signal emerging from the readout resonator passes through a directional coupler (see Chapter 5) and reaches the input of the TWPA. Simultaneously, the pump signal originating from room temperature, as shown in Figure 7.2(a), is applied through the directional coupler to the TWPA's input. It is crucial to remember that parametric amplifiers have an extremely low P1dB, typically in the range of −100 to −80 dBm, so caution must be exercised to prevent saturating them.

7.4.4.2 Cryogenic HEMT Amplifier

After passing through the parametric amplifier, the signal undergoes further amplification within the fridge using a HEMT amplifier, followed by one or two room-temperature LNAs, as depicted in Figure 7.2(a). (See Chapter 5 for a discussion of these amplifiers.)

The HEMT amplifiers require a stable and precisely adjustable voltage supply to ensure optimal performance. It is crucial to avoid exceeding the maximum ratings, as this can easily destroy the HEMT amplifier. In qubit experiments, it is unlikely to reach the RF input drive levels that damage the amplifier. However, it is necessary to exercise extra caution when setting the bias values for the drain-source voltage (V_{ds}), current (I_{ds}), and gate-source voltage (V_{gs}). Special attention should also be given to these sensitive devices' power-up and power-down processes.

7.4.4.3 SLUG Amplifier

In Section 7.4.3.2, we discussed using circulators to isolate the quantum circuit from noisy downstream measurement stages. However, circulators present challenges in terms of size and scalability. The SLUG amplifier, which combines near quantum-limited noise performance with higher reverse isolation than commercial cryogenic isolators, offers a compelling solution to address these issues [11]. Additionally, the SLUG amplifier provides a large instantaneous bandwidth and dynamic range, enabling multiplexed single-shot qubit readout.

While the noise level of a SLUG amplifier will never match that of an optimized JPA, it is possible to achieve added noise of approximately one quantum at frequencies approaching 10 GHz [11]. Unlike the JPA, the SLUG amplifier does not require a separate strong microwave pump tone, which means that the qubit does not need to be protected by multiple stages of cryogenic isolation. The wire-up and operation of the SLUG amplifier are particularly straightforward, requiring only two DC current biases and no microwave pump tones. The pulsed-mode operation of the SLUG amplifier enables the characterization of transmon qubits without the need for cryogenic circulators, and it does not result in any measurable degradation of qubit performance in the measurement chain [11]. Please refer to Chapter 5 for a detailed discussion of the SLUG amplifier.

7.4.5 Further Considerations for the Cryogenic Setup

An RF switch can be employed within the fridge to enhance connectivity and reconfigurability. This allows for efficient switching between different samples during measurements. In scenarios where multiple samples are being tested, the process typically involves warming up the fridge, changing connections, and then cooling it down again, which can be very time-consuming. However, using an RF switch, as illustrated in Figure 7.16, rapid switching between samples inside the fridge becomes possible without needing warmup time or connection changes.

A DC source at room temperature is utilized to control the RF switch. However, it is essential to wire the switch-control circuitry inside the fridge. Mechanical RF switches consist of coils and mechanical contacts manipulated by electrically energized coils. These coils exhibit different resistances at 4K compared to ambient temperature, necessitating specialized control electronics for low temperatures. These electronics must be current-driven instead of voltage-driven as originally intended by the manufacturer. Additionally, it is crucial to isolate the electronics from the PC ground that controls the switch to prevent ground loops. Caution should be exercised with the RF switch since switching operations can introduce heat into the fridge, and it may take a long time to return to the base temperature.

Moreover, switches can also be employed at room temperature to enhance the reconfigurability of the setup, as shown in Figure 7.16(a–c). A cost-effective multifunction DAQ (data acquisition) device like the USB-6009 from National Instruments can be used to control the switch from the PC. Some RF switches may require higher voltages, such as 24V. In such cases, the 5-V output of the DAQ needs to be boosted. A DC-DC boost converter, as depicted in Figure 7.16(d), can be employed to accomplish this.

7.4.6 Vibrational Damping and Decoupling

Ensuring the stability of the construction before installing the fridge is paramount in minimizing vibrations [12, 13]. Additionally, it is crucial to position the compressors and vacuum pumps at a sufficient distance from the fridge to prevent vibrations. These devices generate significant noise and produce dirt, making it preferable to install them outside the building or provide acoustic isolation. Note that longer pumping lines for ^{3}He and ^{4}He on the still and 1k-pot lines can lead to reduced pumping speed, thus affecting the cooling power and base temperature. Therefore, compensating for the increased length using larger tube diameters is necessary to avoid significantly reducing pumping speed.

The vibrations can be measured using geophone sensors. The sensor's generated voltage can be fed into an FFT analyzer to display the noise spectrum. Once

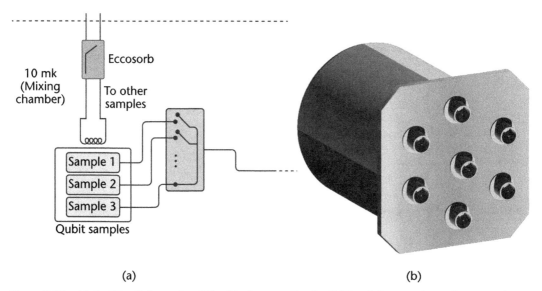

Figure 7.16 (a) An RF switch can be utilized to increase the flexibility of the experimental setup and eliminate the need for multiple connections, disconnections, and warm-up and cool-down cycles of the fridge. (b) A specific type of RF switch, the SP6T, has one input and six outputs. (c) Using switches to allow reconfigurability of the experiment to minimize the number of connects and disconnects of the cables. (d) The DAQ from National Instrument can be used to control the switch from the PC—a DC-DC converter to boost the 5V to 24V.

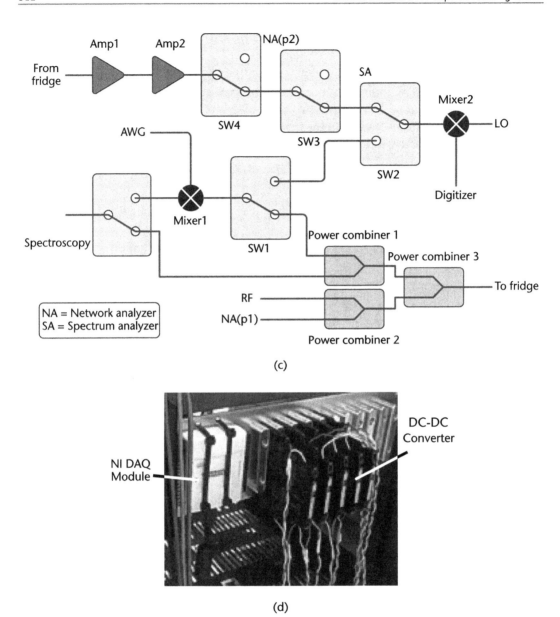

Figure 7.16 *(Continued)*

the installation is complete and the system has cooled down, the temperature inside the fridge can be measured using a nuclear orientation thermometer or the fridge's internal thermometer.

Successful decoupling and damping of vibrations enable the achievement of a stable base temperature of 10–20 mK. However, residual vibrations within the fridge, primarily caused by the pulse tube cooler, can introduce noise coupling to the qubit, thereby reducing its coherence time.

7.4 Cryogenic Setup

Table 7.9 Vibrational Damping and Decoupling

Vibrational Damping and Decoupling	Why Suppress Vibrations?	How to Do It?
Fridge	Reach stable low base temperature of 10–20 mk.	This can be achieved by carefully installing the fridge, including the construction stability, vacuum pumps, compressors, and so on.
Qubit	Avoid inductive coupling of the magnetic field due to the pulse tube cooler. Avoid coupling of mechanically induced electrical noise (due to triboelectric, microphonic, or piezoelectric effect) to the qubit.	Use twisted pairs for DC wiring to minimize the loop area and avoid the induced current in a loop moving in a magnetic field, where the movement comes from the pulse tube cooler. Vacuum insulated cables. Jacketing cables. Flattening cables.

Vibrational damping and decoupling play a critical role in properly operating the fridge and the qubit, as summarized in Table 7.9.

A stable 4-K environment is essential for the operation of a dilution fridge. Traditionally, wet fridges have utilized liquid helium baths to achieve this temperature. However, dry fridges now employ pulse-tube technology, which offers a simpler and more cost-effective solution. These dry fridges can achieve temperatures below 4K without the need for liquid helium. One advantage of pulse-tube technology is its lower vibration levels compared to other cryocoolers, typically ranging from 5 to 10 microns in the cold stages. To further mitigate vibrations, copper braids are employed to decouple the pulse tube stages from other experimental stages. By implementing a very rigid support structure, vibrations can be reduced to levels below 100 nm. While wet fridges based on liquid helium are still utilized, they are primarily reserved for highly sensitive applications such as scanning tunneling microscopes (STMs).

While pulse-tube coolers generally exhibit lower vibrations compared to other cryocoolers, it has been reported that they can generate electrical noise due to the triboelectric effect, especially at temperatures below 4K [12]. Vibration-induced noise can arise from triboelectric, microphonic, and piezoelectric mechanisms. In the case of cables, vibrations cause the dielectric and inner conductor to move relative to the outer conductor, leading to variations in the effective capacitance of the line, which is known as the microphonic effect. Additionally, vibrations can cause the contracted dielectric to rub against the inner or outer conductor, resulting in charge fluctuations and electric noise, known as the triboelectric effect.

To mitigate this type of noise, two strategies are commonly employed: jacketing and flattening the cables [12]. Jacketing involves applying a protective layer to flexible cables, which squeezes the outer conductor onto the dielectric, preventing movement within the cable when cooled. Similarly, flattening a semirigid cable restricts the movement of the dielectric. Alternatively, the use of vacuum-insulated cables can effectively suppress vibration-induced noise. Implementing these measures can significantly reduce pulse-tube noise by more than an order of magnitude and should be considered if vibration-induced issues are detected in measurements.

7.5 Future Directions for the Control Hardware

As highlighted in Chapter 2, scaling represents the most significant hurdle in unlocking the full computational power of quantum computers. To address this crucial challenge, this section provides a brief overview of future directions to tackle control hardware challenges to pave the way for the scalable realization of quantum computers.

7.5.1 Integrated Hardware

This chapter explored the utilization of commercial off-the-shelf (COTS) instruments and components for the control and readout of qubits. The initial step toward improvement involves integrating these room-temperature instruments and components to reduce size and cost while enhancing performance and scalability to accommodate multiple qubits. For instance, digital-signal generation and detection techniques are preferred over traditional analog solutions due to their higher level of integration and lower cost per channel. Figure 7.7 showed the world's first fully digital quantum control hardware developed by Keysight Technologies. A few companies have made significant progress in fully integrating and modularizing room-temperature hardware. Furthermore, open-source hardware platforms, such as QICK developed by Fermilab [14], are available for researchers and companies to customize and modify.

The next logical advancement is the implementation of integrated and scalable cryogenic control systems within the refrigerator. Cryogenic complementary metal–oxide–semiconductor (CMOS) chips hold promise as a viable candidate for this purpose. By leveraging CMOS technology at cryogenic temperatures, we can potentially achieve improved control and readout capabilities for qubits.

7.5.2 Cryogenic CMOS Chips

Cryo-CMOS circuits, leveraging an established electronic design automation (EDA) infrastructure and delivering high-speed and low-power performance, hold promise for co-integrating qubits and enabling large-scale quantum computing. However, cryogenic electronics necessitates ultralow power dissipation and high signal quality to ensure quantum-coherent operations [15, 16]. Typically, cryogenic CMOS chips operate above 1K, benefiting from a few watts of cooling power. Nevertheless, significantly enhancing the cooling power is essential to cater to the demands of thousands of qubits [15].

Although cryo-CMOS chips offer various advantages, the wiring bottleneck persists, as the electronics are physically distant from the qubits operating at the lowest temperature stage. Several potential solutions to this challenge are being explored, including the following:

1. Utilizing a multiplexer board that operates in close proximity to the qubits, satisfying the power requirements at the millikelvin cryogenic stage (typically < 1mW at 100 mK) [15]. Connecting the classical and quantum boards

poses ongoing challenges, and 3D packaging techniques like the room-temperature system-in-package (SiP) method can facilitate the integration of qubits and electronics fabricated on separate dies [16].
2. Exploring the use of hot qubits that operate at the same temperature as electronics. Encouraging results have been observed for spin qubits operating above 1K [17]. However, the existence of hot superconducting qubits remains unproven.
3. Pursuing monolithic cointegration, which presents challenges such as electromagnetic and thermal crosstalk between the electronics and qubits, as well as cryogenic self-heating from CMOS transistors, thereby limiting the integration density of such a quantum/classical system [18].

These approaches represent possible paths toward addressing cryogenic quantum computing systems' wiring and integration challenges. By pushing the boundaries of cryo-CMOS technology and exploring innovative packaging methods, there is the potential to realize scalable and efficient quantum computing architectures.

7.6 Automation and Control of the Experiment

Controlling the instruments and automating qubit operations through a centralized software system are crucial aspects of a qubit setup. Commonly used software packages include Python and MATLAB, while there are also specialized software packages tailored for qubit measurements, such as QCoDeS, pyCQED, qKIT, and Labber. Additionally, many labs develop in-house software platforms catering to their specific measurement requirements.

These software platforms play a vital role in communicating with instruments like signal generators, network analyzers, and digitizers, which typically utilize Ethernet, USB, and GPIB ports. To interact with these instruments effectively, becoming familiar with Standard Commands for Programmable Instruments (SCPI) commands is essential. SCPI is an ASCII-based programming language that facilitates communication with test and measurement instruments. While SCPI commands may not be universally consistent, they exhibit significant similarities across many instruments.

When connecting instruments to a PC, it is crucial to be mindful of grounding loops. Directly connecting an instrument to the PC also connects their respective grounds. To mitigate potential issues, isolators are employed to decouple the PC from the measurement instrument, ensuring proper grounding. Additionally, it is worth noting that the USB specification imposes limits on cable length. The cable length should not exceed 5m for full-speed devices, while for low-speed devices, the limit is set at 3m. These limitations should be considered when utilizing lengthy USB cables in practice.

By understanding the intricacies of software communication, SCPI commands, and the importance of proper grounding, researchers and engineers can effectively control and automate their experimental setups, facilitating accurate and reliable control and measurements of superconducting quantum computers.

References

[1] Krinner, S., "Realizing Repeated Quantum Error Correction in a Distance-Three Surface Code," *Nature*, Vol. 605, 2022, pp. 669–674, https://doi.org/10.1038/s41586-022-04566-8.

[2] Wallraff, A., et al., "Strong Coupling of a Single Photon to a Superconducting Qubit Using Circuit Quantum Electrodynamics," *Nature*, Vol. 431, 2004, pp. 162–167, https://doi.org/10.1038/nature02851

[3] Blais, A., et al., "Cavity Quantum Electrodynamics for Superconducting Electrical Circuits: An Architecture for Quantum Computation," *Phys. Rev. A*, Vol. 69, June 29, 2004.

[4] Krinner, S., et al., "Engineering Cryogenic Setups for 100-Qubit Scale Superconducting Circuit Systems," *EPJ Quantum Technol.*, Vol. 6, No. 2, 2019, https://doi.org/10.1140/epjqt/s40507-019-0072-0.

[5] Raftery, J. J., et al., "Direct Digital Synthesis o Microwave Waveforms for Quantum Computing," *Quantum Physics*, 2017, *arXiv:1703.00942*.

[6] Kalfus, W. D., et al., "High-Fidelity Control of Superconducting Qubits Using Direct Microwave Synthesis in Higher Nyquist Zones," *IEEE Transactions on Quantum Engineering*, Vol. 1, Art no. 6002612, 2020, pp. 1–12, 2020, doi: 10.1109/TQE.2020.3042895.

[7] Schuster, D. I., "Circuit Quantum Electrodynamics," PhD thesis, Yale University, 2007.

[8] Ball, H., W. D. Oliver, and M. J. Biercuk, "The Role of Master Clock Stability in Scalable Quantum Information Processing," *Quantum Physics*, 2016, *arXiv:1602.04551*.

[9] Mueller, F., et al., "Printed Circuit Board Metal Powder Filters for Low Electron Temperatures," *The Review of Scientific Instruments*, Vol. 84, No. 4, 2013, p. 044706.

[10] Boissonneault, M., et al., "Improved Superconducting Qubit Readout by Qubit-Induced Nonlinearities," *Physical Review Letters*, Vol. 105, No. 10, 2010, p. 100504.

[11] Thorbeck, T., et al., "Reverse Isolation and Backaction of the SLUG Microwave Amplifier," *Physical Review Applied*, Vol. 8, 2017, p. 054007.

[12] Rachpon, K., et al., "Vibration-Induced Electrical Noise In A Cryogen-Free Dilution Refrigerator: Characterization, Mitigation, and Impact on Qubit Coherence," *The Review of Scientific Instruments*, Vol. 87, No. 7, 2016, p. 073905.

[13] Kono, S., et al., "Mechanically Induced Correlated Errors on Superconducting Qubits with Relaxation Times Exceeding 0.4 Milliseconds," 2023, arXiv:2305.02591.

[14] Stefanazzi, L., et al., "The QICK (Quantum Instrumentation Control Kit): Readout and control for qubits and detectors," *The Review of Scientific Instruments*, Vol. 93, No. 4, 2021, p. 044709.

[15] Pellerano, S., et al., "Cryogenic CMOS for Qubit Control and Readout," *2022 IEEE Custom Integrated Circuits Conference (CICC)*, Newport Beach, CA, 2022, pp. 01–08.

[16] Pauka, S. J., et al., "A Cryogenic CMOS Chip for Generating Control Signals For Multiple Qubits," *Nature Electronics*, Vol. 4, Jan. 2021, pp. 64–70.

[17] Vandersypen, L. M. K., et al., "Interfacing Spin Qubits in Quantum Dots and Donors—Hot Dense and Coherent," *npj Quantum Information*, Vol. 3, No. 34, 2017.

[18] Ardizzi, A. J., et al. "Self-Heating of Cryogenic High Electron-Mobility Transistor Amplifiers and the Limits of Microwave Noise Performance," *Journal of Applied Physics*, 2022.

CHAPTER 8
Principles of Cryogenics

> Books and minds only work when they are open.
> —*James Dewar*

This chapter explores the fundamentals of cryogenics, delving into various refrigeration techniques that encompass cryogens, pumped-helium systems, mechanical coolers, and dilution refrigeration. The chapter thoroughly analyzes the dilution-fridge operation, offering a comprehensive understanding of its functioning. Additionally, we cover cryogenic thermometry and discuss different aspects of cryogenic-sensor installation, including shielding and thermal anchoring. The chapter concludes by discussing the materials commonly employed in cryogenic environments.

8.1 Introduction

The word cryogenics originates from the Greek words "kryos," meaning frost, and "genic," meaning to produce. Cryogenics studies the creation and effects of extremely low temperatures. The most significant achievement of cryogenic experiments was the discovery of superconductivity by Kamerlingh Onnes in 1911. He cooled a metallic mercury sample to temperatures below 4.2K, using liquid helium-4 (^{4}He), and observed a sudden drop in electrical resistivity to an essentially zero-resistance state. This phenomenon was named superconductivity, also known as charged superfluidity. By 1922, Onnes reached temperatures as low as 0.83K. Table 8.1 shows the key milestones in the field of cryogenics [1–3]. Over the last two centuries, scientists have developed techniques to reach temperatures down to picokelvin [4–6]. Discoveries such as superconductivity and the new phase of matter called Bose-Einstein condensate (BEC) opened avenues to new technologies such as superconducting quantum computers, magnetic resonance imaging systems, and ultrasensitive sensors.

Cryogenic environments are employed wherever superconductors are involved, such as in the superconducting magnets of the Large Hadron Collider (LHC) and superconducting quantum computers [7]. These environments are also utilized to minimize the impact of noise in detectors, including particle detectors and those used in radio astronomy. As discussed in Chapter 4, the noise power level is proportional to the temperature. Consequently, cryogenic environments are employed to maintain ultrasensitive receivers, such as those utilized in radio astronomy, at

Table 8.1 Milestones in Cryogenic Techniques

Year	Accomplishment
1845	Faraday liquified most known gases.
1877	Louis Cailletet and Raoul Pictet liquified oxygen and nitrogen. They were able to reach 80K.
1898	Liquefaction and solidification of hydrogen by James Dewar. He reached 13K.
1908	Kamerlingh Onnes liquified helium. He reached 0.83K in 1922.
1930s	The lower millikelvin range is made possible by adiabatic demagnetization.
1960s	Development of dilution refrigerators (DRs) with millikelvin-level cooling capability. The lowest temperature that current DRs may attain is $T = 5\text{--}20$ mK (record: 2 mK).
1970s and 1980s	Nuclear demagnetization makes microkelvin temperatures available for experiments (single-shot cooling). Today: $T < 100$ μK (record 1.5 μK). Laser cooling of atoms allowed reaching temperatures as low as 3 μK for cesium atoms.
1990s	In 1995, the BEC was created by Eric Cornell and Carl Wieman of the University of Colorado at Boulder using rubidium atoms; later that year, Wolfgang Ketterle of MIT produced a BEC using sodium atoms. Nanokelvin temperatures were achieved (record: 38 pK).

extremely low temperatures. This approach ensures that the noise floor remains minimal, enhancing the receiver's sensitivity.

Additionally, advanced gravitational-wave detectors employ dilution-refrigeration techniques to mitigate the impact of noise and vibrations. These detectors are designed to detect minute changes in diameter, as low as 10^{-20}m (roughly the diameter of a nucleus is 10^{-15}m), within a spherical resonator. Consequently, cooling these detectors is crucial to prevent thermal vibrations from surpassing the detectable vibration threshold [8].

Cryogenic techniques have resulted in several fascinating discoveries, and they will play even more critical roles in the future of physics and engineering. Especially in quantum computing, several qubit platforms, such as superconducting, spin, and topological qubits, depend on cryogenic environments.

8.2 An Overview of Cooling Techniques

Figure 8.1 illustrates various techniques employed to achieve low temperatures; they will be discussed in greater detail in Sections 8.3 to 8.6. Our primary focus is on the dilution-refrigeration technique for cooling superconducting qubits.

As depicted in Figure 8.1, liquid cryogens and mechanical coolers can achieve temperatures as low as 2K. Temperatures below 2K can be reached by implementing pumped systems. Pumped ^{4}He systems can achieve temperatures as low as 1.2K, while pumped ^{3}He systems can reach 300 mK. To cool below 300 mK, the dilution-refrigeration technique is employed, leveraging the cooling effect of a mixture of ^{3}He and ^{4}He liquids. Adiabatic demagnetization can also reach the lower millikelvin range, while nuclear demagnetization enables cooling to microkelvin temperatures.

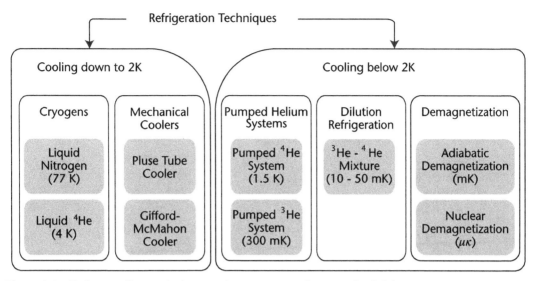

Figure 8.1 Various cooling methods to reach temperatures down to microkelvin.

To achieve temperatures below the microkelvin range, a combination of laser and evaporative cooling techniques enables cooling to nanokelvin and picokelvin temperatures, leading to the emergence of a new phase of matter known as the BEC [6]. In this state, all particles reach the ground state and form a coherent superposition of wavefunctions, exhibiting a macroscopic quantum effect.

8.3 Cryogens

Helium-4 (^{4}He) and nitrogen (N) are the primary cryogens extensively utilized in research and industrial applications. Table 8.2 displays the boiling point and latent heat values for these cryogens. In addition to ^{4}He, obtained from oil and gas wells, helium-3 (^{3}He) is also employed in cryogenics. However, ^{3}He is exceptionally expensive and scarce as it does not occur naturally; it is a by-product of nuclear reactions resulting from the decay of tritium.

8.3.1 Cooling Mechanisms

Cryogens facilitate cooling through two mechanisms: latent heat and enthalpy [3]. When an object cools down in a cryogen, the heat from the object is absorbed,

Table 8.2 Properties of ^{3}He, ^{4}He, and Nitrogen Cryogens [3]

Cryogen	Boiling Point	Latent Heat	Cost
^{3}He	3.19 K	0.026 kJ/mol	Very expensive
^{4}He	4.23 K	2.6 kJ/l	Expensive
Nitrogen	77 K	161 kJ/l	Cheap

causing the cryogen's liquid phase to evaporate. This process relies on the utilization of the cryogen's latent heat. As the cooling continues, the cryogen will eventually completely boil away, ensuring the utilization of all the available latent heat.

Latent heat refers to the energy required per unit mass for a phase change from liquid to gas. As heat is applied to a liquid, its temperature rises until it reaches boiling. At this stage, the liquid remains at the same temperature while undergoing a phase change from liquid to gas. The energy absorbed is not used to increase the temperature but rather to facilitate the transition of molecules from a liquid to a gaseous state, where gas molecules possess higher energy than liquid molecules.

The well-known equation $Q = mc\Delta\theta$ is commonly used to calculate the heat required for a temperature change. However, during a phase transition, there is no temperature change. Instead, the amount of heat Q that needs to be added or removed to facilitate a phase change for an object with mass m is determined by its latent heat L, as given by $Q = mL$.

The greater the latent heat of a cryogen, the more heat it can absorb before undergoing complete evaporation. As evident from Table 8.2, liquid nitrogen possesses a significantly higher latent heat than helium, resulting in a substantially lower amount of nitrogen evaporating for the same temperature reduction. Additionally, nitrogen is approximately a thousand times more cost-effective than liquid helium. However, nitrogen has a limited lowest-achievable temperature of 77K.

The second cooling mechanism is a result of the enthalpy change. The object's heat warms the gas from boiling to room temperature. In this case, gas enthalpy is used. ^{4}He has very low latent heat, but the enthalpy between 4.2 and 300K is substantially higher, at around 200 kJ/l. Using the enthalpy, we may, therefore, further reduce the amount of cryogens we use. The challenge lies in fully utilizing the enthalpy, necessitating the gas to exit the cryostat at room temperature. One approach to maximizing enthalpy gain is decelerating the gas as it exits the cryostat, achieved through baffles or a medium such as cotton wool [3].

One way to cool down a system below 77K is to use ^{4}He directly. However, a combination of nitrogen and ^{4}He would be much more efficient, such that we first cool to 77K with liquid nitrogen and then introduce ^{4}He. Recall that nitrogen has a higher latent heat and is much cheaper than ^{4}He. This cooling strategy is very effective since components' heat capacity (the ability to store heat) in a cryostat drops sharply with temperature. Consequently, by the time the system has reached 77K, most of the heat has already been removed. So, from this point, with a low heat capacity, the ^{4}He needed for cooling down will be much less. Table 8.3 compares the two strategies for cooling 1 kg of copper. Finally, we must remove all the nitrogen before filling the cryostat with helium; otherwise, a large amount of helium will be wasted to solidify the nitrogen as it has a relatively large latent heat of melting.

Table 8.3 Comparison of Two Strategies to Cool Down to 4.2K [3]

Cooling Strategy	Temperature Change	Consumption
Only ^{4}He	300–4.2K	32 liters
Nitrogen	300–77K	0.46 liters
^{4}He	77–4.2K	2.2 liters

8.3.2 Storage and Transportation of Cryogens

Liquid cryogens, such as liquid nitrogen or liquid helium, are stored and transported in dewars named after James Dewar. These cryogenic storage dewars are designed as vacuum flasks specifically for this purpose [1, 2]. They are constructed to maintain a high vacuum between the inner and outer walls, which provides crucial thermal insulation to significantly slow down the rate at which the cryogen boils away due to its low boiling temperature compared to the ambient temperature.

In Figure 8.2(a), you can observe that a dewar has walls made of multiple layers, with a high vacuum maintained between them. This vacuum insulation greatly

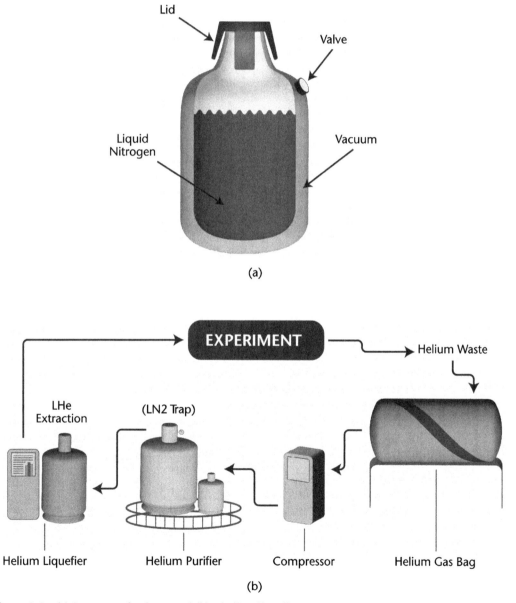

Figure 8.2 (a) Structure of a dewar and (b) a helium liquefier system.

extends the holding time of the dewar, ranging from a few hours to a few weeks, depending on the size and construction of the container.

Technological advancements have led to the development of pressurized super-insulated vacuum containers, which enable even longer storage and transportation periods for liquid nitrogen. These containers have reduced the losses to 2% per day or even less, further enhancing the efficiency of cryogen storage and minimizing wastage.

In order to minimize the risk of an explosion, it is crucial to carefully handle the gas produced when the liquid slowly boils. The gas can escape through a loosely fitted stopper or an open top in a basic design. However, more advanced dewars have high-pressure containment systems that trap the gas above the liquid. This configuration raises the boiling point of the liquid, allowing it to be stored for longer periods. Safety valves are incorporated to automatically release pressure to prevent excessive vapor pressure buildup.

Table 8.4 outlines the primary hazards associated with the handling of liquid cryogens. It is important to note that Table 8.4 is not intended to serve as a comprehensive safety guide. For a comprehensive set of guidelines and precautions, it is recommended to consult the code of practice specifically designed to handle liquid cryogens in a laboratory setting safely.

Recycling helium enables cost savings and ensures the availability of pure, high-pressure helium. Figure 8.2(b) illustrates the utilization of a helium liquefier system to accomplish this task.

8.4 Mechanical Refrigerators

Mechanical coolers, such as Gifford-McMahon (GM) or pulse-tube refrigerators (PTRs), offer an alternative to liquid cryogens and can achieve temperatures as low as 2K. Mechanical cooling systems provide a safer and less complex operation, eliminating the need for a liquefier.

A PTR utilizes a closed-cycle refrigeration technique based on gas compression, expansion, and displacement [1–3]. Figure 8.3(a) illustrates a PTR's components, including a room-temperature compressor and an expansion chamber located within the cryostat. The compressor supplies high-pressure helium (18–22 bar) as the working gas. This high-pressure helium is then directed into the cryostat, where the expansion occurs, resulting in a cooling effect. The gas passes through a high heat capacity "regenerator," which cools it as it enters the cryostat. Furthermore, the gas is further cooled during the expansion phase as it exits. Figure 8.3(b) showcases a real-world example of a PTR.

Achieving temperatures of 2K typically requires a two-stage system. The first stage typically operates at temperatures ranging from 40 to 80K, providing substantial cooling power.

It is worth noting that PTRs consume a significant amount of electrical power. For instance, it may require 11 kW of electrical power to produce 1.5W of cooling at 4.2K. The typical cooling power at 45K is around 40W, while at 4K, it ranges from

Table 8.4 Dangers Associated with Liquid Cryogens

Danger	Description	Precautions
Burns	Contact with liquid cryogens, cold gas generated by evaporating liquid cryogens, and pipework holding cold gas can cause severe skin damage similar to that caused by high heat. In extreme cases, frostbite may occur.	Always use eye protection and gloves when using or carrying cryogens.
Explosion	Volume in the liquid and gas phases varies greatly (by a ratio of roughly 700), which might cause an explosion if cryogenic liquids cannot escape from a volume.	Every cryostat needs safety valves. Make sure the correct valves are open. A dewar explodes when the evaporated gases have nowhere to go due to the accumulation of ice and frozen air in the outlet. To avoid ingress of air, dewars and cryostats must have a small positive pressure. Do not leave the dewars open to the air for any longer than necessary.
Asphyxiation	Oxygen being replaced by evaporating cryogens causes asphyxiation.	Labs must be well-ventilated. Route exhaust gases outside the building. Labs must have oxygen depletion alarms in the lab. Avoid taking a nitrogen or helium dewar in the elevator, and make sure no one else goes into the elevator while you are rushing up or down the stairs.
Combustion hazard	Since the boiling point of nitrogen is lower than oxygen, it allows oxygen to condense from the air and be concentrated. Liquid oxygen can build up to levels that may cause violent reactions with organic materials (i.e., a severe clothing fire could result).	Keep the lid on buckets of nitrogen. Keep cold surfaces that may condense air (such as pipes carrying helium boil-off gas) clean (no grease, oil, or anything else that is readily combustible).

0.5 to 1.5W. Additionally, pulse-tube coolers operate optimally in a vertical orientation, with the second stage pointing downward, which differs from GM coolers.

The primary drawback of mechanical coolers is the vibrations they generate, which can introduce noise that may couple with the device. Despite the absence of moving parts and the utilization of acoustic pulses in PTRs, they can still produce vibrations, although to a lesser extent than GM coolers. Factors such as the electricity supply frequency (e.g., 60 versus 50 Hz) can influence the performance of mechanical refrigerators. Therefore, optimizing and ensuring compatibility between the system and the laboratory's supply frequency is crucial. Minimizing vibrations requires meticulous design and careful attention to various other factors.

Mechanical coolers offer particular advantages when used in cryostats located in remote areas, such as dilution refrigerators. Dry refrigerators, which utilize

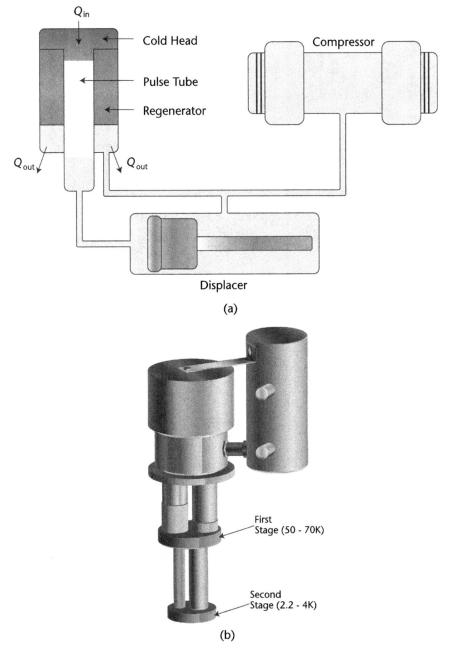

Figure 8.3 (a) Schematic of a pulse-tube cooler and (b) a real-world pulse tube cooler.

mechanical coolers, are employed in contrast to wet-dilution fridges that rely on liquid cryogens. Wet-dilution refrigerators remain suitable for applications highly sensitive to vibrations, such as STMs [9].

Caution must be exercised when working with a mechanically cooled cryostat, as premature opening can result in burns before it has warmed up.

8.5 Pumped-Helium Refrigerators

Evaporative cooling is a familiar concept we encounter when cooling down our soup or coffee by blowing away high-energy particles from the liquid's surface. By reducing the pressure above the liquid-air boundary, more high-energy particles can transition from the liquid to the air, cooling the liquid. In achieving nanokelvin temperatures and forming BEC [4–6], an advanced form of evaporative cooling involving RF signals and magnetic fields is pivotal in removing high-energy atoms from an atomic ensemble. As the gas undergoes cooling, the particle distribution gradually shifts toward lower velocities with decreasing temperature.

As we know, a liquid's boiling point decreases as the pressure decreases. Consequently, achieving lower temperatures by employing pumps to reduce the pressure becomes feasible. Using pumped cooling, temperatures as low as 1K and 250 mK can be achieved for ^{4}He, and ^{3}He, respectively. However, there is a limitation to pumped cooling, as the vaporization rate diminishes as the liquid continues to cool. More precisely, the vapor pressure P_{vap} is determined by $P_{vap} \propto e^{-L/RT}$, where L is the latent heat, T is the temperature, and R is the ideal gas constant. The cooling power $\dot{Q}$ is directly proportional to the vapor pressure; therefore, it decreases exponentially with decreasing temperature. The equation $\dot{Q} = \dot{n}_g L$ can express the cooling power, where n_g is the number of gas particles, and the dot shows the time rate of change. At low pressures, insufficient molecules are available to cool the system effectively.

Figure 8.4(a) illustrates that ^{3}He exhibits a higher vapor pressure than ^{4}He. This is because ^{3}He is lighter than ^{4}He, leading to a lower binding energy between its atoms in the liquid state. Consequently, ^{3}He possesses a lower latent heat of evaporation, resulting in higher vapor pressure [9].

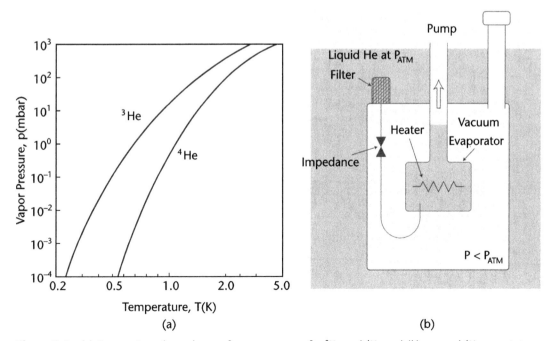

Figure 8.4 (a) Temperature dependence of vapor pressure for ^{3}He and ^{4}He and (b) pumped ^{4}He cryostat.

Figure 8.4(b) depicts a ^{4}He cryostat configuration comprising several components: a main bath operating at 4.2K and an atmospheric pressure of 1 bar, a smaller container known as the 1-K pot, a vacuum chamber, and pumps connected to the vacuum chamber and the 1-K pot. Nitrogen or air impurities can cause blockages in the impedance capillary. To prevent this from occurring, a filter is employed.

The vacuum space is crucial in thermally isolating the 1-K pot from the main 4.2-K bath. The liquid from the 4.2-K bath is pumped and directed through a flow impedance, which causes it to undergo an isenthalpic expansion. Subsequently, the liquid flows into the 1-K pot, lowering its temperature. Half of the evaporation heat is used to cool the liquid itself, while the remaining half can be used to cool the experimental setup.

The ^{4}He heat of evaporation is about 83 J/mol, and a typical flow rate that can be obtained with a midsized mechanical pump is $\dot{V} = 10^{-4}$ mol/s = 0.13 sL/mn. So, a cooling power of around 5 mW is easily achievable.

8.6 Dilution Refrigerator

The dilution refrigerator was initially proposed by Heinz London in 1951 and was experimentally realized in 1964 at Leiden University. In 1967, Oxford Instruments produced the first commercial dilution refrigerator.

Figure 8.5 illustrates a dilution fridge. The fridge consists of four temperature stages, each progressively reducing the temperature until it eventually reaches the millikelvin base temperature. Modern dilution fridges employ a combination of two cooling techniques to reach millikelvin temperatures: mechanical cooling (dry fridge) or a pumped-helium system (wet fridge) to reach kelvin temperatures and dilution refrigeration to reach millikelvin temperatures.

We are already familiar with mechanical cooling and pumped ^{4}He systems. This section explores the dilution refrigeration technique.

Table 8.5 presents the temperatures of different stages and their corresponding cooling powers for a Bluefors dilution fridge. It is evident from Table 8.5 that the cooling power decreases notably as the temperature decreases. As discussed in Section 2.4.1.3, achieving adequate cooling power is one of the challenges when scaling up. Section 8.6.1.2 discusses the factors influencing cooling power and why enhancing it poses a significant challenge.

Table 8.5 Temperatures and Available Cooling Powers of the Indicated Stages of a Bluefors XLD400 Dilution Fridge

Stage Name	Temperature (K)	Cooling Power (W)	Cable Length (mm)
50K	35	30 (at 45K)	200
4K	2.85	1.5 (at 4.2K)	290
Still	882×10^{-3}	40×10^{-3} (at 1.2K)	250
CP	82×10^{-3}	200×10^{-6} (at 140 mK)	170
MXC	6×10^{-3}	19×10^{-6} (at 20 mK)	140

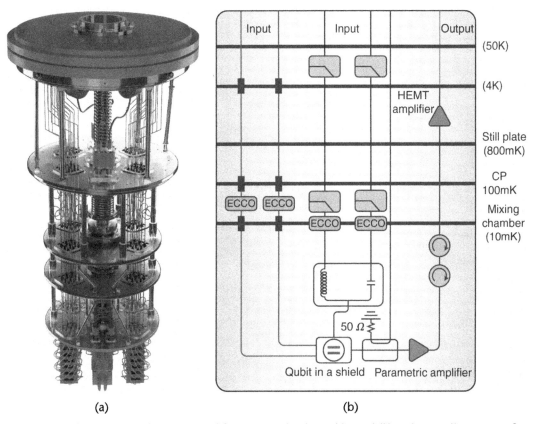

Figure 8.5 (a) A dilution refrigerator used for superconducting qubits and (b) various cooling stages of an Oxford Instruments dilution fridge.

8.6.1 Principle of Dilution Refrigeration

Recall that, in evaporative cooling, the heat applied to the liquid dQ at the constant temperature T causes phase transition, where the particles go from the more ordered liquid phase to a less ordered gaseous phase. As a result, the entropy increases, where the change in entropy dS is given by the second law of thermodynamics $dS = dQ/T$. This is shown in Figure 8.6.

The lowest achievable temperature using evaporative cooling is limited due to the exponentially decaying vapor pressure as the temperature decreases. As we will see shortly, the idea of dilution refrigeration is similar to an upside-down evaporative cooler with the difference that the equivalent of the "vapor pressure," which is the concentration of ^{3}He in ^{4}He, is a constant value of 6.4% and is independent of temperature. The lowest temperature that may be reached by the dilution refrigeration method will be limited by a much weaker temperature dependence (i.e., T^2 compared to the exponential temperature dependence of the Fermi liquid's enthalpy) [9–12], as shown in Figure 8.8(c).

First, we explain the idea of dilution refrigeration in Figure 8.6 with an analogy to evaporative cooling. To simplify the analogy, the dilution process shown

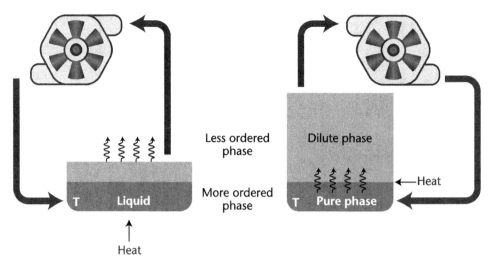

Figure 8.6 Comparison of evaporative cooling and dilution refrigeration.

in Figure 8.6 is upside-down. In the actual case, pure ^{3}He is on top of the diluted mixture [see Figure 8.10(b)]. The equivalent parameters are given in Table 8.6.

The idea of dilution refrigeration is as follows. Use a mixture of ^{3}He and ^{4}He and cool it down until this mixture separates into a pure and dilute phase [see Figure 8.7(b)] [9]. ^{3}He atoms in the pure phase (the more ordered phase) diffuse to the dilute phase (the less ordered phase). This diffusion at the interface of the two phases absorbs energy and therefore has a cooling effect. So, we can thermally connect the object we want to cool down with the interface of the two phases, as shown in Figure 8.6. This happens in a part of the fridge called the mixing chamber.

The diffusion process stops when the ^{3}He molar concentration reaches 6.4%. Therefore, bringing the mixture out of equilibrium is essential to sustain the migration of ^{3}He atoms from the pure to the dilute phase to provide a continuous cooling cycle. This can be done using the combination of osmotic pressure and distillation of ^{3}He.

Now we need to answer the following questions.

Table 8.6 Analogy of Evaporative Cooling and Dilution Refrigeration

Evaporative Cooling	Dilution Refrigeration	Description
Liquid	^{3}He in the pure phase	More ordered phase
Gas	^{3}He in the diluted phase (^{3}He in ^{4}He mixture)	Less ordered phase
Vapor pressure (drops exponentially with temperature)	^{3}He concentration [Constant (6.4%)]	The much weaker T^2 dependence limits the lowest achievable temperature
The liquid absorbs heat	The boundary between the pure and dilute phases absorbs heat	The migration of ^{3}He atoms from the more ordered phase to the less ordered phase absorbs energy

8.6 Dilution Refrigerator

- How does the phase separation happen?
- Why is the concentration of ^{3}He finite and equal to 6.4%?
- How can we create a continuous cooling process in this system?

Sections 8.6.1.1 and 8.6.1.2 answer these questions.

8.6.1.1 Mixture of ^{3}He and ^{4}He

The phase diagram shown in Figure 8.7(a) helps to understand how the ^{3}He-^{4}He mixture behaves at different temperatures. The y-axis shows the temperature,

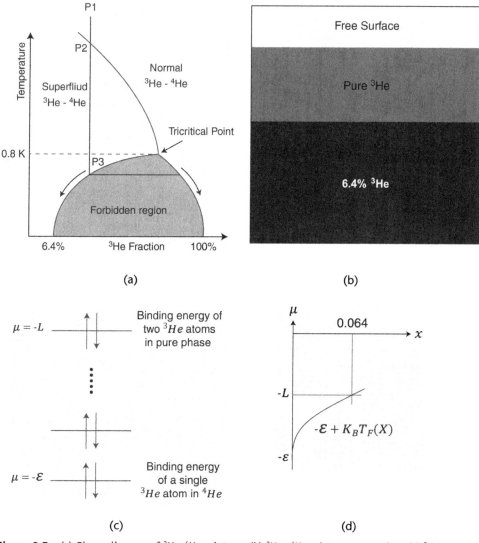

Figure 8.7 (a) Phase diagram of ^{3}He-^{4}He mixture; (b) ^{3}He -^{4}He mixture separation; (c) ^{3}He atoms with opposite spins occupy energy levels until they reach the energy level of latent evaporation of ^{3}He atoms, where the equilibrium is reached at 6.4% molar concentration of ^{3}He; and (d) the change in chemical potential as a function of ^{3}He concentration x in the dilute phase. The $k_B T_F(x)$ is the Fermi energy, which is a function of ^{3}He concentration x.

and the x-axis the molar fraction of ^{3}He. The diagram consists of three regions: the normal, superfluid, and forbidden region that starts at temperatures below the tricritical point at 870 mK [9–12]. The boundaries that separate this region from adjacent ones are referred to as lambda lines due to their resemblance to the Greek letter lambda (λ) sign.

Assume that we start somewhere around point A, where the mixture is in the normal state. As we cool down the mixture, it arrives in the superfluid region. Cooling the mixture further will spontaneously separate it into two different phases. The ^{3}He-rich phase with a higher concentration of ^{3}He is a normal fluid and corresponds to the point on the right side of the lambda line. The ^{4}He-rich phase, also called the dilute phase as it is diluted by ^{3}He molecules, corresponds to the point on the left side of the lambda line and will be a superfluid. ^{3}He-rich phase floats on the mixture as the normal liquid is less dense than the superfluid, as shown in Figure 8.7(b). Lowering the temperature further, the normal fluid approaches a state of being 100% ^{3}He, and the dilute superfluid phase reaches a 6.4% ^{3}He concentration, as shown in Figure 8.7(a).

Now, we want to understand the ^{3}He finite solubility of 6.4%. The ^{3}He atoms are lighter than the ^{4}He atoms, occupying a larger volume as their vibration amplitude is larger. This results in the ^{3}He atoms being closer to ^{4}He atoms than the ^{3}He atoms. Moreover, the attractive force between ^{3}He atoms is a weak Van der Waals force. Being closer to the ^{4}He atoms and having a weak attractive force to other ^{3}He atoms results in the diffusion of ^{3}He atoms from the pure phase to the dilute phase until an equilibrium is reached at a 6.4% molar concentration of ^{3}He in the dilute phase. Now, let us explain this number.

The binding energy between two ^{3}He atoms in the pure phase equals the latent heat of evaporation L. The binding energy ε between one ^{3}He and one ^{4}He atom is larger than between two ^{3}He atoms. As mentioned earlier, ^{3}He atoms get closer to ^{4}He atoms than ^{3}He atoms and are more strongly attracted to ^{4}He atoms than ^{3}He atoms. As more ^{3}He atoms diffuse in liquid ^{4}He, the binding energy between the ^{3}He and ^{4}He atoms decreases until it reaches the binding energy between two ^{3}He atoms, as shown in Figure 8.7(c), where μ is the chemical potential associated with energy change of adding particles to the system. This is due to the Pauli exclusion principle. Since ^{3}He is a fermion, it obeys the Fermi statistics and Pauli exclusion principle, where each energy state is occupied by two ^{3}He atoms with opposite spins, as shown in Figure 8.7(c). As more and more ^{3}He atoms diffuse in the ^{4}He, the binding energy reduces until it reaches the latent heat of evaporation of ^{3}He, and this is where the equilibrium is reached at 6.4% molar concentration of ^{3}He, and the diffusion stops, as shown in Figure 8. 7(d) [9].

8.6.1.2 Circulation of ^{3}He

To sustain a continuous cooling cycle, it is necessary to disturb the equilibrium of the mixture. Without this disturbance, the diffusion of ^{3}He atoms from the pure phase into the mixed phase ceases after reaching equilibrium, causing the cooling process to stop. To address this, a distiller (referred to as a still) is connected to the mixing chamber, as depicted in Figure 8.8(a). The still operates by applying heat, causing the ^{3}He atoms with a lower boiling point to evaporate from the mixture.

Increasing the rate of ^{3}He evaporation by supplying additional heat to the still enhances the cooling power. However, there is a limitation to how much the temperature can rise, as the fraction of ^{3}He in the vapor diminishes rapidly, as shown in Figure 8.8(b). Maintaining a ^{3}He fraction of approximately 90% in the circulated gas is crucial to uphold the efficiency of the dilution process. This optimal still temperature is typically in the range of 0.7 to 0.8K [9–12].

The question arises regarding how the mixture moves from the mixing chamber to the still. The concentration of ^{3}He in the liquid within the still decreases as ^{3}He vapor is pumped out. This difference in ^{3}He concentration establishes an osmotic pressure gradient along the tube connecting the mixing chamber and the still, causing ^{3}He to be drawn from the mixing chamber. Let us now explore how adequate osmotic pressure is generated.

The Fermi nature of ^{3}He allows it to traverse the superfluid medium of the mixture with minimal interaction with the surrounding ^{4}He. This phenomenon can be likened to a Fermi gas traveling through the massive vacuum of superfluid ^{4}He. Consequently, the ideal gas law (Hoff's law) can be employed. According to this law, which applies to a substance with n moles at temperature T in a solution with volume V, the osmotic pressure Π can be determined by: $\Pi = nRT/V$.

In a dilution refrigerator, the total volume V can be replaced with the molar volume $V_{m,4}$ of ^{4}He and n, with the fraction of ^{3}He found within the solution. The osmotic pressure difference between the phase boundary (having pressure Π_{mc}) and elsewhere along the dilute side (having pressure Π_{st}) is given by

$$\Delta \Pi \cong \frac{(x_{mc} T_{mc} - x_{st} T_{st}) R}{V_{m,4}} \quad (8.1)$$

Here, x_{mc} is 6.4%, T_{mc} to be estimated at 20–50 mK, and T_{st} at around 0.7K for optimized partial pressure. This gives the maximum osmotic pressure of almost 20 mbar, equivalent to the hydrostatic pressure of 1m of liquid helium. So, even if the mixing chamber and still are separated by a vertical distance of 1m, this osmotic pressure difference is sufficient to suck ^{3}He from the mixing chamber.

Let us look at the cooling power provided by dilution refrigeration. The amount of absorbed heat equals the difference in the molar enthalpy of ^{3}He in the dilute phase $H_d(T)$, and the concentrated phase $H_c(T)$. Both of these enthalpies are a function of temperature. For n_3 moles that pass through the phase boundary, the absorbed heat is given by $Q = n_3[H_d(T) - H_c(T)]$. The rate of heat absorption or cooling power is given by $\dot{Q} = \dot{n}_3 \Delta H$, where $\dot{n}_3$ is the ^{3}He molar circulation rate, and the enthalpy difference is given by [9]

$$\Delta H = T(S_d - S_p) = 84 T^2 \frac{\text{J}}{\text{mol}} \quad (8.2)$$

where S_d and S_p are the entropy of the dilute and pure phases, respectively.

Figure 8.8(c) compares the cooling power using dilution refrigeration and evaporative cooling. As can be seen, the lowest achievable temperature using dilution refrigeration is limited by the much weaker temperature dependence of the enthalpy of Fermi liquids (T^2).

Due to imperfect heat exchangers, the returning ^{3}He is always slightly warmer than the outgoing ^{3}He. Therefore, a more accurate cooling power expression accounts for both the temperature of the final heat exchanger, T_{ex} and the temperature of the mixing chamber, T_{mc} [9]

$$\dot{Q} = \dot{n}_3 \left(95 \, T_{mc}^2 - 11 \, T_{ex}^2\right) \tag{8.3}$$

For $\dot{n}_3 = 100 \, \mu$mol/s, to reach the temperature of $T = 10$ mK, the cooling power will be around $\dot{Q} = 1 \, \mu$W.

8.6.2 Components of a Dilution Refrigerator

This section investigates some of the essential components of a dilution refrigerator. Regarding various shielding used in the fridge, see Section 7.4.3.

8.6.2.1 Pumps

As mentioned in Section 8.6.1.2, the circulation of ^{3}He allows for sustaining the cooling cycle, where the ^{3}He vapor needs to get pumped. There are several pumps from which to pick. When choosing a pump, it is essential to note that it needs to be clean to avoid dilution fridge contamination and provide high enough pressure to reduce the condensing time.

As shown in Figure 8.9(a), the evaporated ^{3}He coming from the still goes to the oil-free turbo pump (roots pump can also be used, but it is not as clean as a turbo pump) backed by two rotary pumps (scroll pump can be used but it generates dust that may cause blockage and damage to other pumps) and a diaphragm pump (scroll pump can be used, but it is less clean and generates lower pressure) at the high-pressure end. The chain of pumps increases the pressure from about 0.01 mbar to 0.5–2.5 bar [9].

In a wet dilution refrigerator, there is also a need for ^{4}He circulation, which can be done using rotary pumps, as shown in Figure 8.9(b).

8.6.2.2 Heat Exchangers

Heat exchangers play a crucial role in establishing thermal coupling between various components within the refrigerator. For instance, as depicted in Figure 8.10(a), heat exchangers enable the cooling of the ^{3}He gas flowing toward the mixing chamber by utilizing the cold outgoing ^{3}He gas flowing toward the still.

Based on (8.9), the cooling process ceases (i.e., $\dot{Q} = 0$) when $T_{ex}/T_{mc} \geq 3$. This condition occurs when the incoming ^{3}He gas has a temperature three times higher than the diluted solution. Therefore, to achieve a mixing chamber temperature of 20 mK, the incoming ^{3}He gas from the still, which has a temperature of 700 mK, must be cooled down to at least 60 mK. Accomplishing this cooling requires highly efficient heat exchangers. In this section, we explore the impact of viscous heating

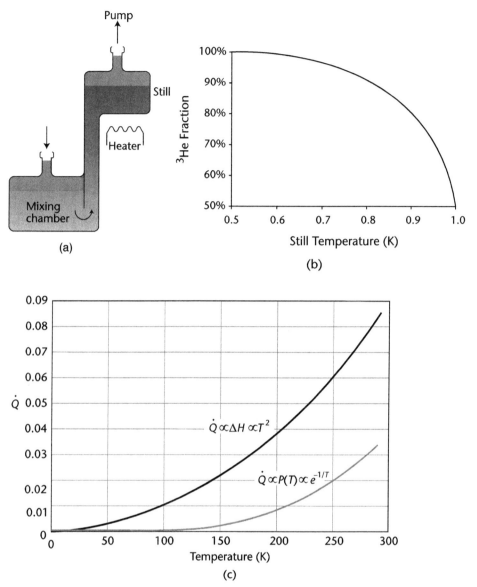

Figure 8.8 (a) Migration of mixture from the mixing chamber to the still, (b) fraction of ^{3}He as a function of still temperature, and (c) comparison of cooling power of dilution refrigeration and evaporative cooling.

and Kapitza resistance on heat exchanger performance and discuss two types of heat exchangers, namely continuous and stepped, as illustrated in Figure 8.10(b).

The thermal contact between two materials is subject to an interfacial thermal resistance, commonly referred to as Kapitza resistance. This resistance arises from the scattering of energy carriers, such as phonons or electrons, at the interface. The differences in electronic and vibrational properties of the materials on either side of the interface contribute to this scattering phenomenon.

Figure 8.10(c) illustrates that the Kapitza resistance is inversely proportional to T^3 and exhibits a rapid increase with decreasing temperature T [9]. One approach to mitigate the Kapitza resistance between helium and metals at lower temperatures is

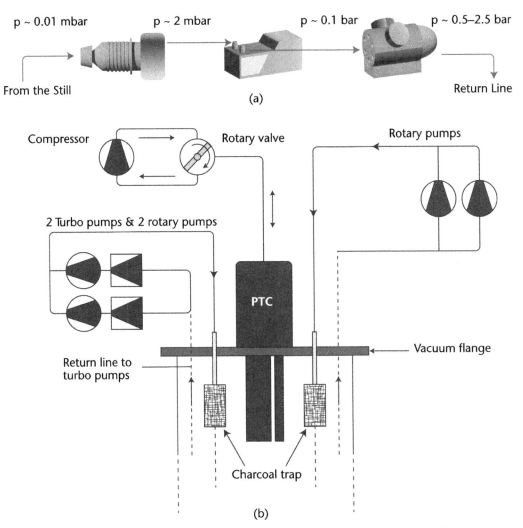

Figure 8.9 (a) Pumps used in a dilution fridge and (b) rotary and turbo pumps in a wet dilution fridge. (*After:* [9].)

to enhance the surface area by employing sintered silver with a specific area of 1m² per gram of silver. Consequently, multiple-step heat exchangers utilizing sintered silver are utilized in the lowest section of the dilution unit. As the temperature decreases, the amount of silver and the size of the flow channels are increased in each step. These stepped heat exchangers are depicted in Figure 8.10(b).

The flow of a viscous fluid through a small channel generates heat due to friction. Figure 8.10(d) demonstrates that the ^{3}He viscosity rises at lower temperatures, approximately eight times higher in the diluted phase than in the concentrated phase. Consequently, the diameter of the flow channels in heat exchangers is enlarged to minimize this effect. However, larger flow channels conduct more heat and necessitate a greater quantity of the expensive ^{3}He [9–12].

Continuous heat exchangers find application at higher temperatures, typically positioned immediately after the still, where the Kapitza resistance and viscosity are considerably lower. A continuous heat exchanger facilitates a gradual and continuous heat exchange along a temperature gradient. It performs effectively down to approximately 30 mK and is ^{3}He-efficient. One common type of continuous heat exchanger is the tube-in-tube configuration, wherein a smaller tube is inserted within a larger tube. The inner tube carries liquid ^{3}He while the cooler dilute phase flows through the annular space between the tubes.

As we have observed, the exponential increase in viscosity and Kapitza resistance at lower temperatures poses significant challenges in achieving extremely low millikelvin temperatures. The current record for the lowest attainable temperature in a dilution fridge stands at 2 mK. However, practical limitations make temperatures below 2 mK unfeasible using dilution refrigeration alone.

8.6.2.3 Impedance

A flow impedance refers to a constriction that restricts the fluid flow and enables the accumulation of a pressure difference. Generally, the impedance Z is defined in relation to the flow through the capillary $\dot{V}$ and the pressure drop across it, ΔP, as follows:

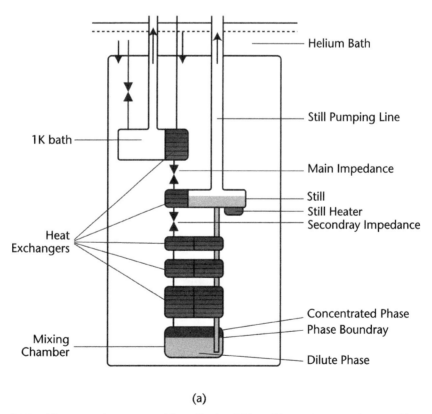

(a)

Figure 8.10 (a) Heat exchangers used in a dilution fridge, (b) continuous and stepped heat exchanger, (c) temperature dependence of Kapitza resistance, and (d) temperature dependence of viscosity.

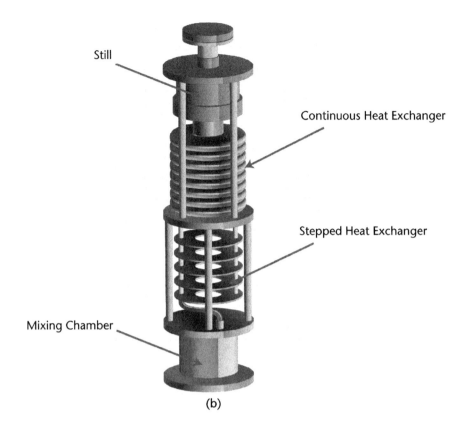

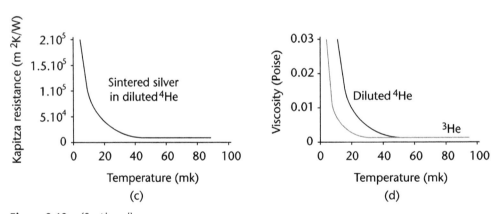

Figure 8.10 *(Continued)*

$$\dot{V} = \frac{\Delta P}{(\eta Z)}$$

where η represents the viscosity of the liquid. Controlling the flow rate is the most common application of an impedance. Impedances within the range of $10^{10} - 10^{11}$ cm^{-3} are typically utilized in various cryogenic applications. The impedance is usually constructed using an appropriate length of fine CuNi capillary tube. To increase the impedance value, a short length of wire, such as a manganin wire, is often inserted inside it [3].

Figure 8.10(a) illustrates the path of gaseous ^{3}He at room temperature as it travels toward the 1-K pot. To liquefy ^{3}He, generating a pressure greater than the vapor pressure of ^{3}He at 1K, which is approximately 300 mbar, is necessary. This objective is accomplished by utilizing the main impedance immediately after the 1-K pot [9–13].

Impedance provides the Joule-Thomson (JT) expansion that contributes to gas condensation. When the highly compressed gas is suddenly allowed to expand, it causes cooling. This is called the JT effect.

When the ^{3}He travels through the still's heat exchangers, it should maintain a pressure significantly higher than its vapor pressure to prevent ^{3}He from re-evaporating and to increase the system's heat load. In order to prevent this, it is usual practice to install a secondary impedance right beneath the still, enabling pressure inside the still to stay above vapor pressure.

8.6.2.4 Heat Switch

Sometimes it is necessary to make thermal contact between fridge parts that must be thermally isolated when cold. A heat switch is a component that turns on and off the thermal contact depending on the temperature. A heat switch can use a tube full of gas that freezes out at the operating temperature, a movable mechanical contact, or materials that go superconducting at the operating temperature (note that superconductors are not good thermal conductors). For example, a gas-gap heat switch is used between the still and the mixing chamber. The switch is closed when the cooldown first begins, and the components are still at room temperature. This allows quick cooldown of the fridge. Later when the temperature drops below a specific value, the switch becomes open to prevent a big heat load on the mixing chamber during steady-state operation.

8.6.2.5 Cold Trap and Filters

The vapors with impurities are condensed in a cold trap to stop them from spreading to other fridge parts. A dilution refrigerator uses two types of cold traps, liquid nitrogen LN2 and liquid helium LHe. Their application is to purify ^{3}He as it goes through an LN2 cold trap before flowing back into the condenser line.

In addition to dilution fridges, cold traps are utilized in pumps for various purposes. One example involves preventing impurities, like nitrogen, from reaching the vacuum pump, as their condensation can result in contamination or blockage. Similarly, certain types of pumps can potentially introduce contaminants into the vacuum system. To address this issue, a cold trap is employed at the inlet of these pumps to capture contaminants and prevent their dispersion.

8.6.3 Dry and Wet Dilution Fridge

Figure 8.11 illustrates both wet and dry dilution fridges. In the case of a wet dilution fridge, a pumped ^{4}He system is employed to achieve a temperature of 1K. Section 8.5 introduced the pumped helium system, and Sections 8.6.2.1 and 8.6.2.2 described the pumps and gas-handling systems.

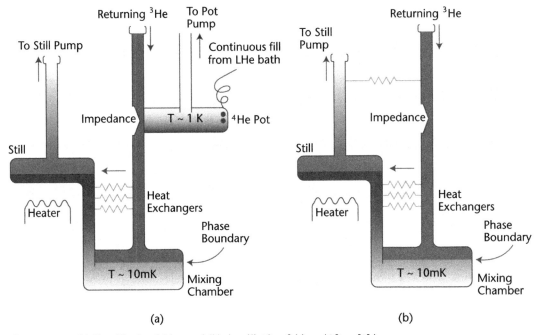

Figure 8.11 (a) Wet dilution fridge and (b) dry dilution fridge. (*After:* [9].)

On the other hand, a dry dilution fridge replaces the pumped ^{4}He system with a pulse-tube cooler. In a dry fridge, an additional heat exchanger, positioned just before the impedance, replaces the 1-K pot. This configuration is shown in Figure 8.11(b). The pulse-tube cooler necessitates using a compressor and a rotary valve, depicted in Figure 8.9(b). As mentioned in Section 8.2, an optimal arrangement involves locating the compressor remotely, which helps reduce vibrations and acoustic noise. This is particularly beneficial when working close to the dilution refrigerator.

8.7 Cryogenic Thermometry

Thermometers are required for temperature monitoring at various points in the refrigerator, which establishes the proper control commands to manage the fridge temperature. Thermometry at cryogenic temperatures necessitates careful attention to several factors since these details impact reliable readings and whether the experiment can achieve the desired temperature.

8.7.1 Cryogenic Temperature Measurements

Any property that is a function of temperature can be used as a thermometer. Table 8.7 shows some widely used thermometers [14, 15].

When choosing thermometers, a trade-off exists between measurement resolution, stability, and resistance to radiation and magnetic fields. The selection of the most suitable thermometer depends on the specific application and the required performance parameters. Figure 8.12(a) shows the resolution and temperature range covered by positive temperature coefficient (PTC) and negative temperature coefficient (NTC) sensors.

The operational temperature range of various temperature sensors is shown in Table 8.8. Examining the parameters listed in Table 8.9 assists in selecting an appropriate temperature sensor.

For certain types of thermometers, the temperature is determined by measuring the sensor's resistance. The four-wire measurement scheme, as depicted in Figure 8.12(c), is a widely recognized method for accurately measuring resistance. In this scheme, two leads are dedicated to the current source, while the other two are used for voltage measurement. This configuration eliminates measurement errors due to voltage drops in the current-carrying wires.

To conduct the measurement, the sensor is subjected to voltage or current excitation, and its response is measured. However, minimizing the heat dissipation in the sensor resulting from the excitation process is crucial. Excessive heat dissipation can lead to self-heating of the sensor, which can ultimately impact the accuracy of temperature measurements. It is, therefore, essential to determine the maximum permissible excitation level in practice by gradually increasing the current or voltage until a discernible change in resistance is observed.

To account for the decrease in power at lower temperatures, it is recommended to use a constant current for exciting thermometers with positive temperature coefficients, such as metallic resistance thermometers. On the other hand, thermometers with negative temperature coefficients, like semiconductor thermometers, are ideally excited with a constant voltage.

Table 8.7 Different Types of Thermometers Used at Cryogenic Temperatures [14]

Type	Examples	Description
Negative-temperature coefficient (NTC) resistors	Cernox Germanium Ruthenium oxide Carbon resistors Thermistors	NTC resistors are more sensitive at low temperatures
Positive-temperature coefficient (PTC) resistors	Platinum Rhodium-iron Platinum-cobalt Thermistors	PTC resistors are more sensitive at high temperature
Diodes	Silicon GaAs or GaAlAs	
Other	Thermocouples Capacitors Cryogenic linear temperature sensor Coulomb blockade thermometer Noise thermometer Paramagnetic salts thermometer	

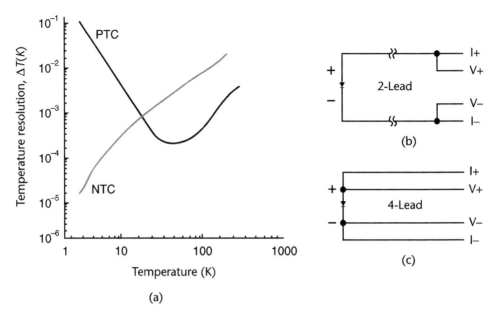

Figure 8.12 (a) Resolution of NTC and PTC resistors, (b) 2-lead resistance measurement, and (c) 4-lead resistance measurement.

Table 8.8 Parameters for Temperature-Sensor Selection [14]

Sensor	Minimum Temperature (K)	Maximum Temperature (K)
Germanium	0.05	100
RhFe	0.3	400
Capacitor	1	250
Si Diode	1	400
Thermocouple	3	1,000
Platinum	10	1,000

Resistance thermometers can be excited using alternating (AC) or direct (DC) currents. In DC measurements, it is necessary to have a signal that is sufficiently larger than any measurement noise level. On the other hand, AC excitation allows for high-precision measurements using a lock-in receiver, which enables the measurement of much smaller signals than simple DC measurements. However, it is worth noting that the cost of instrumentation for DC measurements is generally lower than precision AC measurements involving lock-in amplifiers.

In addition to the potential self-heating caused by DC signals, the accuracy of DC measurements can also be affected by thermal electromotive force (EMF). The Seebeck effect explains that thermal EMF is generated when dissimilar metals come into contact, such as the transition from the device terminal to the device itself. This thermal EMF generates a voltage typically in the microvolt range, which can introduce noise during electrical measurements. Thermal EMF becomes a critical consideration when dealing with small resistor values. To improve the measurement accuracy with DC excitation, it is possible to address thermal EMFs by

8.7 Cryogenic Thermometry

Table 8.9 Parameters for Temperature-Sensor Selection [14]

Parameter	Questions to Ask	Description
Temperature range	Minimum useful temperature? Maximum survivable temperature?	
Magnetism	Is the sensor going to be used in a magnetic field?	Field sensitivity is temperature-dependent. Some sensors work very well at some temperatures but very poorly at different temperatures. In general, offsets are worse at higher fields and lower temperatures. Recommended: Cernox above 1K; Rox below 1K
Radiation	How does ionizing radiation such as neutron and gamma radiation affect the offsets?	The general rule of thumb is that amorphous bulk materials such as cernox, rhodium-iron, germanium, and platinum have less susceptibility to ionizing radiation than more complex structures such as silicon diode and GaAlAs transistors, which are more sensitive to ionizing radiation.
Vacuum	What is the vacuum level (HV, UHV)?	At UHV, bare chips or devices in metal-ceramic packages with metal-to-ceramic seals perform the best.
Performance	What are the sensitivity, response time, and mass?	
Packaging	Size, wiring, and fixturing method?	Packaging affects the response time.
Price	Maximum available budget?	

reversing the current and obtaining the average resistance from the two current directions. This approach helps to eliminate the effects of thermal EMFs on the measurement, allowing for more accurate results.

AC bridge networks equipped with null detectors offer the highest level of accuracy by eliminating thermal EMFs resulting from temperature gradients. This approach enables the attainment of a remarkable resistance resolution of 1 $\mu\Omega$, which corresponds to a temperature resolution of 1 μK in a 25-Ω standard platinum resistance thermometer (SPRT) [14]. A commonly employed low frequency of approximately 30 Hz is utilized to circumvent circuit reactance issues.

8.7.2 Installation Considerations

To achieve a highly accurate temperature readout, it is crucial to pay careful attention to the sensor installation. Sections 8.7.2.1–8.7.2.4 discuss placement, wiring, heat sinking, and shielding factors.

8.7.2.1 Wiring

Cryogenic wires with low thermal conductivity are employed for sensor connections to minimize heat leakage. The most commonly used type of cryogenic wire is phosphor bronze. Other cryogenic wires are made of manganin, constantan, stainless steel, and superconducting materials. Moreover, nichrome heater wire and copper are utilized for heater connections. For the wire insulation, thin film formvar or polyimide is typically employed. The typical gauge for cryogenic wires ranges from 32 to 40 AWG[1] [14].

As discussed in Chapter 7, using twisted-pair wires helps reduce noise interference. Therefore, it is advisable to employ twisted-pair wires for sensor wiring. Color-coded and bundled cables allow for easy tracing of fridge wires, simplifying the heat-sinking process. When dealing with a small number of sensors, twisted-pair wires can be utilized. However, looms are more practical in cases involving numerous sensors as they provide access to multiple wires and have a flat design.

To maximize measurement accuracy in cryogenic wiring, it is crucial to adhere to the following rules:

- Keep signal leads separate to prevent any coupling between them.
- Keep signal away from heat loads, such as the control heater.
- Shield signal leads.
- Utilize the same wire throughout the entire experiment, from top to bottom, to minimize thermal EMF.

8.7.2.2 Sensor Placement and Shielding

To ensure accurate temperature measurement, it is crucial to position the sensor near the area where it will be measured. Furthermore, keeping the sensor sufficiently distant from any sources of optical radiation, heat emission, or control heaters is important.

The sensor should be enclosed within a shielded and reflective package to provide optimal protection. Additionally, it is necessary to install adequate radiation shielding at each stage of the dilution refrigerator.

The heat transfer between surfaces at different temperatures, T_{Hot} and T_{Cold}, can be determined by *heat flow* $\propto T_{\text{Hot}}^4 - T_{\text{Cold}}^4$ [15].

When a temperature sensor is close to a surface at room temperature, it can reach radiation power levels of tens of milliwatts. This radiation power poses a risk of undesired sensor heating, which can overwhelm the sensor reading. It is crucial to employ measures that effectively block optical radiation to address this issue.

One effective approach is to incorporate radiation shielding around each refrigerator stage and between the various stages within the fridge. Figure 8.13(a) illustrates the utilization of baffles, which efficiently block and reflect optical radiation. These measures ensure the necessary protection of the temperature sensor. Various sensitive components, including the sensor, can be shielded using metalized tape, a shielded reflective package, or super insulation, as shown in Figure 8.13(b). Extra care must be taken when using tapes or any materials that could adversely affect the fridge's vacuum, as maintaining vacuum cleanliness is crucial for the proper operation of the fridge.

---

1. The American Wire Gauge (AWG) is the most commonly used standard for measuring wire gauge.

8.7 Cryogenic Thermometry

Implementing such protective measures can minimize the adverse impact of optical radiation on the temperature sensor, leading to more accurate and reliable temperature measurements.

In addition to radiation, external noise sources have the potential to couple to the sensor through the leads. This leaked noise into the fridge that couples to the sensor contribute to the self-heating effect of the sensor. Therefore, it is crucial to employ meticulous installation techniques to minimize the impact of external noise reaching the sensor. Some noise-suppression techniques are shown in Table 8.10.

8.7.2.3 Thermal Anchoring

The temperature of the thermometer can exceed that of the sample due to heat leakage to the thermometer through the electrical leads and self-heating. To prevent this, certain measures can be taken, such as the following:

- Establishing effective thermal contact between the thermometer and the sample to ensure good heat transfer. Thermometer manufacturers ensure

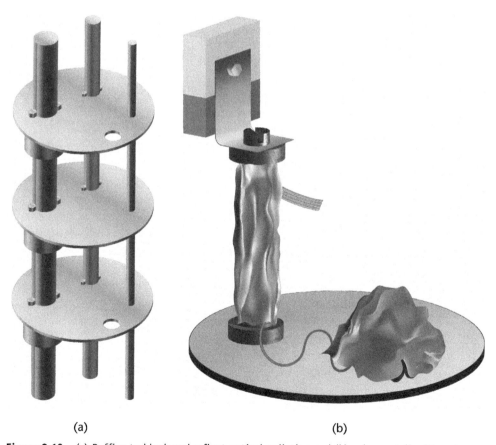

Figure 8.13 (a) Baffles to block and reflect optical radiation and (b) using metalized tape to block and reflect optical radiation.

Table 8.10 Noise-Suppression Methods for Electric, Magnetic, and Vibrational Noise

Type of Noise	Cause	Remedy
E-field noise	Capacitive voltage coupling	Use a "grounded" metal shield such as a Faraday cage
H-field noise	Current coupling induced by the magnetic field	Use twisted-pair wiring and ferrous metals
Vibrational noise	Mechanically coupled into sensor cables, generating triboelectric signals due to insulating material friction of the shielded cable	Secure the cables

good thermal contact between the thermometer element and its packaging, so achieving proper thermal contact with the sample primarily involves ensuring sufficient thermal contact between the package and the sample. We will see how this can be implemented in Section 8.7.2.4.
- Minimizing heat conduction into the thermometer.

Now, let us delve into the methods for minimizing heat conduction. Heat is conducted through electrical wires following the heat conduction equation $\dot{Q} = -\kappa A (dT/dx)$ [15, 16], where $\dot{Q}$ is the rate of energy flow, κ the thermal conductivity, A the heat transfer area, x the position, and T the temperature. So, reducing the heat transfer needs a low thermal conductivity coefficient and a small area. Therefore, cryogenic wires are made of very thin and low–thermal conductivity materials such as manganin and phosphor bronze.

As mentioned in Chapter 7, thermally anchoring wires at each stage is crucial to minimize heat loads on lower stages. This thermal anchoring reduces heat leakage from the upper stages to the sensor. Figure 8.14 illustrates the proper thermal anchoring of sensor wires at each temperature stage.

For a Cernox sensor, the recommended heat dissipation rates at 4K and 300 mK are approximately 0.1 μW and 0.1 pW, respectively [14]. It is imperative to dissipate this heat rapidly; otherwise, it can overwhelm the sensor reading. With a 1-m length of a 4-wire 32 AWG phosphor bronze wire bundle, the power transferred from the 77-K stage to the sensor at the 4-K stage amounts to around 120 μW. Hence, ensuring appropriate heat sinking becomes essential.

Wires with larger gauge sizes have smaller cross-sections. As a result, the necessary length for adequate heat sinking increases with the cross-sectional area corresponding to smaller gauge wires, as demonstrated by Table 8.11. This is why copper is not the preferred choice for cryogenic applications. It requires a significantly longer length to achieve efficient heat sinking than materials like stainless steel, manganin, and phosphor bronze [14].

Once an adequate length of wire is wound around a post, applying GE varnish can further enhance thermal conductivity by serving as an overcoat. In situations where looms are utilized, they can be thermally connected by sandwiching them between flat high-thermal conductivity metallic sheets.

8.7 Cryogenic Thermometry

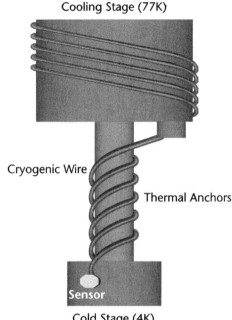

Figure 8.14 Thermal anchoring.

Table 8.11 The Necessary Heat-Sinking Length in Millimeters for Various Types of Wires to Bring the Room-Temperature Down to Liquid Nitrogen (300–7K) Temperature

Material	Wire Gauge	
	40 AWG	24 AWG
Manganin	3 mm	22 mm
Phosphor Bronze	5 mm	32 mm
Copper	19 mm	160 mm

8.7.2.4 Sensor Mounting

Figure 8.15(a) displays sensors with various packaging options. Depending on the package type, mounting techniques such as soldering, clamping, insertion into a hole, or screwing down can be employed [14]. The primary objective of mounting the sensor is to establish optimal thermal contact with the measured entity. To achieve this, it is beneficial to adhere to the following principles when mounting thermal sensors in a cryogenic environment:

- Utilize an interface material like thermal grease to maximize the contact area [refer to Figure 8.15(b)].
- Apply some force to secure the sensor package in place.
- As a general guideline, ensure that the overall contraction difference between two materials does not exceed two times. Excessive differential contraction,

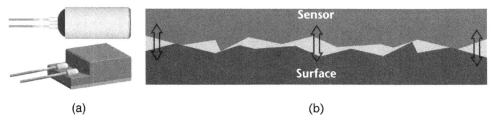

Figure 8.15 (a) Various mounting packages for thermometers and (b) stycast or grease can be used for mounting sensors with a flat surface. Stycast is permanent, while grease is easily movable and needs some other mechanical support.

exceeding this limit, can lead to thermally induced mechanical stress, causing the materials to fracture or separate.

8.8 Materials in Cryogenic Environments

According to the heat conduction eqation, thermal energy flow is directly proportional to thermal conductivity. Hence, comprehending how the thermal conductivity of different materials varies with temperature is crucial for effectively managing the thermal effects within the fridge.

The ability of heat to permeate into a material is related to its thermal conductivity κ, measured in units of $W \cdot m^{-1} \cdot K^{-1}$. The thermal conductivity of a material is influenced by its phononic (related to the crystal structure) and electronic properties. The phononic contribution to thermal conductivity is denoted as κ_{ph}, while the electronic contribution is denoted as κ_e. At low temperatures, the electronic thermal conductivity is proportional to temperature ($\kappa_e \propto T$), while the phononic contribution (κ_{ph}) is proportional to the cube of temperature ($\kappa_{ph} \propto T^3$) [16].

Table 8.12 provides information on the temperature dependence of thermal conductivity for different materials, illustrating how thermal conductivity varies with changing temperatures.

As discussed in Section, superconductors can serve as heat switches due to their sharp decline in thermal conductivity when in the superconducting state. This characteristic enables their use as the open state of the switch.

Table 8.13 presents the average thermal conductivities of three different materials at various temperatures: phosphorus-deoxidized copper (commonly employed for bars, rods, and pipes), electrolytic tough pitch copper (typically used in wire spools), and stainless steel (18/8) [1].

Let us now explore some commonly utilized materials in cryogenic environments. Sapphire is widely employed in cryogenic environments due to its properties as a dielectric with good thermal conductivity. This characteristic remains effective down to temperatures as low as 4K. However, at extremely low temperatures, sapphire's thermal conductivity sharply declines. Specifically, between 4K and 100 mK, sapphire's thermal conductivity undergoes a substantial drop, decreasing by over 60,000. In contrast, metals experience a relatively modest decrease, with their thermal conductivity reducing by three orders of magnitude smaller.

Table 8.12 Thermal Conductivity Temperature Dependence for Various Materials at Low Temperatures

Materials	Condition	Temperature Dependence of κ
Pure metals	$\kappa_e \gg \kappa_{ph}$	$\kappa_m \propto T$
Insulators	$\kappa_{ph} \gg \kappa_e$	$\kappa_i \propto T^3$
Nonpure metals or metallic alloys	κ_e is comparable with κ_{ph}	$\kappa_a \propto T$
Superconductors	$T > T_c$ (Normal state) similar behavior to metals	$\kappa_{sn} \propto T$
	$T < T_c$ (Superconducting state)	$\kappa_{ss} \propto T^3$
		$\kappa_{ss} \ll \kappa_{sn}$

Table 8.13 Temperature Dependence of the Mean Thermal Conductivity for Phosphorus Deoxidized Copper, Electrolytic Tough Pitch Copper, and Stainless Steel

Temperature(K)	300–77 K W/(m·K)	300–4 K W/(m·K)	77–4 K W/(m·K)	4–1 K W/(m·K)	1–0.1K W/(m·K)
Electrolytic copper	410	570	980	200	40
Deoxidized copper	190	160	80	5	1
Stainless steel	12.3	10.3	4.5	0.2	0.06

In addition to electrical and thermal properties, cryostat materials must have specific mechanical properties. When metals are cooled, some become brittle. Any metal with face-centered cubic (FCC) crystal structure, such as copper, brass, and stainless steel, is reliable[2] for cryogenic use [3]. When choosing materials, it is also essential to pay attention to machinability. For example, brass can be easily machined, which is not the case for stainless steel. When shaping metals, note that the extruded bar can have longitudinal cracks that might lead to leaks; hence plate construction is recommended for anything that has to be leak-proof.

Recall that another widely utilized material in low-temperature laboratories is OFHC copper. OFHC not only enables efficient heat transfer but also possesses good machinability. It finds application in various uses, such as constructing clamps for thermal anchoring or even serving as 3D waveguide cavities for superconducting qubit experiments.

A handful of well-established materials are commonly employed in low-temperature laboratories for most applications. Table 8.14 presents these materials, providing an overview of their usage.

Materials such as Apiezon grease enhance heat conduction at high-vacuum cryogenic temperatures [14, 15]. This material offers several advantageous properties, including chemical inertness, rapid cooling to cryogenic temperatures, and resistance to cracking at low temperatures. As previously observed, Apiezon

---

2. The materials with an FCC structure have more slip planes compared to body-centered cubic (BCC) materials, which allows more deformation for the relieving stresses.

Table 8.14 List of Materials Widely Used in Cryogenic Applications [14]

Category	Materials
Thermal media	Apiezon greases
	Silicon grease (Dow Corning)
	Cry-Con grease (copper loaded)
	Indium foil
Adhesive materials	IMI 7031 varnish
	Stycast epoxies
	Conductive epoxies
	Ceramic potting epoxies (some are metal-based and conduct!)
Fastening materials	Dacron fiber (dental floss)
	Nylon (fishing line)
	Small metal leaf spring
	Kapton tape
	Masking tape
	3M Scotch electrical tape with polyester film (Type 56)
Wire materials	Copper
	Phosphor bronze
	Manganin
	Constantan
	Nichrome
	Stainless steel
	Superconducting
	Coaxial (single strand, dual lead, quad lead, dual twist, quad twist)
Miscellaneous	Superinsulation
	Fiberglass and Teflon sleeving
	Heat-shrink tubing

grease can effectively increase the contact area for temperature sensors, thereby improving thermal coupling across the entire contact surface.

Additionally, different types of epoxies, such as Stycast, serve as adhesive materials for securing components within the refrigerator. Chapter 7 discussed their application in creating copper powder filters. Fishing lines or dental floss are commonly employed as fastening materials in cryogenic setups. A small metal leaf spring is considered one of the best components for fastening. When using tapes like Kapton tape, it is important to exercise caution as they can impact the cleanliness of the environment, which in turn may adversely affect the vacuum conditions inside the fridge. We have already explored cryogenic wires and their uses.

References

[1] White, G. K, and P. J. Meeson, *Experimental Techniques in Low-Temperature Physics* (Fourth ed.), Oxford: Clarendon Press, 2002.

[2] Ekin, J. W., *Experimental Techniques for Low-Temperature Measurements*, Oxford University Press, 2006.

[3] Woodcraft, A. L., "An Introduction to Cryogenics," *European Organization for Nuclear Research Cern/At*, 2007.

[4] Wieman, C., and E. Cornell, "Several Years Later: The Creation of a Bose-Einstein Condensate in an Ultracold Gas," Lorentz Medal, October 12, 2021, https://www.nist.gov/publications/several-years-later-creation-bose-einstein-condensate-ultracold-gas.

[5] Seo, B., "Apparatus for Producing a 168Er168Er Bose–Einstein Condensate," *Journal of the Korean Physical Society*, Vol. 82, pp. 901–906.

[6] Ketterle, W., "Bose–Einstein Condensation Of Ultracold Atomic Gases," *Physica Scripta*, No. T66, 1996, p. 31.

[7] Tavian, L., "Latest Developments in Cryogenics at CERN," Tech. rep. CERN-AT2005-011-ACR., July 2005, http://cds.cern.ch/record/851586.

[8] Bortoli, F. S., "On the Dilution Refrigerator Thermal Connection for the SCHENBERG Gravitational Wave Detector," *Brazilian Journal of Physics*, Vol. 50, 2020, pp. 541–547.

[9] Batey, G., and G. Teleberg, *Principles of Dilution Refrigeration: A Technology Guide*, Oxford, UK: Oxford Instruments, 1998.

[10] Kakaç, S., H. F. Smirnov, and M. R. Avelino, *Low Temperature and Cryogenic Refrigeration*, Springer, 2003.

[11] Balshaw, N. H., "Practical Cryogenics: An Introduction to Laboratory Cryogenics," Oxford Instruments, 1996.

[12] Rapp, R. E., R. T. Johnson, and J. C. Wheatley, "Principles and Methods of Dilution Refrigeration. II," *Journal of Low Temperature Physics*, Vol. 4, No. 1, 1971, pp. 1–39.

[13] Richardson, R. C., and E. N. Smith, *Experimental Techniques in Condensed Matter Physics at Low Temperatures*, Addison-Wesley, 1988.

[14] Lakeshore Cryotronics, "Cryogenic Sensors: Installation Techniques for Success," https://www.lakeshorecryotronics.com/sensors-install-webinar.

[15] Pobell, F., *Matter and Methods at Low Temperatures*, Springer-Verlag, 1994.

[16] Duthil, P., "Material Properties at Low Temperature," Institut de Physique Nucléaire d'Orsay, IN2P3-CNRS/Université de Paris Sud, Orsay, France.

About the Author

Alan Salari is an experimental quantum physicist and microwave engineer actively involved in developing cutting-edge technologies. With extensive experience gained from working at leading industrial entities such as Hughes Network Systems, Quantum Circuits Inc., and Fraunhofer Institute, he has contributed to diverse fields such as: superconducting qubits, ultracold atomic gases, pixel detectors for the Large Hadron Collider, microwave plasma generators, and radar and satellite communication systems.

Alan is a senior member of IEEE, the Quantum Economic Development Consortium (QED-C), and the Mid-Atlantic Quantum Alliance. Notably, he has been honored with numerous awards from esteemed organizations, including the National Science Foundation (NSF); Discovery Partners Institute (DPI), Chicago; the Alexander von Humboldt Foundation; the German Ministry of Economic Affairs and Energy (EXIST Gründerstipendium); and Siemens AG.

Index

A

Absorption loss, 263
ac-Stark shift
  about, 91, 109
  Hamiltonian, 106–9
  photon number calibration with, 109–12
  use of, 91
Active load, 290–91
Amplification, 247–48
Amplifiers
  about, 210–11
  comparison, 213
  cryogenic HEMT, 309–10
  parametric, 215–19, 308–9
  quantum-limited, 211–15
  SLUG, 219, 220, 310
  TWPA, 216–19, 308–9
Amplitude manipulation
  amplifiers, 210–19
  attenuators, 219–21
  bias-tree, 228–29, 230
  circulator/isolator, 223–26
  DC block, 228
  directional coupler, 222–23
  microwave switch, 226–28
  power combiner, 221–22
  signal addition/subtraction, 221–29
  techniques, 209, 210
  See also Signal processing
AM-PM conversion, 159–60
Analog signal generation
  about, 177–78
  room-temperature setup, 279–83
Analog-to-digital converter (ADC), 148, 253–54
Analytical methods
  about, 168
  circuit theory, 171–72
  distributed-element regime, 171
  lumped-element regime, 169–71
  Maxwell's equations, 167, 168–69
  optical regime, 171
Anharmonicity, 76
Ansatz, 87
Arbitrary wave generators (AWGs)
  about, 92, 179–80
  block diagram, 181
  figures of merit, 180–83
  flux bias pulses, 116
Artificial atoms, 69
Atom-field coupling, 99
Atomic model, 2
Attenuation
  about, 131
  sources, 131, 192
  for specific transmission line length, 191
Attenuation constant
  about, 191–92
  conductive losses, 192–93
  dielectric losses, 193
  mismatch losses, 194
  radiation losses, 194
Attenuators, 219–21, 296–99
Automation/control of the experiment, 315

B

Babinet's principle, 267
Balme series transitions, 70
Bandpass filters, 231, 232
Band-stop filters, 231
Bandwidth
  AWG, 181–82
  signal generator, 178
  transmission line, 197–99
Bare states, 99
Bell state, 28, 37
Bias-tree, 228–29, 230
Bit-flip code, 46
Bit-flip gate, 35
Bloch sphere, 16–17, 82
Breakout box, 286
Broad-bandwidth DDS, 285

C

Capacitive coupling, 118, 119
Cavity quantum electrodynamics (CQED)
  about, 86, 93, 94–95
  atom sizes, 96
  dispersive regime, 120
  element comparison, 95, 96
  fast readout, 105–6
  generic system, 95, 96
  high SNR, 104–5

353

Cavity quantum electrodynamics (CQED) (Cont.)
  illustrated, 96
  Jaynes-Cummings model, 95
  JC Hamiltonian, 97–100
  quantization of EM field, 95–97
  qubit readout requirements, 104
  strong coupling, 101–2
  strong dispersive regime, 102–4
  vacuum Rabi oscillations, 100–101
Charge dispersion, 76
Circuit-analysis techniques, 172
Circuit modeling
  about, 172
  numerical methods, 176–77
  scattering parameters, 172–75
  S-parameters, 175–76
Circuit QED (cQED), 86
Circuit theory, 171–72
Circulator/isolator, 223–26
Classical mechanics, 1, 3–4
Coaxial cable, 200–201
Coherence time, 39–40
Coherent superposition, 24–25
Cold trap and filters, dilution fridge, 337
Commutation relations, 22–23
Conducted emission, 260
Conductive losses, 192–93
Control hardware
  about, 277
  cryogenic CMOS chips, 314–15
  cryogenic setup, 282, 286, 290–313
  future directions for, 314–15
  high-level description of setup, 277–79
  integrated, 314
  low-level description of setup, 279
  room-temperature setup, 279–90
  schematic, 280
Controlled-NOT (CNOT) gate, 38
Cooling mechanisms, 319–20
Cooling techniques, 318–19
Cooper pair box (CPB)
  about, 71
  charge fluctuations, 73
  eigenenergies, 71, 73, 74–75
  Hamiltonian of, 71–72
  illustrated, 72
Cooper pairs
  about, 61
  behavior of, 62
  density of, 64
  electrons, 62
  energy gap, 62–64
  formation of, 61
  Josephson junction, 66

See also Superconductivity
Coplanar waveguide (CPW)
  applications of, 203
  asymmetrical discontinuities, 204–5
  grounded (GCPW), 203, 204
  resonators, 240, 242–43
  structure, 203
  symmetric, 203–4
Copper powder filters, 299–303
Coupling
  about, 176
  atom-field, 99
  capacitive, 118, 119
  inductive, 118
  microwave resonators, 239–42
  qubit-cavity, 113–15
  strong, 101–2
CPHASE gate, 118
Critical temperature, 57
Cross-modulation, 161
Crosstalk, 52, 261
Cryogenic CMOS chips, 314–15
Cryogenic engineering, 54
Cryogenic HEMT amplifiers, 309–10
Cryogenic microwave cabling, 293–95
Cryogenic mu-metal shield, 271
Cryogenics
  cooling techniques, 318–19
  cryogens, 319–22, 323
  dilution refrigerator, 326–38
  environments, materials in, 346–48
  introduction to, 317–18
  mechanical refrigerators, 322–24
  milestones in techniques, 318
  pump-helium refrigerators, 325–26
  thermometry, 338–46
Cryogenic setup
  DC wiring, 290
  description, 282
  filtering, 299–304
  heat loads, 290–91
  low-pass filters in, 286
  noise-suppression techniques, 291–307
  shielding, 304–7
  signal amplification, 307–10
  switches, 310–11
  vibrational damping and decoupling, 311–13
  See also Control hardware
Cryogenic thermometry
  installation considerations, 341–46
  measurements, 338–41
  sensor mounting, 345–46
  sensor placement and shielding, 342–43
  thermal anchoring, 343–45
  wiring, 341–42

Cryogens
    about, 319
    cooling mechanisms, 319–20
    liquid, dangers associated with, 323
    storage and transportation of, 321–22
CW spectroscopy, 113

D

Data acquisition, 55
DC block, 228
DC wiring, 290
de Broglie, Louis, 2
Density matrix, 27
Derivative Removal by Adiabatic Gate (DRAG) pulse, 92–93
Deutsch-Jozsa algorithm, 45
Dewars, 321
Dielectric losses, 193
Digital phase shifters, 245–46
Digital predistortion (DPD), 135
Digital signal generation, 283–85
Digital-to-analog converter (DAC), 148
Digitization/demodulation, 250
Dilution refrigerator
    about, 326
    circulation of ^{3}He, 330–32
    cold trap and filters, 337
    components of, 332–37
    dry, 337–38
    evaporative cooling comparison, 328
    heat exchangers, 332–36
    heat switch, 337
    illustrated, 327
    impedance, 335–37
    mixture of ^{3}He and ^{4}He, 329–30
    principle of, 327–32
    pumps, 332, 334
    temperature and cooling powers, 326
    wet, 337–38
    See also Cryogenics
Dirac, Paul, 2–3
Direct digital synthesis (DDS), 177
Directional coupler, 222–23
Direct RF sampling, 251–55
Dispersion
    about, 195
    charge, 76
    as distortion source, 134–35
    effect in transmission lines, 195
    material, 195
    waveguide, 195
Dispersive JC Hamiltonian, 109, 110
Dispersive two-qubit interactions, 120–21
Distortion
    about, 132

    deliberate introduction of, 132
    digital predistortion (DPD), 135
    dispersion and, 134–35
    intermodulation (IMD), 157, 236
    nonflat frequency response and, 134
    nonlinear effects and, 134
    sources of, 134–35
Distributed-element filters, 231
Distributed-element regime, 171
Distributed-element transmission lines, 184, 185
DiVencenzo criteria, 32–33
Double-p matric, 40
Double-pole, double-throw (DPDT) switch, 226, 227
Downconversion, 233–36, 249–50
Dry dilution fridge, 337–38
Dynamic range
    about, 161–62
    DAC, 184
    illustrated, 162
    SLUG amplifiers, 219, 310
    spurious-free (SFDR), 161, 162
    TWPA, 216

E

Effective number of bits (ENOB), 180, 183
Ehrenfest's theorem, 19
Electromagnetic compatibility (EMC)
    about, 257
    conducted emission, 260
    crosstalk, 261
    electronic system interaction, 259–60
    interference sources, 260–61
    primary levels of approach, 259
    radiated emission, 260
    requirements, 258–59
Electromagnetic shielding
    about, 261–62
    effectiveness, 262–65
    grounding effect, 267–68
    illustrated use, 262
    onboard electronic components, 270
    penetration, 265–67
    radiated path and, 304–5
    for source types, 263
    techniques for qubits, 268–71
    See also Shielding
Electron beam lithography (EBL), 78
Energy gap, 62–64
Energy-level splitting, 99–100
Energy uncertainty, quantum coherence and, 25–26
Equivalent circuit
    of capacitor, 172
    g-factor and, 114

Equivalent circuit *(Cont.)*
    illustrated, 116–17
    of qubit-cavity coupling, 113–15
Equivalent noise temperature, 138–39, 143
Error vector magnitude (EVM)
    about, 162
    bathtub curve, 164
    expression of, 163
    importance of, 162–63
    requirements, 164
Expectation value, 18–19

F

Figures of merit
    AWG, 180–83
    signal generator, 178–79
    transmission lines, 199–200
Filtering
    absorptive, 229
    cryogenic setup, 299–304
    electromagnetic compatibility (EMC) and, 271–72
    microwave signal processing, 248–49
    reflective, 229
    room-temperature setup, 271, 285–87
Filters
    about, 229
    bandpass, 231, 232
    band-stop, 231
    categorization, 229–31
    copper powder, 299–303
    dilution refrigerator, 337
    distributed-element, 231
    high-pass, 231
    infrared, 303–4
    low-pass, 231–32, 286
    lumped-element, 231
    mechanisms, 229
    Pi, 232
    T, 232
    use of, 271
    See also Frequency manipulation
Finite-difference time domain (FDTD), 177
Finite-element method (FEM), 177
Flicker noise, 136
Flux bias gate, 90
Flux quantization, 59
Fourier transform
    quantum, 45
    scaling property, 22
    uncertainty principle and, 21–22
Free particles, Schrödinger equation and, 8–9
Frequency manipulation
    about, 229
    filters, 229–32

microwave resonators, 237–44
techniques, 231
up-/downconversion, 233–36
See also Signal processing
Frequency range, 178

G

Gain, 176, 212
Gain compression, 153–54
Gates
    bit-flip, 35
    CNOT, 38
    CPHASE, 118
    fidelity, 39–41
    flux bias, 90
    Hadamard, 35–36, 37
    iSWAP, 38
    p/8-gate, 36
    phase-flip, 36
    single-qubit, 34–37
    three-qubit, 38
    two-qubit, 37–39
G-factor, 114
Ginzburg-Landau theory, 63
Grounded CPW (GCPW), 203, 204
Grounding
    about, 272
    effect on shields, 267–68
    improvement, 274
    room-temperature setup, 289–90
    strategy, 272–74
    of wires, 274–75
Ground plane, 272
Grover's algorithm, 44
GSM standard, 141

H

Hadamard gate, 35–36, 37
Hahn echo experiment, 81, 82
Hamiltonians
    atom and field interactions, 86–87
    Bloch sphere, 16–17
    of Cooper pair box (CPB), 71–72
    free particle, 8–9
    harmonic oscillator, 11–13
    JC, 97–100
    ladder operators, 13–14
    potential well, 9–11
    quantum LC oscillator, 14–15
    qubit, 16
    transmon qubit, 115
Harmonic distortion, 150–53, 158
Harmonic oscillators, 11–13
Harmonics, 179
Heat exchangers, dilution fridge, 332–36

Heat flow, 342
Heat loads, 290–91
Heat switch, dilution fridge, 337
Heisenberg's uncertainty principle. *See* Uncertainty principle
Helium, 319, 322
Hemispheric emissivity, 307
Hermite polynomials, 11
Hertz, Heinrich, 125, 126
High-pass filters, 231
Hollow waveguides, 206–7
Homodyne detection, 250–51

I

Image frequency, 233–36
Impedance, dilution fridge, 335–37
Impedance matching, 189–90
Inductive coupling, 118
Infrared filters, 303–4
Input/output (IO) management, 52
Insertion loss, 175, 176, 200, 201
Interference, 136
Interference sources, 260–61
Intermodulation
    calculation of, 155–56
    defined, 154
    distortion, 157
    frequencies, extracting, 156
    products, 154–55
Intermodulation distortion (IMD), 157, 236
Intraresonator photon number, 111
Ions, 47, 48
Isolation, 176, 236
Isolators, 223–26
iSWAP, 38

J

Jaynes-Cummings model, 86, 95
JC Hamiltonian
    dispersive, 109, 110
    interaction Hamiltonian, 97
    ladder, 100
    solutions, 98–100
    terms, 97
    See also Hamiltonians
Josephson energy, 67
Josephson inductance, 67
Josephson junctions
    about, 66
    cooling, 69
    Cooper pairs, 66
    illustrated, 67
    LC circuits with, 70
    nonlinearity, 68
    parameters, 68

    See also Superconductivity

K

Kerr-type terms, 110
Kirchhoff's voltage law (KVL), 170
Kirchoff's current law (KCL), 170–71

L

Ladder operators, 13–14
Lamb shift, 106–12
Laws of physics, 3–4
Linearity, 132–34
Lithography techniques, 78
Longitudinal relaxation, 78–80
Lorentz shape, 25
Lowering operator, 13, 16
Low-noise amplifier (LNA), 247–49
Low-pass filters, 231–32, 286
Lumped-element filters, 231
Lumped-element regime, 169–71
Lumped-sum transmission lines, 184, 185

M

Marconi, Guglielmo, 125–26
Material dispersion, 195
Materials
    in cryogenic applications, 346–48
    ferromagnetic, 264–65
    hemispheric emissivity for, 307
    permeable ferromagnetic, 224
    thermal contact between, 333
Matrix mechanics, 2
Maximum noise power, 138
Maxwell's equations, 167, 168–69
Measurement concept, 4
Mechanical refrigerators, 322–24
Meissner effect, 57–58
Memory size, 180
Method of moments (MoM), 177
Microstrip line, 201–3
Microwave cabling, 292–95
Microwave components
    analysis, 167–77
    analytical methods, 168–72
    circular modeling, 172–76
    numerical methods, 176–77
    performance specifications, 209
    signal detection, 247–55
    signal generation, 177–83
    signal processing, 209–47
    signal transmission, 183–209
    tools for analysis of, 167–77
    See also specific components
Microwave connectors, 295
Microwave controllers, 207–9

Microwave engineering
  about, 127–28
  brief history of, 125–27
  as engineer skillset, 54–55
  role in shaping modern world, 127
  technical areas related to, 128
Microwave links
  components, 129
  link budget analysis, 130
  for quantum computer, 129
  received signal level, 131
Microwave resonators
  about, 237–38
  behavior, 237–39
  coupling, 239–42
  CPW, 240, 242–43
  critically coupled, 242
  overcoupled, 241
  rectangular cavity, 245
  transmission-line open-ended, 240
  waveguide, 243–44
Microwave signal generator, 281
Microwave switch, 226–28
Microwave systems
  AM-PM conversion, 159–60
  analysis, 128–64
  cross-modulation, 161
  dynamic range, 161–62
  error vector magnitude, 162–64
  gain compression, 153–54
  harmonic distortion, 150–53, 158
  intermodulation, 154–57
  microwave link, 129–31
  nonlinear effects in, 149–61
  signal degradation factors, 131–48
  third-order intercept points (IP3), 157–59
Minimum detectable signal (MDS), 140–41, 247
Mismatch losses, 194
Mixers, 233, 234
Mixing chamber (MXC), 290–91
Multilayer insulation (MLI) pads, 306

N

Nanofabrication, 55
Network analyzers, 174–75
Neutral atoms, 47–48
Nitrogen, 319, 320, 322
Noise
  flicker, 136
  phase, 144–45, 179
  quantization, 145–48, 149
  quantum, 136
  shot, 136
  thermal, 136–43, 149
Noise figure (NF)
  about, 141
  equivalent noise temperature and, 143
  of passive device, 142
Noise-suppression techniques
  about, 291–92
  with attenuators, 296–99
  with circulator, 306–7
  for conducted path, 292–304
  for electric, magnetic, vibrational noise, 344
  microwave cabling and, 292–95
  microwave connectors and, 295–96
  for radiated path, shielding and, 304–7
  summary, 293
  See also Cryogenic setup
Noise temperature, 212
Noise voltage, 137
Noisy intermediate-scale quantum (NISQ), 46
Non-Abelian anyons, 51
Nonflat frequency response, 134
Nonlinear effects
  about, 149
  AM-PM conversion, 159–60
  cross-modulation, 161
  distortion and, 134
  gain compression, 153–54
  harmonic distortion, 150–53, 158
  inputs and, 150
  intermodulation, 154–57
  overview, 149–50
  third-order intercept points (IP3), 157–59
  See also Microwave systems
Non-TEM transmission lines, 185, 196–200
Nuclear magnetic resonance (NMR), 78
Numerical methods, circuit modeling, 176–77

O

Optical regime, 171
Organization, this book, xii
Otocoupler, 274
Oxygen-free high-conductivity (OFHC) enclosure, 269–70

P

Parametric amplifiers, 215–19, 308–9
Parametric mixing, 213
Passive load, 290
Permeable ferromagnetic materials, 264–65
Phase constant, 194–96
Phase-flip code, 46
Phase-flip gate, 36
Phase manipulation, 244–47
Phase noise, 144–45, 146, 179
Phase shifters, 244–47
Photon number calibration, 109–12
Pi filters, 232

Potential wells, 9–11
Power combiner, 221–22
Power-handling capability, 200
Power leakage, 236
Power spectral density, 148
Propagation constants, 190–91
Propagation modes, 196
Pulse broadening, 196
Pulsed spectroscopy, 113
Pulse shaping, 91–93
Pulse-tube refrigerators (PTRs), 322–24
Pumped-helium refrigerators, 325–26
Pumps, dilution fridge, 332, 334
Purcell effect, 94
Purcell filter, 106, 108
Purcell relaxation rate, 103
Pure dephasing, 80–81

Q

Quality factor, 239
Quantization error, 147–48
Quantization noise
    about, 145–47
    deviations, 147
    effect, 147
    thermal noise comparison, 149
Quantum algorithms
    about, 42–43
    Deutsch-Jozsa algorithm, 45
    development of, 44
    Grover's algorithm, 44
    quantum Fourier transform, 45
    Shor's algorithm, 44–45
    structure, 43
Quantum bus, 117
Quantum circuits, 41–42
Quantum coherence
    about, 24–25
    coherent superposition and, 24–25
    density matrix and, 27
    energy uncertainty and, 25–26
    process illustration, 24
Quantum computing
    about, *xi*, 31
    applications of, 33–34
    in banking and finance, 34
    challenges and opportunities in, 51–55
    in chemistry, material discovery, and drug
        development, 34
    in cybersecurity, 33–34
    DiVencenzo criteria and, 32–33
    hardware engineer skillsets, 54–55
    heat and power dissipation, 52–53
    information processing, 34–47
    input/output (IO) management, 52
    introduction to, 31–55
    ions, 47, 48
    neutral atoms, 47–48
    noise and crosstalk, 52
    platforms, 47–51
    power of, 31–32
    semiconductor qubits, 48–51
    size of the system, 53
    slew rate, rise time, and bandwidth, 53–54
    technical challenges of scaling, 52–54
Quantum electrodynamics (QED), 3, 112–13
Quantum entanglement, 27–28
Quantum error correction, 46, 78
Quantum Fourier transform, 45
Quantum harmonic oscillator, 11–13, 14
Quantum information processing
    about, 34
    algorithm, 42–45
    circuits, 41–42
    error correction, 46
    gate fidelity, 39–41
    single-qubit gates, 34–37
    two-qubit gates, 37–39
Quantum key distribution (QKD), 34
Quantum LC oscillator, 14–15
Quantum-limited amplification, 211–15
Quantum measurement
    about, 17
    coherence and, 24–27
    collapse of wave function and, 17–18
    expectation value and, 18–19
    Heisenberg's picture and, 23
    uncertainty principle and, 20–23
    variance or uncertainty and, 19–20
Quantum mechanics
    brief history of, 1–3
    calculations, 5–6
    classical mechanics versus, 1, 3–4
    concepts, 1
    entanglement, 27–29
    introduction to, xii
    measurement, 4, 17–27
    observables in, 5–6
    quantization, 69
    Schrödinger equation and, 4–17
    vacuum and, 79
Quantum noise, 136
Quantum nondemolition (QND)
    about, 93
    measurement, 102
    photon population of resonator and, 109
    protective readout, 102
Quantum-processing unit (QPU), 278
Quantum processor, 78, 122
Quantum-process tomography (QPT), 40, 41

Quantum supremacy, 46–47
Quantum tunneling, 11
Quasiparticles, 84–85
Qubit-cavity coupling, 113–15
Qubit coherence time scales
  about, 78
  coherence improvement strategies and, 85
  longitudinal relaxation, 78–80
  sources of qubit decoherence and, 83–85
  transverse relaxation, 80–82
  See also Superconducting qubits
Qubit control
  about, 86
  drive pulse shaping, 91–93
  in practice, 115–16
  Rabi oscillations, 86–89
  rotation about X- and Y-axes, 90
  rotation around Z-axis, 90–91
  rotations, 89–90
  See also Superconducting qubits
Qubit decoherence
  improvement strategies, 85
  quasiparticles, 84–85
  source effect on coherence times, 83
  sources, 83–85
  time improvement, 85
  times, 83
Qubit drive pulse shaping, 91–93
Qubit readout
  about, 93
  ac-Stark shift and Lamb shift, 106–12
  CQED, 94–106
  fast, 105–6
  in practice, 115–16
  Purcell effect, 94
  requirements, 104
  See also Superconducting qubits
Qubits
  about, 16
  crosstalk and, 52
  energy levels, 25–26
  naturally occurring, 48
  noise and, 52
  nonideal properties in, 25
  semiconductor, 48–51
  shielding techniques for, 268–71
  spin, 49
  state space of, 16
  superconducting, 49–50, 69–122
  topological, 50–51
  transmon, 73–78
  unentangled, 28
Qubit states
  controlling, 89
  distribution, 107
  single, 28
  two, 28, 37

R

Rabi oscillations, 86–89
Radiated emission, 260
Radiation losses, 194
Radiative load, 291
Raising operator, 13, 16
Ramsey experiment, 82
Randomized benchmarking, 40–41, 42
Reciprocity theorem, 225
Reflection loss, 263
Reflections, multiple, 265
Refrigerators
  dilution, 326–38
  pulse-tube (PTRs), 322–24
  pumped-helium, 325–26
  See also Cryogenics
Relaxation
  longitudinal, 78–80
  time measurement, 80
  transverse, 80–82
Residual gas load, 291
Resonator coupling, 239–42
Return loss, 175, 200, 201
Reversibility concept, 3, 4
RF sampling, 251–55
RF switches, 310–11
Rivest-Shamir-Adleman (RSA), 33–34
RMS vacuum voltage, 115
Room-temperature filtering, 271
Room-temperature microwave cabling, 292
Room-temperature setup
  description, 282
  filtering, 285–87
  further considerations, 288–90
  grounding, 289–90
  low-pass filters in, 286
  signal detection, 287–88
  signal generation, 279–85
  signal processing, 285–88
  synchronization, 289
  See also Control hardware
Rotating-wave approximation (RWA), 88, 98
Rotations, qubits, 89–91
Ruhmkorff's coils, 125

S

Sample rate, 280
Scalar signal generators, 178
Scanning tunneling microscope (STM), 11, 313
Scattering parameters, 172–75
Schrödinger, Erwin, 2
Schrödinger equation

about, 4–5
Ansatz in solving, 87
Bloch sphere and, 16–17
free particle and, 8–9
harmonic oscillator and, 11–13
ladder operators and, 13–14
potential function, 7
potential well and, 9–11
quantum LC oscillator and, 14–15
qubit and, 16
solving, 6–17
standard Hamiltonians, 8–17
time-dependent, 6, 7–8
wave function and, 5
Self-mixing effect, 250
Semiconductor qubits
about, 48
spin, 49
superconducting, 49–50, 69–122
topological, 50–51
See also Qubits
Sensor mounting, 345–46
Shielding
cryogenic, 342–43
effectiveness, 262–65
electromagnetic, 261–71, 304–5
noise suppression and, 304–7
thermal, 306–7
Shor's algorithm, 44–45
Shot noise, 136
Signal addition/subtraction, 221–29
Signal amplification
about, 307–8
cryogenic HEMT amplifier, 309–10
parametric amplifiers, 308–9
SLUG amplifier, 310
in superconducting qubit experiment, 308
See also Cryogenic setup
Signal degradation factors
about, 131
attenuation, 131
distortion, 132
distortion sources, 134–35
interference, 136
linearity, 132–34
phase noise, 144–45, 146
quantization noise, 145–48
thermal noise, 136
See also Microwave links
Signal detection
about, 247
amplification, 247–48
digitization/demodulation, 250
direct RF sampling, 251–55
downconversion, 249–50

filtering, 248–49
homodyne detection, 250–51
room-temperature setup, 287–88
superheterodyne detection, 251, 252, 253–54
See also Microwave components
Signal flow, 247
Signal generation
about, 177
analog signal generation, 177–78, 279–83
digital methods for, 179–83
digital signal generation, 283–85
room-temperature setup, 279–85
signal generator figures of merit, 178–79
See also Microwave components
Signal integrity, 257–58
Signal processing
about, 209
amplitude manipulation, 209–29
frequency manipulation, 229–44
performance specifications, 209
phase manipulation, 244–47
room-temperature setup, 285–88
See also Microwave components
Signal-to-noise ratio (SNR), 104–5, 139–40
Signal transmission
about, 183–85
microwave connectors, 207–9
non-TEM transmission lines, 196–200
TEM-mode transmission lines, 185–96
transmission line types, 200–207
See also Microwave components
Single-pole double-throw (SPDT) switch, 226, 227
Single-qubit gates, 34–37
SLUG amplifiers, 219, 220, 310
S-parameters
definition of, 173
forward, 174
matrix, 174
measurement of, 174–75
terminology, 175–76
Spectroscopic qubit measurements, 112–13
Spin qubits, 49
Spurious-free dynamic range (SFDR), 161, 162, 183
Standard Commands for Programmable Instruments (SCPI), 315
Standard deviation, 19
Standard platinum resistance thermometer (SPRT), 341
Standing wave ratio (SWR), 188
State-preparation-and measurement (SPAM) errors, 40–41
Stern-Gerlach experiment, 2
Stokes theorem, 60

Strong coupling, 101–2
Strong dispersive regime, 102–4
Superconducting quantum interference device (SQUID), 61, 68, 74–75, 77, 216, 278
Superconducting qubits
  about, 49–50
  artificial atom, 69–71
  cable types used in setup, 294
  calibration of single-qubit operations, 121–22
  coherence time scales, 78–85
  control, 86–93
  control hardware, 277–315
  Cooper pair box, 71–73
  decoherence sources, 83–85
  parameter adjustments and, 71
  transmon, 73–78
  two-qubit system, 116–21
Superconductivity
  Josephson junction, 66–69
  type-II superconductors, 65, 66
  type-I superconductors, 65–66
Superconductors
  canonical momentum, 59–60
  Cooper pairs, 61–64
  hole in sample, 60
  introduction to, 57–69
  properties, 57–59
Superheterodyne detection, 251, 252, 253–54
Sweet spots, 73
Switches, 310–11
Switching speed, 179
Synchronization, room-temperature setup, 289

T

T-bate, 36
TEM transmission lines
  about, 185–87
  attenuation constant, 191–94
  circuit parameters, 185–86
  impedance matching, 189–90
  phase constant, 194–96
  propagation constants, 190–91
  terminated lossless, 187–89
  See also Transmission lines
Terminated lossless transmission lines, 187–89
T filters, 232
Thermal anchoring, 343–45
Thermal electromotive force (EMF), 340–41
Thermal energy, 71
Thermal noise
  about, 136
  equivalent noise temperature, 138–39, 143
  maximum noise power, 138
  minimum detectable signal, 140–41
  noise figure, 141–43
  noise voltage, 137
  quantization noise comparison, 149
  SNR, 139
  See also Signal degradation factors
Thermal shielding, 306
Thermometer measurements, cryogenic, 338–41
Third-order intercept points (IP3), 157–59
Three-qubit gates, 38
Three-wave mixing, 211–13
Threshold theorem, 46
Time-dependent Schrödinger equation, 6, 7–8
Topological qubits, 50–51
Transistor-to-transistor (TTL) signal, 227
Transmission lines
  about, 183
  bandwidth, 197–99
  coaxial cable, 200–201
  CPW, 203–6
  dispersion effect in, 195
  distributed-element, 184, 185
  figures of merit, 199–200
  hollow waveguides, 206–7
  load impedance, 188
  lumped-sum, 184, 185
  microstrip line, 201–3
  non-TEM, 196–200
  propagation modes, 196
  TEM-mode, 185–96
  terminated lossless, 187–89
  types of, 200–207
Transmon qubits
  about, 73
  anharmonicity, 76
  capacitance, building, 77
  charge dispersion, 76
  design formulas summary, 77
  Hamiltonian parameters, 115
  regime, operating in, 75
  transition frequency, 75
  See also Superconducting qubits
Transverse electromagnetic (TEM) propagation mode, 171
Transverse relaxation, 80–82
Traveling-wave parametric amplifier (TWPA), 216–19, 308–9
Tunable phase shifters, 246–47
Tunneling, 11
Two-dimensional electron gas (2DEG), 49
Two-qubit gates, 37–39
Two-qubit system
  about, 116–17
  capacitive coupling, 118, 119
  dispersive interaction, 120–21
Two-tone measurement, 154, 155

U

Uncertainties, 19–20
Uncertainty principle
  about, 20–21
  commutation relations and, 22–23
  Fourier transform and, 21–22
  implications, 21
  in qubit energy levels and, 25
Upconversion, 233–36

V

Vacuum modes, 79
Vacuum Rabi oscillations, 100–101
Variance, 19
Vector-signal generators, 178
Vertical resolution, 180
Vibrational damping and decoupling, 311–13
Voltage-controlled oscillators (VCOs), 177
Voltage standing-wave ratio (VSWR), 188

W

Wave functions
  about, 4
  coherent superposition of, 25
  collapse of, 17
  momentum-space, 21–22
  position-space, 21–22
  quantum harmonic oscillator, 12
  relative phase, 7
Waveguide dispersion, 195
Waveguide resonators, 243–44
Wave packets, 10, 20
Wet dilution fridge, 337–38
Wilkinson combiner, 221, 222
Wires, grounding of, 274–75

X

X-gate, 35, 117

Y

YIG-tuned oscillators, 177
Ytterbium ion, 47

Z

Zero resistance, 57, 58
Z-gate, 36, 117

Artech House Microwave Library

Behavioral Modeling and Linearization of RF Power Amplifiers, John Wood

Chipless RFID Reader Architecture, Nemai Chandra Karmakar, Prasanna Kalansuriya, Randika Koswatta, and Rubayet E-Azim

Chipless RFID Systems Using Advanced Artificial Intelligence, Larry M. Arjomandi and Nemai Chandra Karmakar

Control Components Using Si, GaAs, and GaN Technologies, Inder J. Bahl

Design of Linear RF Outphasing Power Amplifiers, Xuejun Zhang, Lawrence E. Larson, and Peter M. Asbeck

Design Methodology for RF CMOS Phase Locked Loops, Carlos Quemada, Guillermo Bistué, and Iñigo Adin

Design of CMOS Operational Amplifiers, Rasoul Dehghani

Design of RF and Microwave Amplifiers and Oscillators, Second Edition, Pieter L. D. Abrie

Digital Filter Design Solutions, Jolyon M. De Freitas

Discrete Oscillator Design Linear, Nonlinear, Transient, and Noise Domains, Randall W. Rhea

Distortion in RF Power Amplifiers, Joel Vuolevi and Timo Rahkonen

Distributed Power Amplifiers for RF and Microwave Communications, Narendra Kumar and Andrei Grebennikov

Electric Circuits: A Primer, J. C. Olivier

Electronics for Microwave Backhaul, Vittorio Camarchia, Roberto Quaglia, and Marco Pirola, editors

An Engineer's Guide to Automated Testing of High-Speed Interfaces, Second Edition, José Moreira and Hubert Werkmann

Envelope Tracking Power Amplifiers for Wireless Communications, Zhancang Wang

Essentials of RF and Microwave Grounding, Eric Holzman

Frequency Measurement Technology, Ignacio Llamas-Garro, Marcos Tavares de Melo, and Jung-Mu Kim

FAST: Fast Amplifier Synthesis Tool—Software and User's Guide, Dale D. Henkes

Feedforward Linear Power Amplifiers, Nick Pothecary

Filter Synthesis Using Genesys S/Filter, Randall W. Rhea

Foundations of Oscillator Circuit Design, Guillermo Gonzalez

Frequency Synthesizers: Concept to Product, Alexander Chenakin

Fundamentals of Nonlinear Behavioral Modeling for RF and Microwave Design, John Wood and David E. Root, editors

Generalized Filter Design by Computer Optimization, Djuradj Budimir

Handbook of Dielectric and Thermal Properties of Materials at Microwave Frequencies, Vyacheslav V. Komarov

Handbook of RF, Microwave, and Millimeter-Wave Components, Leonid A. Belov, Sergey M. Smolskiy, and Victor N. Kochemasov

High-Efficiency Load Modulation Power Amplifiers for Wireless Communications, Zhancang Wang

High-Linearity RF Amplifier Design, Peter B. Kenington

High-Speed Circuit Board Signal Integrity, Second Edition, Stephen C. Thierauf

Integrated Microwave Front-Ends with Avionics Applications, Leo G. Maloratsky

Intermodulation Distortion in Microwave and Wireless Circuits, José Carlos Pedro and Nuno Borges Carvalho

Introduction to Modeling HBTs, Matthias Rudolph

An Introduction to Packet Microwave Systems and Technologies, Paolo Volpato

Introduction to RF Design Using EM Simulators, Hiroaki Kogure, Yoshie Kogure, and James C. Rautio

Introduction to RF and Microwave Passive Components, Richard Wallace and Krister Andreasson

Klystrons, Traveling Wave Tubes, Magnetrons, Crossed-Field Amplifiers, and Gyrotrons, A. S. Gilmour, Jr.

Lumped Element Quadrature Hybrids, David Andrews

Lumped Elements for RF and Microwave Circuits, Second Edition, Inder J. Bahl

Microstrip Lines and Slotlines, Third Edition, Ramesh Garg, Inder Bahl, and Maurizio Bozzi

Microwave Component Mechanics, Harri Eskelinen and Pekka Eskelinen

Microwave Differential Circuit Design Using Mixed-Mode S-Parameters, William R. Eisenstadt, Robert Stengel, and Bruce M. Thompson

Microwave Engineers' Handbook, Two Volumes, Theodore Saad, editor

Microwave Filters, Impedance-Matching Networks, and Coupling Structures, George L. Matthaei, Leo Young, and E. M. T. Jones

Microwave Imaging Methods and Applications, Matteo Pastorino and Andrea Randazzo

Microwave Material Applications: Device Miniaturization and Integration, David B. Cruickshank

Microwave Materials and Fabrication Techniques, Second Edition, Thomas S. Laverghetta

Microwave Materials for Wireless Applications, David B. Cruickshank

Microwave Mixer Technology and Applications, Bert Henderson and Edmar Camargo

Microwave Mixers, Second Edition, Stephen A. Maas

Microwave Network Design Using the Scattering Matrix, Janusz A. Dobrowolski

Microwave Power Amplifier Design with MMIC Modules, Howard Hausman

Microwave Radio Transmission Design Guide, Second Edition, Trevor Manning

Microwave and RF Semiconductor Control Device Modeling, Robert H. Caverly

Microwave Transmission Line Circuits, William T. Joines, W. Devereux Palmer, and Jennifer T. Bernhard

Microwave Techniques in Superconducting Quantum Computers, Alan Salari

Microwaves and Wireless Simplified, Third Edition, Thomas S. Laverghetta

Millimeter-Wave GaN Power Amplifier Design, Edmar Camargo

Modern Microwave Circuits, Noyan Kinayman and M. I. Aksun

Modern Microwave Measurements and Techniques, Second Edition, Thomas S. Laverghetta

Modern RF and Microwave Filter Design, Protap Pramanick and Prakash Bhartia

Neural Networks for RF and Microwave Design, Q. J. Zhang and K. C. Gupta

Noise in Linear and Nonlinear Circuits, Stephen A. Maas

Nonlinear Design: FETs and HEMTs, Peter H. Ladbrooke

Nonlinear Microwave and RF Circuits, Second Edition, Stephen A. Maas

On-Wafer Microwave Measurements and De-Embedding, Errikos Lourandakis

Parameter Extraction and Complex Nonlinear Transistor Models, Günter Kompa

Passive RF Component Technology: Materials, Techniques, and Applications, Guoan Wang and Bo Pan, editors

PCB Design Guide to Via and Trace Currents and Temperatures, Douglas Brooks with Johannes Adam

Practical Analog and Digital Filter Design, Les Thede

Practical Microstrip Design and Applications, Günter Kompa

Practical Microwave Circuits, Stephen Maas

Practical RF Circuit Design for Modern Wireless Systems, Volume I: Passive Circuits and Systems, Les Besser and Rowan Gilmore

Practical RF Circuit Design for Modern Wireless Systems, Volume II: Active Circuits and Systems, Rowan Gilmore and Les Besser

Principles of RF and Microwave Design, Matthew A. Morgan

Production Testing of RF and System-on-a-Chip Devices for Wireless Communications, Keith B. Schaub and Joe Kelly

Q Factor Measurements Using MATLAB, Darko Kajfez

Radio Frequency Integrated Circuit Design, Second Edition, John W. M. Rogers and Calvin Plett

Reflectionless Filters, Matthew A. Morgan

RF Bulk Acoustic Wave Filters for Communications, Ken-ya Hashimoto

RF Circuits and Applications for Practicing Engineers, Mouqun Dong

RF Design Guide: Systems, Circuits, and Equations, Peter Vizmuller

RF Linear Accelerators for Medical and Industrial Applications, Samy Hanna

RF Measurements of Die and Packages, Scott A. Wartenberg

The RF and Microwave Circuit Design Handbook, Stephen A. Maas

RF and Microwave Coupled-Line Circuits, Rajesh Mongia, Inder Bahl, and Prakash Bhartia

RF and Microwave Oscillator Design, Michal Odyniec, editor

RF Power Amplifiers for Wireless Communications, Second Edition, Steve C. Cripps

RF Systems, Components, and Circuits Handbook, Ferril A. Losee

Scattering Parameters in RF and Microwave Circuit Analysis and Design, Janusz A. Dobrowolski

The Six-Port Technique with Microwave and Wireless Applications, Fadhel M. Ghannouchi and Abbas Mohammadi

Solid-State Microwave High-Power Amplifiers, Franco Sechi and Marina Bujatti

Stability Analysis of Nonlinear Microwave Circuits, Almudena Suárez and Raymond Quéré

Substrate Noise Coupling in Analog/RF Circuits, Stephane Bronckers, Geert Van der Plas, Gerd Vandersteen, and Yves Rolain

System-in-Package RF Design and Applications, Michael P. Gaynor

Technologies for RF Systems, Terry Edwards

Terahertz Metrology, Mira Naftaly, editor

Understanding Quartz Crystals and Oscillators, Ramón M. Cerda

Vertical GaN and SiC Power Devices, Kazuhiro Mochizuki

The VNA Applications Handbook, Gregory Bonaguide and Neil Jarvis

Wideband Microwave Materials Characterization, John W. Schultz

Wired and Wireless Seamless Access Systems for Public Infrastructure, Tetsuya Kawanishi

For further information on these and other Artech House titles, including previously considered out-of-print books now available through our In-Print-Forever® (IPF®) program, contact:

Artech House
685 Canton Street
Norwood, MA 02062
Phone: 781-769-9750
Fax: 781-769-6334
e-mail: artech@artechhouse.com

Artech House
16 Sussex Street
London SW1V 4RW UK
Phone: +44 (0)20 7596 8750
Fax: +44 (0)20 7630 0166
e-mail: artech-uk@artechhouse.com

Find us on the World Wide Web at: www.artechhouse.com